Gene Regulation by Steroid Hormones II

Edited by
A.K. Roy and J.H. Clark

With 149 Figures

Springer-Verlag
New York Heidelberg Berlin

ARUN K. ROY
Professor of Biological Sciences, Oakland University, Rochester, Michigan 48063, USA

JAMES H. CLARK
Professor of Cell Biology, Baylor College of Medicine, Houston, Texas 77030, USA

Cover Picture: Electronmicrograph of a hybrid molecule between ovalbumin gene and ovalbumin mRNA. The intervening sequences are shown as loops. Courtesy of Dr. Eugene Lai, Baylor College of Medicine.

Library of Congress Cataloging in Publication Data
Main entry under title:
Gene regulation by steroid hormones II.
  Bibliography: p.
  Includes index.
  1. Steroid hormones—Physiological effect—Congresses. 2. Genetic regulation—Congresses. 3. Gene expression—Congresses. I. Roy, A.K. (Arun K.) II. Clark, James H. [DNLM: 1. Gene expression regulation—Congresses. 2. Steroids—Physiology—Congresses. WK 150 G3261]
QP572.S7G46   1983   591.19'27   82-19250

Typeset by Bi-Comp, Incorporated, York, Pennsylvania.
Printed and bound by Halliday Lithograph, West Hanover, Massachusetts.
Printed in the United States of America.

9 8 7 6 5 4 3 2 1

ISBN 0-387-90784-X Springer-Verlag New York Heidelberg Berlin
ISBN 3-540-90784-X Springer-Verlag Berlin Heidelberg New York

Foreword

Receptors and Gene Expression

It is now more than three years since the last Meadow Brook Conference on Hormones, and a great deal has happened in the meantime. We have become comfortable with the totally unanticipated fact that the coding sequences of genes are in discontinuous arrangements and that the RNA transcribed from them must be extensively processed to form messenger RNA. We have also learned about the strategy of ''mixing and matching'' of genetic segments so that a small amount of DNA can go a long way in producing a huge variety of different proteins, as in the immunoglobulin system. The explosive effort directed toward DNA sequence analysis has led us to the conclusion that there are signals within the DNA that specify sites of transcription initiation and possibly sites for interacting with regulatory molecules such as hormones and their receptors. The current intense interest in the structure of chromatin beyond the nucleosome—that is, the superstructural characteristics of the genetic material—is finally yielding meaningful results that give promise for understanding the regulation of gene activity.

ROBERT F. GOLDBERGER

Preface

Research on the molecular mechanism of steroid hormone action continues at an extraordinary pace and a great deal of progress has been made. Steroid hormones have been localized on target genes providing the long awaited evidence for the concept of a direct effect of the steroid-receptor complex on gene regulation. Purified steroid receptors have been dissected to identify different functional domains. Monoclonal antibodies directed against different antigenic determinants of steroid receptors have also provided precise tools for the exploration of the molecular biology of receptor action. Parallel investigation on the structure of the target genes, especially the non-coding segments seems to be leading to a tantalizing model reminiscent of prokaryotic operon. As always, these important developments are overshadowed by a long list of both previously unresolved and new questions. Comprehensive explanation of the molecular mechanism of steroid hormone action needs to take into consideration the multiplicity of specific steroid binding proteins, the role of cellular architecture such as nuclear matrix and cytoskeleton and the interacting influence of the non-steroidal hormones in the steroidal regulation of gene expression. The Second Meadow Brook Conference was organized to bring together leading researchers with a view to providing a critical appraisal of the past achievements and a sense for the future direction of this important area of regulatory biology.

The conference was made possible through generous financial supports from Oakland University, Merck & Company, G.D. Searle & Company, E.R. Squibb & Sons and Upjohn Company. Particular thanks are expressed to other members of the program committee—John D. Baxter, Donald S. Coffey and Shutsung Liao. Finally, we must add that excellent cooperation from the authors and the editorial staff of Springer-Verlag, especially Dr. Philip Manor, has made the job of editing this volume a delightful experience.

Spring 1983

ARUN K. ROY
JAMES H. CLARK

Table of Contents

List of Contributors

J.N. ANDERSON; Department of Biological Sciences, Purdue University, West Lafayette, IN 47907, USA

C.W. BARDIN, The Population Council, The Rockefeller University, 1230 York Avenue, New York, NY 10021, USA

E.R. BARRACK, The James Buchanan Brady Urological Institute, Department of Urology, The Johns Hopkins Oncology Center, Department of Pharmacology and Experimental Therapeutics, The Johns Hopkins University School of Medicine, Baltimore, MD 21205, USA

J.D. BAXTER, Howard Hughes Medical Institute Laboratories, Metabolic Research Unit and The Departments of Medicine, Biochemistry and Biophysics, and Pharmaceutical Chemistry, University of California, San Francisco, CA 94143, USA

K.S. BLOOM, Department of Biological Sciences, Purdue University, West Lafayette, IN 47907, USA

M.L. BROCK, Department of Biochemistry, University of Illinois, 1209 W. California, Urbana, IL 61801, USA

J.R. BROOKS, Merck Sharp and Dohme Research Laboratories, P.O. Box 2000, Rahway, NJ 07065, USA

J. CARLSTEDT-DUKE, Department of Medical Nutrition, Karolinska Institute, Huddinge University Hospital F69, S-141 86 Huddinge, Sweden

G. CATHALA, Howard Hughes Medical Institute Laboratories, Metabolic Research Unit and The Departments of Medicine, Biochemistry and Biophysics, and Pharmaceutical Chemistry, University of California, San Francisco, CA 94143, USA

B. CHATTERJEE, Department of Chemistry, Oakland University, Rochester, MI 48063, USA

J.H. CLARK, Department of Cell Biology, Baylor College of Medicine, Houston, TX 77030, USA

D.S. COFFEY, The James Buchanan Brady Urological Institute, Department of Urology, The Johns Hopkins Oncology Center, Department of Pharmacology and Experimental Therapeutics, The Johns Hopkins University School of Medicine, Baltimore, MD 21205, USA

W.F. DEMYAN, Department of Biological Sciences, Oakland University, Rochester, MI 48063, USA

B. DWORNICZAK, Ruhr-Universität, Department of Biochemistry, 463 Bochum, Postfach 2148, West Germany

N. EBERHARDT, Howard Hughes Medical Institute Laboratories, Metabolic Research Unit and The Departments of Medicine, Biochemistry and Biophysics, and Pharmaceutical Chemistry, University of California, San Francisco, CA 94143, USA

B.J. GERMAIN, Department of Biological Sciences, Purdue University, West Lafayette, IN 47907, USA

R.F. GOLDBERGER, Office of the Provost and Vice-President, Columbia University, New York, NY 10027, USA

G. L. GREENE, The University of Chicago, Ben May Laboratory for Cancer Research, 950 E 59th Street, Chicago, IL 60637, USA

J.-Å. GUSTAFSSON, Department of Medical Nutrition, Karolinska Institute, Huddinge University Hospital F69, S-141 86 Huddinge, Sweden

M.A. HAYWARD, Department of Biochemistry, University of Illinois, 1209 W. California, Urbana, IL 61801, USA

V.V. ISOMAA, The Population Council, The Rockefeller University, 1230 York Avenue, New York, NY 10021, USA

O.A. JÄNNE, The Population Council, The Rockefeller University, 1230 York Avenue, New York, NY 10021, USA

S.M. JUDGE, The University of Chicago, Ben May Laboratory for Cancer Research, 950 E. 59th Street, Chicago, IL 60637, USA

M. KARIN, Department of Microbiology, School of Medicine, University of Southern California, 2025 Zonal Avenue, Los Angeles, CA 90033, USA

K.L. KELNER, Department of Cell Biology, Baylor College of Medicine, Houston, TX 77030, USA

S. KOBUS, Ruhr-Universität, Department of Biochemistry, 463 Bochum, Postfach 2148, West Germany

P.A. KOLLMAN, Department of Pharmaceutical Chemistry, University of California, San Francisco, CA 94143, USA

J.L. KOSTYO, Department of Physiology, University of Michigan Medical School, 1335 E. Catherine, Ann Arbor, MI 48109, USA

N.C. LAN, Howard Hughes Medical Institute Laboratories, Metabolic Research Unit and The Departments of Medicine, Biochemistry and Biophysics, and Pharmaceutical Chemistry, University of California, San Francisco, CA 94143, USA

T. LIANG, Merck Sharp and Dohme Research Laboratories, P.O. Box 2000, Rahway, NJ 07065, USA

S. LIAO, The University of Chicago, Ben May Laboratory for Cancer Research, 950 E. 59th Street, Chicago, IL 60637, USA

G. LITWACK, Fels Research Institute and Department of Biochemistry, Temple University School of Medicine, Philadelphia, PA 19140, USA

B.M. MARKAVERICH, Department of Cell Biology, Baylor College of Medicine, Houston, TX 77030, USA

M. MAYER, Fels Research Institute and Department of Biochemistry, Temple University School of Medicine, Philadelphia, PA 19140, USA

A.R. MEANS, Department of Cell Biology, Baylor College of Medicine, Houston, TX 77030, USA

S. MELLON, Howard Hughes Medical Institute Laboratories, Metabolic Research Unit and The Departments of Medicine, Biochemistry and Biophysics, and Pharmaceutical Chemistry, University of California, San Francisco, CA 94143, USA

P.P. MINGHETTI, Department of Cell Biology, Baylor College of Medicine, Houston, TX 77030, USA

N.M. MOTWANI, Department of Biological Sciences, Oakland University, Rochester, MI 48063, USA

T.S. NATH, Department of Biological Sciences, Oakland University, Rochester, MI 48063, USA

T. NGUYEN, Howard Hughes Medical Institute Laboratories, Metabolic Research Unit and The Departments of Medicine, Biochemistry and Biophysics, and Pharmaceutical Chemistry, University of California, San Francisco, CA 94143, USA

T.D. NGUYEN, Howard Hughes Medical Institute Laboratories, Metabolic Research Unit and The Departments of Medicine, Biochemistry and Biophysics, and Pharmaceutical Chemistry, University of California, San Francisco, CA 94143, USA

S.K. NORDEEN, Howard Hughes Medical Institute Laboratories, Metabolic Research Unit and The Departments of Medicine, Biochemistry and Biophysics, and Pharmaceutical Chemistry, University of California, San Francisco, CA 94143, USA

J.D. O'CONNOR, Department of Biology, University of California, Los Angeles, CA 90024, USA

V. OHL, Fels Research Institute and Department of Biochemistry, Temple University School of Medicine, Philadelphia, PA 19140, USA

S. OKRET, Department of Medical Nutrition, Karolinska Institute, Huddinge University, Hospital F69, S-141 86 Hiddinge, Sweden

B.W. O'MALLEY, Department of Cell Biology, Baylor College of Medicine, Houston, TX 77030, USA

A.E.I. PAJUNEN, The Population Council, The Rockefeller University, 1230 York Avenue, New York, NY 10021, USA

E.J. PECK, JR., Department of Biochemistry, University of Arkansas Medical School, Little Rock, AR 72205, USA

O. PONGS, Ruhr-Universität, Department of Biochemistry, 463 Bochum, Postfach 2148, West Germany

G.H. RASMUSSON, Merck Sharp and Dohme Research Laboratories, P.O. Box 2000, Rahway, NJ 07065, USA

A.K. ROY, Department of Biological Sciences, Oakland University, Rochester, MI 48063, USA

A.G. SALTZMAN, University of Chicago, Ben May Laboratory for Cancer Research, 950 E. 59th Street, Chicago, IL 60637, USA

K. SCHALTMANN-EITELJÖRGE, Ruhr-Universität, Department of Biochemistry, 463 Bochum, Postfach 2148, West Germany

W.T. SCHRADER, Department of Cell Biology, Baylor College of Medicine, Houston, TX 77030, USA

B. SEKULA, Fels Research Institute and Department of Biochemistry, Temple University School of Medicine, Philadelphia, PA 19140, USA

D.J. SHAPIRO, Department of Biochemistry, University of Illinois, 1209 W. California, Urbana, IL 61801, USA

J.P. STEIN, Department of Endocrinology, University of Texas Medical School, Houston, TX 77025, USA

B. STEVENS, Department of Biology, University of California, Los Angeles, CA 90024, USA

J. STEVENS, American Cancer Society, 777 Third Avenue, New York, NY 10017, USA

Y.-W. STEVENS, Memorial Sloan-Kettering Cancer Center, New York, NY 10021, USA

J.S. STROBL, National Cancer Institute, NIH, Bethesda, MD 20205, USA

E.B. THOMPSON, National Cancer Institute, NIH, Bethesda, MD 20205, USA

J.N. VANDERBILT, Department of Biological Sciences, Purdue University, West Lafayette, IN 47907, USA

N.L. WEIGEL, Department of Cell Biology, Baylor College of Medicine, Houston, TX 77030, USA

M.E. WOLFF, Department of Pharmaceutical Chemistry, University of California, San Francisco, CA 94143, USA

Ö. WRANGE, Department of Medical Nutrition, Karolinska Institute, Huddinge University Hospital F69, S-141 86 Huddinge, Sweden

W.W. WRIGHT, The Population Council, The Rockefeller University, 1230 York Avenue, New York, NY 10021, USA

Chapter 1

Gene Structure and Evolution

BERT W. O'MALLEY, ANTHONY R. MEANS, AND
JOSEPH P. STEIN

I. Introduction

The organization of eukaryotic genes has been an area of acute interest to us
for several years. One particularly attractive model system for studying the
structural organization of functionally related genes in a single tissue has
been the hen oviduct (O'Malley et al. 1969). The chicken ovalbumin gene has
been cloned and sequenced (Rosen et al. 1975; Monahan et al. 1976; McReynolds et al. 1977; Woo et al. 1978; Woo et al. 1981). More recently, the
chicken ovomucoid gene was cloned and partially sequenced (Stein et al.
1978; Lai et al. 1979; Catterall et al. 1979; Stein et al. 1980; Catterall et al.
1980). We discovered that both of these oviduct genes exhibited a surprisingly complex structure; each gene contained seven nontranslated regions of
DNA sequence (intervening sequences or introns) interspersed among eight
genomic DNA regions (exons) that code for the mature polypeptide chains.
A number of laboratories in addition to our own have identified the existence
of these intervening sequences in diverse eukaryotic genes (Breathnach et
al. 1977; Weinstock et al. 1978; Garapin et al. 1978; Mandel et al. 1978;
Lindenmaier et al. 1979; Cochet et al. 1979; Brack and Tonegawa 1977; Maki
et al. 1980; Early et al. 1980; Jeffreys and Flavell 1977; Tilghman et al. 1978;
Valenzuela et al. 1978; Lomedico et al. 1979; Bell et al. 1980; Nunberg et al.
1980; Gorin and Tilgham 1980; Fyrberg et al. 1980). Even though the existence of these intervening sequences seems to be the rule rather than the
exception in eukaryotic genes, their function and origin remain an intriguing
mystery.

The ovomucoid gene afforded us a unique opportunity to examine the
raison d'être of intervening sequences. We have sequenced most of the
chicken ovomucoid gene, including all of the intron–exon junctions. During
the course of our investigation, the sequence of the secreted chicken ovomucoid protein was published (Kato et al. 1978). We are thus able to analyze the
structure of the ovomucoid gene in light of the protein sequence. The results
of this analysis, which we present here, have allowed us to propose a theory
of the function of intervening sequences in the organization of the ovomu-

coid gene. This in turn has led us to propose a mechanism for the evolution of complex eukaryotic genes.

II. Isolation and Structural Analysis of the Ovomucoid Gene

The nucleotide sequence of ovomucoid mRNA was determined from two recombinant DNA plasmids, pOM100 and pOM502. (These recombinant plasmids were derived from the parent bacterial plasmid, pBR322; the OM refers to the fact that ovomucoid DNA sequences were inserted, and the three-digit numbers simply refer to the chronological isolation sequence.) Preparation of the recombinant plasmid pOM100, which contains cDNA synthesized from a partially purified $mRNA_{om}$, has been described (Stein et al. 1978). This plasmid contained a 650 base pair (bp) DNA_{om} insert but lacked ~150 bp corresponding to the 5' end of $mRNA_{om}$. To obtain cDNA sequences containing the 5' end of the mRNA sequence, a 204 bp fragment removed from the middle of the ovomucoid cDNA sequence in pOM100 by cutting with the restriction endonuclease Pst I was purified and hybridized with a partially purified preparation of $mRNA_{om}$. DNA complementary to the 5' end of $mRNA_{om}$ was synthesized by reverse transcriptase in the absence of an oligodeoxythymidilate primer. The use of a sequence-specific primer prevented the polymerization of cDNA from mRNAs other than $mRNA_{om}$. Double-stranded $cDNA_{om}$ was prepared and cloned in the Pst 1 site of pBR322 (Catterall et al. 1980). The largest resulting recombinant plasmid was designated pOM502. This plasmid contained a DNA_{om} insert of 538 nucleotides, excluding the GC tails at either end of the $cDNA_{om}$ sequence.

These two plasmids were sequenced, using the method of Maxam and Gilbert (1977). Together, they contained all 821 nucleotides of the $cDNA_{om}$.

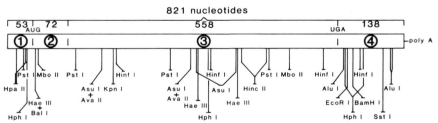

Fig. 1. The ovomucoid structural gene sequence. The lengths of various regions of the mRNA, in nucleotides, are given at the top of the figure. Region 1, the 5' leader sequence, refers to the portion of the mRNA at the 5' end that occurs before the AUG initiation signal and hence is not translated into a protein sequence. Region 2, the signal peptide sequence, is that portion of the mRNA that is translated into the signal peptide, the N-terminal portion of the protein, which is clipped off during the secretion of the protein. Region 3, the ovomucoid protein sequence, is that portion of the mRNA that codes for the secreted protein sequence. Region 4, the 3' noncoding sequence, is that portion of the mRNA that follows the UGA termination signal and hence is not translated into a protein sequence.

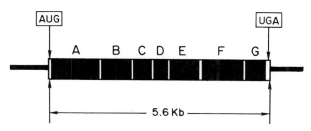

Fig. 2. Structural organization of the chicken ovomucoid gene. The ovomucoid exon sequences are shown as open bars, the intron sequences as filled bars, and the flanking genomic DNA as a solid line. The seven introns are labeled A through G. The overall length of the ovomucoid gene is 5.6 kb.

The ovomucoid structural gene sequence determined from these plasmids is schematically presented in Fig. 1. The 821 nucleotides that represent the mature ovomucoid mRNA can be divided into four regions. There is a 53 nucleotide 5' leader sequence prior to the AUG initiator codon. This is followed by a 72 nucleotide signal peptide sequence, which is translated but removed from the secreted polypeptide during the secretion process (Blobel and Dobberstein 1975). The 558 nucleotides that code for the secreted ovomucoid are next, followed by the 138 nucleotides of 3' noncoding sequence that are located after the UGA translation termination signal. A poly(A) tail of unspecified length follows this sequence.

The cloned ovomucoid structural gene sequences were nick-translated with labeled nucleotides and used to isolate the ovomucoid gene from a library of chicken oviduct DNA fragments. The structural organization of the ovomucoid native gene was then determined by restriction endonuclease, electron microscopic mapping, and limited DNA sequence analysis. This physical map of the entire ovomucoid gene is shown in Fig. 2. The gene is 5.6 kilobases (kb) in length, with seven intervening sequences of various lengths interspersed among the eight coding (exon) segments. Remarkably, only 15% of the ovomucoid gene is represented in the mature mRNA.

III. Sequence Analysis of the Ovomucoid Gene

The DNA sequencing method of Maxam and Gilbert (1977) was used to sequence more than 70% of the chicken ovomucoid gene, including all the exons, the sequences surrounding all 14 intron–exon junctions, and several complete introns. Because the ovomucoid mRNA was sequenced, we were able to determine exactly where in the coding sequence the introns are inserted. The DNA sequences surrounding these 14 intron–exon junctions are listed in Fig. 3. These junctions demonstrate two different types of splicing sites: those with a redundant sequence at the adjacent junctions and those with no redundancy. There is an extensive terminally redundant sequence surrounding introns A and B and a very limited redundancy surrounding introns F and G. The splice point could occur at any position

Junction	Sequence	Exon	Intron	Length
I/A	GCTTCCT CCCAG GTGAGTAACTCC	I		111
A/II	TTTTTCC CCCAG ATGCTGCCTTTG		A	~900
II/B	TTTGGGGCT GAG GTG AGAAAGAGA	II		20
B/III	AACTTTGTC GAG GTG GACTGCAGT		B	725
III/C	TGCCTACAGCATGTGTGTACTGCA	III		137
C/IV	TTCCCTCTTCAG AGAATTTGGAAC		C	~450
IV/D	GAAACTGTTCCT GTAAGTGAAACC	IV		58
D/V	CTCCTTCCACAG ATGAACTGCAGT		D	264
V/E	TGCCCACAAAGT GTTATTGTACCG	V		137
E/VI	TTTTCCTTTCAG AGAGCAGGGGGC		E	760
VI/F	GCTGCTGTGAGT GT GAGTAGCACA	VI		67
F/VII	GTGCTTTTGCAG GT TGACTGCAGT		F	~1,150
VII/G	CAATGCAGTCGT GT ACGTACAGCC	VII		110
G/VIII	CCTCGCTTTCAG GG AAAGCAACGG		G	~475
	↑	VIII		181

Fig. 3. DNA sequences at all 14 intron–exon junctions of the ovomucoid gene. The DNA sequences at each of the intron–exon junctions are listed at the left, with exon sequences underlined. The sequences are aligned so that the putative splice points are above the arrow. Terminally redundant sequences at either end of a particular intron are shown in boxes. The actual length of each intron and exon, in nucleotide pairs, is shown at the right.

within the boxes and still yield the correct mature mRNA sequence. Terminally redundant sequences such as these have been found at all splicing sites examined prior to ovomucoid, although the redundant length varied from one to five nucleotides (Feeney and Allison 1969). However, the junctions at either end of three ovomucoid introns, C, D, and E, have no redundant sequences. Since the mRNA sequence is known (Catterall et al. 1980) (underlined in Fig. 1), there is only one dinucleotide bond at each junction that could be cleaved and joined with the adjacent exon sequence to yield the correct mature mRNA sequence. In each case, splicing at the indicated position yields an intron whose sequence begins with a GT and ends with an AG dinucleotide.

The sizes of the exon sequences are listed in the columns at the right of Fig. 3. These coding segments vary widely in length, the largest, exon VIII, being nine times longer than the shortest, exon II. Exon II is remarkable in that it is only 20 nucleotides in length. This is among the shortest structural gene segments (exons) described to date; only the D gene segment of the immunoglobulin heavy chain gene, which might consist of as few as 13 nucleotides, is as small as exon II (Early et al. 1980). The size of such short exons would seem to put a severe limitation on any model of RNA splicing that involves a specific protein and/or RNA interaction with the coding (exon) sequence.

IV. The Sequence of the Secreted Hen Ovomucoid

Avian ovomucoids are a family of gycoproteins that account for about 10% of the protein content of all bird egg whites (Feeney and Allison 1969; Feeney 1971). Chicken ovomucoid has been shown to be responsible for most of the trypsin inhibitory activity of chicken egg white (Lineweaver and Murray 1947). Laskowski and co-workers have for several years been investigating the comparative biochemistry of these avian ovomucoids. While we were completing the sequence analysis of the ovomucoid gene, this group published the complete amino acid sequence of the secreted chicken ovomucoid (Kato et al. 1978). Furthermore, these investigators noted a general sequence homology among ovomucoids and several mammalian submandibular and seminal plasma trypsin inhibitors. Specifically, the sequence of chicken ovomucoid shown in Fig. 4 could be divided into three structural domains, I, II, and III. Each domain is apparently capable of binding one molecule of trypsin or other serine protease, so there is a good correlation of functional and structural domains. Whereas all three domains contain three disulfide bonds in almost identical positions in each domain, the sequence homology between domains I and II is far stronger than that between domains I and III or II and III. Kato et al. (1978) used these observations to propose that distinct intragenic doubling events of a primordial gene resulted in the present ovomucoid gene. This hypothesis can now be examined more critically by comparing the ovomucoid genomic DNA sequence with the amino acid sequence of the secreted polypeptide.

V. Ovomucoid Intervening Sequences Specify Functional Domains

The illustration of the amino acid sequence in Fig. 4 is depicted so that the functional, trypsin-binding domains of Kato et al. (1978) are clearly delineated. They are depicted (in one dimension) as globular regions held together by disulfide bonds and separated by short connecting peptides. An analysis of the positions of the intervening sequences within the ovomucoid gene determined that several introns interrupted the portions of the gene sequence that code for the connecting peptides. We used the locations of the introns to redefine the three functional domains of chicken ovomucoid.

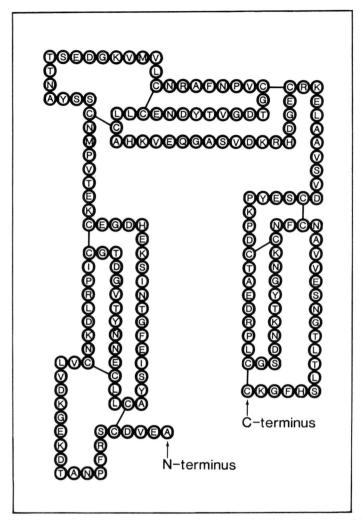

Fig. 4. Amino acid sequence of the secreted chicken ovomucoid. The single-letter amino acid code by Dayhoff is used. The drawing is meant to depict three similar structural regions held together by disulfide bonds (solid lines) separated by short connecting peptides.

These domains are depicted separately in Fig. 5. They are aligned to show the structural similarities as well as the similar positions of the introns within the domains. Each domain is separated from the next by one intervening sequence. All domains also contain one internal intervening sequence at identical positions within the three domains. As the lower part of Fig. 5 shows, 46% of the amino acids of domains I and II are identical. Sixty-six percent of the mRNA sequences coding for domains I and II are identical. The homology of domains I and II with domain III is less extensive but still

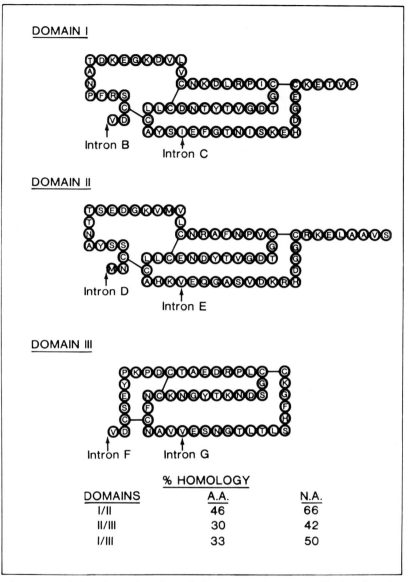

DOMAIN I

DOMAIN II

DOMAIN III

Intron B Intron C

Intron D Intron E

Intron F Intron G

DOMAINS	% HOMOLOGY A.A.	N.A.
I/II	46	66
II/III	30	42
I/III	33	50

Fig. 5. Alignment of the three domains of the secreted ovomucoid showing the similar positions of introns. The top portion shows the secondary structure of the three domains of the secreted ovomucoid polypeptide chain, aligned to emphasize the structural similarities. Disulfide bonds are indicated by the solid lines. The positions of the six introns that interrupt the DNA sequence coding for the three domains are shown with arrows. The homology between the amino acid sequence of the domains and the mRNA sequence coding for these domains is shown at the bottom. The amino acid homologies were calculated by comparing the amino acids at each position of the three domains, beginning with the first amino acid following introns B, D, and F, as shown. The nucleic acid homologies were calculated in the same manner, by comparing each nucleotide that codes for the amino acids shown. (Therefore, there are three times as many comparisons as for the amino acid homology.) A nine amino acid deletion after the ninth position of domain III is assumed to maximize homology with domains I and II.

significant. This suggests that the domains are related, a point that will be discussed in more detail later.

Although introns B, D, and F apparently separate the genomic domains coding for the functional peptide domains shown in Fig. 5, what about the other three introns, C, E, and G? These three introns occur at identical positions within the three domains, I, II and III, shown in Fig. 5. These "internal" introns divide the subdomains that code for the trypsin-binding sites, the N-terminal halves of the peptide domains in Fig. 5, from the C-terminal subdomains. It is quite possible that a primordial ovomucoid domain consisted only of the 5' subdomain and that the internal introns were created in the course of the addition of the 3' subdomain. Adding another disulfide bond to the primordial ovomucoid polypeptide (a single domain) may have resulted in increased stability and increased resistance to digestion by the protease bound by the ovomucoid protein. If these subdomains were separated and the disulfide bridge broken, the functional integrity of the inhibitor would probably be destroyed.

Thus, a strong argument can be made that the intervening sequences of the ovomucoid gene divide the gene into structural domains. These genomic domains code for structural domains of the ovomucoid polypeptide, which appear to correspond as well to functional polypeptide domains. We conclude, therefore, that the ovomucoid intervening sequences specify functional domains of the protein and suggest that this is likely a general concept of eukaryotic gene evolution.

VI. Evolution of the Ovomucoid Gene

The hypothesis that the present ovomucoid is the result of two intragenic doubling events is quite tenable in light of our investigations of the structural organization of the ovomucoid gene. All three functional polypeptide domains can be aligned so that the secondary structures are remarkably similar. All three domains contain three disulfide bonds in virtually identical positions. Three introns, B, D, and F, delineate the three functional domains of the secreted ovomucoid polypeptide as well as a separate domain that contains the signal peptide. All three domains contain an internal intron at identical positions within the genomic coding sequence. A more detailed analysis of the DNA sequences surrounding the six splice points indicated in Fig. 5 is even more revealing. The splice points (B, D, and F) that separate each domain occur *between two codons* of the genomic DNA sequence. The internal splice points (C, E, and G) within each domain, however, occur *between the second and third nucleotides of a codon* of the genomic DNA sequence. Thus, not only are the positions of the introns within the amino acid sequence the same for all three domains, but the pattern of intron interruption of the codon sequence is also identical. These observations are only plausible if the present ovomucoid gene is the result of two intragenic duplications of a primordial ovomucoid gene that itself contained introns.

The primordial ovomucoid gene probably consisted only of the 5' subdomain of domain III. Intron G would have been created in the course of addition of the 3' subdomain, which would have added stability to the primordial protease inhibitor, as discussed earlier. This recombinational event would have resulted in the creation of the original ovomucoid domain, represented as domain III in Fig. 5. Two genomic duplications, with the retention of noncoding sequence within and at the 5' end of the primordial gene, would subsequently have led to the present ovomucoid gene. After the initial duplication event, in which domain III gave rise to domain II, there must have been a deletion of 27 nucleotides (coding for nine amino acids, see Fig. 5) in the 5' region of domain III. The duplication of domain II to form domain I must have occurred much more recently on the evolutionary scale, as judged by the significant homology between the nucleotides at corresponding positions of the genomic sequences coding for domains I and II.

An approximate time when this second duplication must have occurred can be estimated by comparing the avian ovomucoids with mammalian protease inhibitors. For example, the dog submandibular protease inhibitor has been sequenced and was shown to consist of two polypeptide domains; the dog domain II (the carboxy terminal domain) shows close homology to the ovomucoid (carboxy terminal) domain III (Kato et al. 1978). Since all avian species examined to date contain ovomucoids with three domains, the duplication to form the newest domain (I) must have occurred after the divergence of birds from mammals (about 300 million years ago). The evolutionary ancestor to birds and mammals must, therefore, have contained an ancestral ovomucoid with two domains, and we can conclude that the intervening sequences must have existed in the ovomucoid gene well before 300 million and probably over 500 million years ago.

Up to this point our discussion has not included the signal peptide because it is not part of the secreted ovomucoid and is, of course, significantly different from the three polypeptide domains. Intron B separates the genomic DNA coding for the bulk of the secreted ovomucoid from the genomic domain that codes for the signal peptide and the first two amino acids of the secreted ovomucoid polypeptide. However, intron A separates the DNA coding for the first 19 amino acids of the signal peptide from the DNA coding for the remainder of the signal peptide and the first two amino acids of the secreted protein. If each exon corresponds to a functional or structural domain, then the first 19 amino acids of the ovomucoid signal peptide should comprise a domain that contains the information necessary for translocation, while the remainder of the signal peptide might comprise a domain that contains the peptidase specificity. In support of this prediction, the initial 14 amino acids that constitute the signal peptide sequence are encoded in exon I of the conalbumin gene (Cochet et al. 1979), as are the first 15 amino acids of the mouse immunoglobin light chain leader sequence (Bernard et al. 1978; Tonegawa et al. 1978). However, this does not appear to be the case for the rat preproinsulin genes (Cordell et al. 1979; Lomedico et al. 1979). In our evolutionary scheme, the genomic ovomucoid domain containing these first two exons must have been added to the 5' end of the primordial ovomucoid gene at the time of the development of the capacity for secretion of the ovomucoid polypeptide.

We can now formulate a theory for the evolution of the ovomucoid gene and its resultant protein structure, as shown in Fig. 6. The top line shows a single exon sequence, which codes for an ancestral protease inhibitor. This inhibitor was probably only moderately successful at its chosen task, the inhibition of proteases, because of digestion of the inhibitor by the protease. However, with the addition of another appropriate exon to the 3' end of the original exon (step 1), the primordial ovomucoid gene was created, which now coded for a stable inhibitor because of the addition of the strategically located third disulfide bond, as discussed earlier. This polypeptide corresponds to the present-day ovomucoid domain III. Step 2 represents the initial duplication of the primordial ovomucoid gene, perhaps by unequal

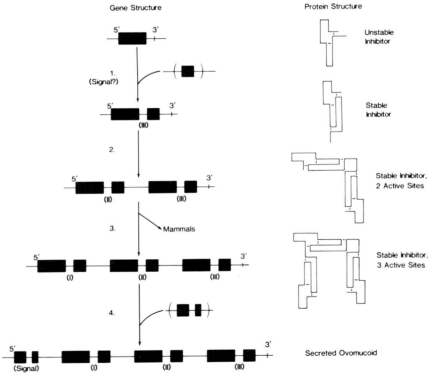

Fig. 6. The proposed evolution of the chicken ovomucoid gene. The column on the left represents the ovomucoid gene structure at various times during its evolution, and the protein structure for which it codes is shown in the column to the right. The 5' and 3' represent the transcription initiation and termination sites, respectively. The solid boxes represent exon sequences. Between exons, the thin lines represent intervening sequences; outside of the exons, the thin lines represent untranslated or flanking sequences.

crossing-over to form an ovomucoid gene coding for a stable inhibitor with two active sites, domains II and III. It was at this point in evolution, about 300 million years ago, that the avian and mammalian lineages diverged. Mammals have retained this two-domain protease inhibitor, while the avian gene evolved further. Step 3 represents the duplication of domain II, again most likely an unequal crossover event, to create a new domain, the present polypeptide domain I. The result was an ovomucoid gene that codes for the three-domain inhibitor that birds have today. Because all avian ovomucoids examined to date have this three-domain structure (Kato et al. 1978), the partial genomic duplication of step 3 must have occurred before the avian speciation, or 80–300 million years ago. Step 4 represents the addition of the exons coding for the signal peptide, which created an ovomucoid that could be secreted by the cells responsible for its synthesis (e.g., tubular gland cells of the oviduct). However, as noted in step 1, this addition could have occurred at any point in the evolution of the gene, and indeed must have occurred before the secretion of the protease inhibitor became a necessity.

VII. Introns and Eukaryotic Genes

In conclusion, our results on the structural organization of the ovomucoid gene have led us to formulate a unified theory of the evolution of the ovomucoid gene. The structure of this gene is consistent with the model proposed by Gilbert and ourselves (Gilbert 1978; Gilbert 1979; O'Malley et al. 1979) that exons may represent protein domains that were "marshaled" together during evolution. Furthermore, we have extended these ideas to make certain predictions concerning the evolution of complex eukaryotic genes in general. Intervening sequences apparently specify structural–functional domains of proteins. This is certainly the case for the ovomucoid, immunoglobin (Sakano et al. 1979), and insulin genes (Lomedicao et al. 1979), and apparently it is also true for the globin gene (Craik et al. 1980).

The origin and function of intervening sequences cannot be proved, but we can make some educated guesses now as to how and why they arose. We have deduced that the ovomucoid introns are very old; in fact, they probably date from the time when the genes were first formed. As shown in Fig. 6, introns were formed when two exon segments, which already had a built-in function, were brought together to form a primordial gene. The recombinational advantage that these intron sequences afforded is shown in more detail in Fig. 7. Column I shows the problem inherent in the assembly of complex genes from diverse exon segments: The recombination must occur at precise sites at either end of the exons. Otherwise, the reading frame of the resulting mRNA would be altered and the functional integrity of the resultant polypeptide possibly destroyed. Since the recombination process itself is random, a requirement for absolute precision in breakage (or excision) of the segments of DNA to be assembled makes the event an occurrence of low probability. Thus, the evolutionary process would be length-

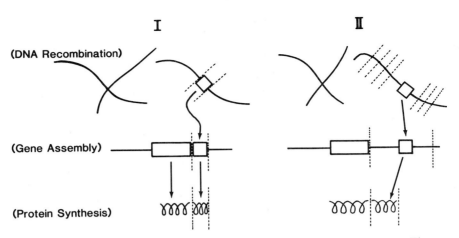

Fig. 7. An evolutionary hypothesis for the origin of intervening sequences. The open boxes represent coding regions of DNA (exons), while the thin solid lines represent either introns or flanking DNA sequences. Column I shows gene assembly from diverse exon elements without benefit of introns, and column II represents the alternative method of gene assembly in which the generation of introns is accounted for.

ened greatly. Column II depicts the same event, as we now propose it actually occurred, leading to the development of modern complex eukaryotic genes. Recombination, to bring together the diverse exon segments, could have occurred at any one of numerous sites in the flanking DNA sequence on either side of the exon sequences. This model would greatly facilitate genetic evolution, since the probability of occurrence would be orders of magnitude more. In this manner, eukaryotes would be afforded the capacity for greater genetic development per unit time. After recombination, an intervening sequence would have been created between the two exons. Provided that an early mechanism existed (or developed) for splicing these introns out of the primary RNA transcripts, this process would allow the rapid development of a diverse functional gene. Recent evidence from a number of laboratories has shown that such a splicing enzyme does exist and that introns do not provide a barrier to functional gene expression (Knapp et al. 1978; Blanchard et al. 1978; Kinniburgh and Ross 1979; Hamer and Leder 1979; Halbreich et al. 1980; Seif et al. 1979). Thus, our experimental data and the line of reasoning described earlier lead us to believe that intervening sequences are intimately related to the origin and evolution of complex eukaryotic genes.

Summary

Two-thirds of the natural chicken ovomucoid gene has been sequenced, including all exons and the intron sequences surrounding all 14 intron–exon junctions. The junction sequences surrounding four of the introns are redundant; however, the sequences surrounding the other three introns contain no redundancies and thus the splicing sites at either end of these three introns are unambiguous. The splicing in all cases conforms to the GT-AG rule. We compare the structural organization of the ovomucoid gene with the ovomucoid protein sequence to examine theories of the evolution of ovomucoids as well as the origin of intervening sequences. This analysis suggests that the present ovomucoid gene evolved from a primordial ovomucoid gene by two separate intragenic duplications. Furthermore, sequence analyses suggest that introns were present in the primordial ovomucoid gene before birds and mammals diverged, abut 300 million years ago. Finally, the positions of the introns within the ovomucoid gene support the theory that introns separate gene segments that code for functional domains of proteins and provide insight on the manner by which eukaryotic genes were constructed during the process of evolution.

References

Bell GI, Pictet RL, Rutter WJ, Cordell B, Tischer E, Goodman HM (1980) Nature 284: 26–33

Bernard O, Hozumi N, Tonegawa S (1978) Cell 15: 1133–1144

Blanchard JM, Weber J, Jelinek W, Darnell JE (1978) Proc Natl Acad Sci USA 75: 5344–5348

Blobel G, Dobberstein B (1975) J Cell Biol 67: 852–862

Brack C, Tonegawa S (1977) Proc Natl Acad Sci USA 74: 5652–5656

Breathnach R, Mandel JL, Chambon P (1977) Nature 270: 314–319

Catterall JR, Stein JP, Lai LC, Woo SLC, Dugaiczyk A, Mace ML, Means AR, O'Malley BW (1979) Nature 278: 323–327

Catterall JR, Stein JP, Kristo P, Means AR, O'Malley BW (1980) J Cell Biol 87: 480–487

Cochet M, Gannon F, Hen R, Maroteaux L, Perrin F, Chambon P (1979) Nature 282: 567–575

Cordell B, Bell G, Tischer E, DeNoto FM, Ullrich A, Pictet R, Rutter WJ, Goodman HM (1979) Cell 18: 533–543

Craik CS, Buckman SR, Beychok S (1980) Proc Natl Acad Sci USA 77: 1384–1388

Early P, Huang H, Davis M, Calame K, Hood L (1980) Cell 19: 981–992

Feeney RE (1971) In: Fritz H and Tschesche H (eds) Proceedings of the International Research Conference on Proteinase Inhibitors. De Gruyter, Berlin, pp 162–168

Feeney RE, Allison RG (1969) Inhibitors of Proteolytic Enzymes. Evolutionary Biochemistry of Proteins. Wiley-Interscience, New York

Fyrberg EA, Kindle KL, Davidson N, Sodja A (1980) Cell 19: 365–378

Garapin AC, Lepennec JP, Roskam W, Perrin F, Cami B, Krust A, Breathnach R, Chambon P, Kouvilsky P (1978) Nature 273: 349–354

Gilbert W (1978) Nature 271: 501

Gilbert W (1979) In: Axel R, Maniatis T, and Fox CF (eds.) ICN-UCLA Symposia on Molecular and Cellular Biology, 14. Academic Press, New York

Gorin MB, Tilgham SM (1980) Proc Natl Acad Sci USA 77: 1351–1355

Halbreich A, Pajot P, Foucher M, Grandchamp C, Slonimski P (1980) Cell 19: 321–329

Hamer DH, Leder P (1979) Cell 18: 1299–1302

Jeffreys AJ, Flavell RA (1977) Cell 12: 1097–1103

Kato I, Kohr WJ, Laskowski MJ (1978) In: Magnusson S, Ottesen M, Foltman B, Dano K, Neurath H (eds.) Proceedings of the 11th FEBS Meeting, 47. Pergamon Press, Oxford pp 197–206

Kinniburgh AJ, Ross J (1979) Cell 17: 915–921

Knapp G, Beckmann JS, Johnson PF, Fuhrman SA, Abelson J (1978) Cell 14: 221–236

Lai EC, Stein JP, Catterall JF, Woo SLC, Mace ML, Means AR, O'Malley BW (1979) Cell 18: 829–842

Lindenmaier W, Nguyen-Huu MC, Lurz R, Stratmann M, Blin N, Wurtz T, Hauser HJ, Sippel AE, Schutz G (1979) Proc Natl Acad Sci USA 76: 6196–6200

Lineweaver H, Murray CW (1947) J Biol Chem 171: 565–572

Lomedicao P, Rosenthal N, Efstratiatis A, Gilbert W, Kolodner R, Tizard R (1979) Cell 18: 545–558

Maki R, Traunecker A, Sakano H, Roader W, Tonegawa S (1980) Proc Natl Acad Sci USA 77: 2138–2142

Mandel JL, Breathnach R, Gerlinger P, LeMeur M, Gannon F, Chambon P (1978) Cell 14: 641–653

Maxam AM, Gilbert W (1977) Proc Natl Acad Sci USA 74: 506–564

McReynolds LA, Catterall JR, O'Malley BW (1977) Gene 2: 217–231

Monahan JJ, McReynolds LA, O'Malley BW (1976) J Biol Chem 251: 7355–7362

Nunberg JN, Kaufman RJ, Chang ACY, Cohen SN, Schimke RT (1980) Cell 19: 355–364

O'Malley BW, McGuire WL, Kohier PO, Korenman SG (1969) Recent Prog Horm Res 25: 105–160

O'Malley BW, Roop DR, Lai EC, Nordstrom JL, Catterall JF, Swaneck GE, Colbert DA, Tsai MJ, Dugaiczyk A, Woo SLC (1979) Recent Prog Horm Res 35: 1

Rosen JM, Woo SLC, Molder JW, Means AR, O'Malley BW (1975) Biochemistry 14: 69–78

Sakano H, Rogers JH, Huppi K, Brack C, Traunecker A, Maki R, Wall R, Tonegawa S (1979) Nature 277: 627–633

Seif I, Khoury G, Dhar R (1979) Nucl Acids Res 6: 3387–3398

Stein JP, Catterall JF, Woo SLC, Means AR, O'Malley BW (1978) Biochemistry 17: 5763–4772

Stein JP, Catterall JF, Kristo P, Means AR, O'Malley BW (1980) Cell 21: 681–687

Tilghman SM, Tiemeier DC, Seidman JF, Peterlin BM, Sullivan M, Maizel JV, Leder P (1978) Proc Natl Acad Sci USA 75: 724–729

Tonegawa S, Maxam AM, Tizard R, Bernard O, Gilbert W (1978) Proc Natl Acad Sci USA 75: 1485–1589

Valenzuela P, Venegas A, Weinberg F, Bishop R, Rutter WJ (1978) Proc Natl Acad Sci USA 75: 190–194

Weinstock R, Sweet R, Weiss M, Cedar H, Axel R (1978) Proc Natl Acad Sci USA 75: 1299–1303

Woo SLC, Dugaiczyk A, Tsai MJ, Lai EC, Catterall JR, O'Malley BW (1978) Proc Natl Acad Sci USA 75: 3688–3692

Woo SLC, Beattie WG, Catterall JR, Dugaiczyk A, Staden R, Brownlee GG, O'Malley BW (1981) Biochemistry 20: 6437–6446.

Discussion of the Paper Presented by B. W. O'Malley

ROY: Your theory that the hormone first causes differentiation and then induction is based on the withdrawal, right? What is the longest time of withdrawal you have used?

O'MALLEY: We have used three weeks withdrawn.

ROY: Would you have expected the same result if the withdrawal period was longer, say months or even a year? The reason I am asking this is that instead of causing differentiation, the hormone may be acting in a two-step process, the initial part of which is loosening of the chromatin structure and the second phase of which is mRNA induction. If you wait long enough is it possible that the chromatin will fold back to its original configuration?

O'MALLEY: I wouldn't have any data that would specifically exclude your point; however, one must keep in mind that the shutdown of transcription occurs within a day and one-half after withdrawal (total shutdown), although most of it is shut down within about 12 hours. After 2 weeks the DNA is still in the "open" conformation in chromatin.

BAXTER: Bert, the finding of the repetitive DNA as the junctions of the DNase sensitivity is quite exciting. My first question: Is that about 10^5 abundancy?

O'MALLEY: This is a middle repetitive DNA. So there are some thousands of copies. As you are probably aware, it is difficult to accurately access the exact number of copies of middle repetitive DNA because of cross-reaction within the families. All I can say is that there are probably more than 1000 copies but there are definitely much less than 100,000 copies.

BAXTER: Do you know that the DNA in that region is expressed?

O'MALLEY: The repetitive DNA in this region appears not be expressed. However, in total cell RNA, there is some hybridization with the CR-1 DNA. Some of these sequences may be transcribed in the cell.

BAXTER: How do you know that the DNA is not expressed?

O'MALLEY: Only because we have the whole clone containing those sequences. So the sequences bordering the 200 base repeat region are not expressed.

BAXTER: But is it still possible that those sequences themselves are expressed.

O'MALLEY: Yes, it is still possible; however, most of the RNA we see after filter transfer is not that small.

BAXTER: Could you speculate for a moment? Because there are a number of copies of these sequences in the DNA, one wonders if this is really something unique for ovalbumin gene expression, or does it reflect something about the structure in the junction that permits these types of sequences to come in? Is the presence of these sequences responsible for differential ovalbumin gene expression?

O'MALLEY: That these repetitive sequences are related to differentiation is a viable hypothesis. At present, none of these considerations can be excluded.

BAXTER: In looking at the expression of the DNA, did you ever just do straight Northern blots with that DNA as a probe?

O'MALLEY: The Northern shows some hybridizable RNA and of course Southerns confirm that there are many, many sequences in the DNA.

O'CONNOR: Is it reasonable if you are talking about differentiation or better yet determination to expect to see the appearance during developmental time of rearrangements involving terminal repeats in hormonally responsive tissue?

O'MALLEY: At the present time, that is a question we must answer and we intend to carry out experiments related to your question. At least two possibilities exist— rearrangement during evolution or rearrangement during chick development.

COFFEY: First, I'd like to congratulate you, Bert; that is beautiful work, as usual. First of all, Dr. Small and Dr. Vogelstein have found similar (middle) repetitive sequences at the base of the loops attached to the matrix you are trying to study, the type of DNA that attaches it down to each of those 100-kb pairs. It would seem that it might be that you have activated the whole loop in this process. Is this sort of what you are implying?

O'MALLEY: All I can say right now is that there is middle repetitive DNA in the matrix, and I think within the next month or two we ought to be able to answer your question, whether it's the Cr-1 subfamily that is enriched in the matrix preparation.

COFFEY: The second thing, some people are looking for enzymes still associated with this 7–8% of the proteins here in the matrix. It is not surprising that a lot of them would get washed off during the process, but some of the alpha-DNA polymerase, of considerable amounts, left on there. Have you looked to see if any of the enzymes involved in processing are part of this system?

O'MALLEY: That is exactly what we are going to do now.

GOLDBERGER: Bert, have you tried to test the DNase sensitivity of the CR sequences themselves or is it too close to count?

O'MALLEY: We have not been able to test only the CR-1 sequences because of technical considerations.

Discussants: A. K. ROY, B. W. O'MALLEY, J. D. BAXTER, J. D. O'CONNOR, D. S. COFFEY, and R. F. GOLDBERGER

Chapter 2

Effects of Steroid Hormones on Chicken Oviduct Chromatin

JOHN N. ANDERSON, JEFF N. VANDERBILT,
KERRY S. BLOOM, AND BONNIE J. GERMAIN

I. Introduction

Our current knowledge of the mechanisms by which steroid hormones regulate transcription has been derived primarily from the analysis of defined gene products in a few well-characterized steroid-responsive systems. One such model system is the chicken oviduct, which synthesizes egg white proteins in response to a variety of steroid hormones (reviewed in Palmiter, 1975; Schimke et al., 1975; O'Malley et al., 1979). The hormonal regulation of ovalbumin, the major egg white protein, has been extensively studied, since steroid hormones induce the synthesis of this protein by stimulating the transcription of the ovalbumin gene in oviduct chromatin. A complete understanding of the mechanisms by which steroid hormones modulate gene expression in this system, therefore, requires a description of the chromatin environment of the ovalbumin gene during the hormone induction process.

Microscopic studies have indicated that the folding of DNA along the chromatin fiber represents an important mechanism in the control of differential gene activity. Condensation of specific chromosomal regions, whole chromosomes, or entire chromosomal complements during the differentiation of certain cells is commonly viewed as a mechanism for the inactivation of genetic expression (Littau et al., 1964; Pelling, 1964; Comings, 1966). The generation of extended chromatin fibers from more compacted regions also occurs shortly before (Foe, 1978; Pruitt and Grainger, 1981) or at the time of (McKnight et al., 1978) transcriptional initiation. The observations that the puffing patterns of dipteran polytene chromosomes are correlated with the appearance of specific transcriptional products during normal development or following hormone or heat-shock treatment (Clever, 1966; Ashburner and Bonner, 1979) provide additional support for the proposal that transcription may be controlled, at least in part, by dynamic changes in the architecture of the chromatin.

Although there is no cytological or electron microscopic evidence relating the structure of the ovalbumin gene in chromatin to its transcriptional activity, nuclease probes can be used to explore certain features of the chromatin conformation of the ovalbumin gene. In the first section of this paper, nucleases were employed to provide a description of the ovalbumin chromatin in the oviduct and to elucidate the changes in this organization that accompany ovalbumin gene activation by steroid hormones. Since the association of proteins with specific nuclear loci may be responsible for chromatin structure changes and gene activation, we have prepared a library of monoclonal antibodies against proteins associated with hen oviduct chromatin. In the second section of this review, these antibodies were used as probes to study several individual proteins associated with chick oviduct chromatin during hormone stimulation and withdrawal. The results presented here demonstrate that steroid hormones induce a variety of very specific changes in the structure and composition of oviduct chromatin.

II. Nucleases as Probes for the Ovalbumin Gene in Oviduct Chromatin

A. Structure of the Ovalbumin Chromatin at the Nucleosome Level

The extraordinary length of DNA in eukaryotic cells is compacted within the interphase nucleus in such a manner as to maintain its functional capacity. The packaging of the DNA into nucleosomes represents the primary level of compaction in which the DNA is periodically wrapped around histone core particles to create the well-known "beads on a string" conformation that can be visualized in electron micrographs (reviewed in Kornberg 1977; Chambon 1978; McGhee and Felsenfeld 1980). The nucleosome core contains 140 base pairs (bp) of DNA complexed with two each of the four major histones (H2A, H2B, H3, and H4), while the linker DNA between adjacent core particles contains 15–100 bp of DNA that is probably associated with histone H1. A consequence of the histone core–linker organization of DNA in chromatin is that the linker DNA is more accessible to attack by micrococcal nuclease than is the DNA in the nucleosomal core particles. The sequential breakdown of chromatin DNA by micrococcal nuclease is shown in Fig. 1. Mild digestion of chromatin with this enzyme generates a series of nucleoprotein particles containing DNA fragments whose molecular weights are multiples of the basic nucleosomal unit (Fig. 1, lanes C–E). As digestion proceeds, the large fragments are broken down into smaller ones until more than 80% of the chromatin-containing DNA is converted into mononucleosomal core particles containing 140 bp of DNA. At the digestion limit, the mononucleosomes are broken down into subnucleosomal particles containing DNA fragments ranging in size from 140 bp down to 40 bp.

Historically, histones have been viewed as inactivators of genetic expression since their addition to nuclei or DNA in vitro suppresses RNA synthesis (Bonner and Huang 1963). Transcriptionally active genes may therefore reside in nucleosome-free stretches of DNA in chromatin. Previous reports, however, have shown that the concentrations of transcribing ovalbumin

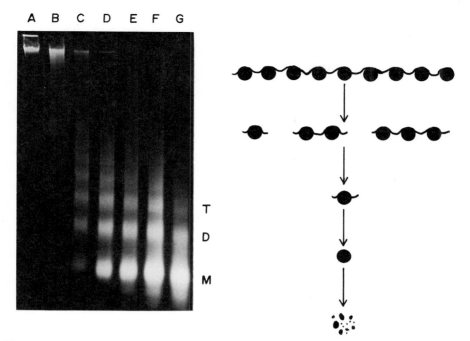

Fig. 1. Fragmentation of nuclear DNA by micrococcal nuclease. Hen oviduct nuclei, purified as described by Bloom and Anderson (1978a), were incubated for 5 min at 37°C with increasing amounts of micrococcal nuclease. Purified DNA from these nuclei was electrophoresed on a 1.8% agarose gel slab and stained with ethidium bromide. Gel lanes A–G correspond to 0, 2, 30, 75, 100, 200, and 350 units of enzyme/ml of nuclear suspension, respectively. Mononucleosome (M), dinucleosome (D), and trinucleosome (T) bands are indicated. On the right is a schematic of the breakdown of chromatin with micrococcal nuclease.

genes, globin genes and genes coding for poly(A) mRNA in rat liver remain unchanged when greater than 80% of the total nuclear DNA is converted to mononucleosomal length fragments by extensive digestion of nuclei with micrococcal nuclease (Garel and Axel 1976; Weintraub and Groudine 1976; Bloom and Anderson 1978b). It was generally concluded from these observations that active genes are complexed with histones that protect these sequences from extensive micrococcal nuclease digestion. These studies, however, do not necessarily prove that active genes are associated with nucleosomes containing 140–160 bp of DNA, since the length of the active sequences following extensive nuclease digestion may be different from the length of the bulk of the inactive DNA. We have therefore attempted to provide more direct evidence for the nucleosomal packaging of the transcriptionally active ovalbumin gene in the chicken oviduct. First, nuclei from hen oviduct and liver were digested with micrococcal nuclease until more than 85% of the DNA was converted to fragments of less than 200 bp. The DNA was denatured and separated into monomer length (140–160 bases) and submonomer length (less than 140 bases) fragments by electrophoresis on alkaline agarose gels. Titration of these DNAs with the ovalbumin cDNA (Fig. 2) revealed that the concentration of the ovalbumin gene in monomer and submonomer length DNA fragments from both oviduct and liver was the same as the concentration of this sequence in nondigested oviduct DNA. To further study the length of the ovalbumin DNA protected from digestion by micrococcal nuclease, DNA fragments produced by extensive micrococcal nuclease digestion of nuclei from ovalbumin-producing and nonproducing tissues were hybridized to full-length ovalbumin cDNA. The resulting hybrids were treated with S1 nuclease, electrophoresed on a polyacrylamide gel, and detected autoradiographically (Fig. 3). Visual inspection of the autoradiogram reveals that the length of the protected monomeric and submonomeric fragments is the same regardless of whether the ovalbumin gene is in the transcriptionally active (Fig. 3; lanes A, C, and D) or inactive (Fig. 3, lanes B and E) state. The results presented in Figs. 2 and 3 suggest, therefore, that the mere presence or absence of nucleosomal core particles

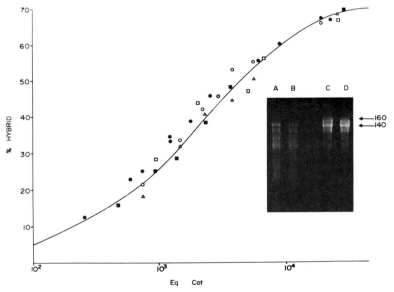

Fig. 2. Concentration of the ovalbumin gene in mononucleosomal and submononucleosomal length DNA fragments following extensive micrococcal nuclease digestion. Hen oviduct and liver nuclei were incubated with micrococcal nuclease (450 U/ml, 15 min) until about 90% of the DNA was less than 200 nucleotides in length. The DNA was denatured and separated into mononucleosome length (140–200 nucleotides) and subnucleosome length (<140 nucleotides) fragments by preparative electrophoresis on a 1.8% alkaline agarose gel. Nondigested oviduct DNA ($\triangle$), mononucleosomal length DNA from oviduct ($\bullet$) and liver ($\bigcirc$), and subnucleosomal length DNA from oviduct ($\blacksquare$) and liver ($\square$) were annealed to [^{32}P]ovalbumin cDNA for quantitation of ovalbumin gene sequences. The insert shows samples of the mononucleosomal length (C and D) and subnucleosomal length (A and B) DNA from oviduct (A and C) and liver (B and D) electrophoresed on a denaturing 7 M urea, 10% polyacrylamide slab gel. [Data from Bloom and Anderson (1981).]

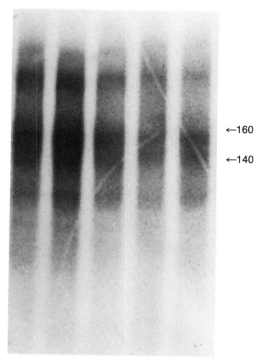

Fig. 3. Length of the ovalbumin coding DNA that is resistant to extensive micrococcal nuclease digestion. Nuclei from hen oviduct (A), hen liver (B), primary DES-stimulated chick oviducts (C), estrogen-withdrawn oviducts (D), and secondary estrogen-stimulated oviducts (E) were incubated with micrococcal nuclease (450 U/ml; 10 min) until about 85% of the DNA was less than 200 nucleotides in length. The purified DNA was annealed to full-length [^{32}P]ovalbumin cDNA for 8 h at 68°C and the mixture treated with S-1 nuclease to destroy single-stranded sequences. The samples were then electrophoresed on an 8% native polyacrylamide gel and visualized by autoradiography. [Data from Bloom and Anderson (1981).]

along the transcriptional unit of the ovalbumin gene in chromatin is not the determining factor in regulating the transcription of this gene.

Although the length of the ovalbumin-coding DNA within the nucleosomal core is apparently constant regardless of the transcriptional status of the ovalbumin gene (Figs. 2 and 3), variations in the distance between nucleosomal core particles could be related to the gene's transcriptional activity. To examine this possibility, nuclei from ovalbumin-producing and nonproducing tissues were digested with increasing amounts of micrococcal nuclease. The DNA fragments obtained were separated according to size by electrophoresis and stained (Fig. 4, top). The DNA was then transferred to nitrocellulose filters, and the nucleosomal spacing pattern of the ovalbumin-and globin-coding DNA was assessed by hybridization of the filter-bound DNA

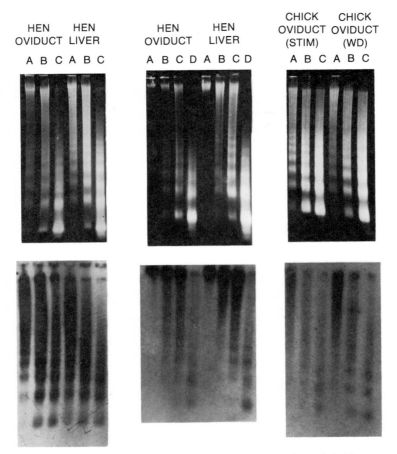

Fig. 4. Nucleosomal spacing pattern of the ovalbumin and globin genes. Nuclei from the indicated tissues were incubated with increasing amounts (from A to C, left and right or A to D, middle) of micrococcal nuclease (10–150 U/ml) for 5 min at 37°C. Top: DNA fragments from the digests were electrophoresed on native 1.8% agarose gels and stained with ethidium bromide. Bottom: The DNA in the gels shown above was blotted onto nitrocellulose filters by the method of Southern (1975) and hybridized to [^{32}P]globin cDNA (left panel) or [^{32}P]ovalbumin cDNA (middle and right panels); the globin or ovalbumin gene sequences were then visualized by autoradiography. The lower-molecular-weight nucleosomal fragments were reduced on the autoradiographs because DNA fragments less than 500 bp are retained very poorly on nitrocellulose (Southern 1975 and Wu et al. 1979). [Data from Bloom and Anderson (1981).]

with labeled ovalbumin and globin cDNA (Fig. 4, bottom). The transcriptionally inactive globin genes in the oviduct and liver and the quiescent ovalbumin genes in liver and estrogen-withdrawn oviducts exhibited nucleosomal repeat patterns that were similar to each other and similar to the bulk of the stained chromatin DNA. The repeat length of the active ovalbumin gene in hen and estrogen-stimulated oviducts also appeared similar to the bulk of the chromatin. However, Fig. 4 shows that the micrococcal nuclease pattern of the ovalbumin-coding DNA in hen and estrogen-stimulated chick oviduct differs from that of bulk chromatin and inactive genes in that the hybridization pattern is smeared and the oligonucleosomal bands are less distinct. The blurring of the nuclease digestion patterns, which is especially apparent in the more minimal digestions, has been consistently observed in eight experiments similar to those described in Fig. 4. Partial loss of resolution of oligonucleosomal length fragments has also been observed upon the transcriptional activation of the heat-shock genes of *Drosophila* (Wu et al. 1979) and the active rRNA-coding sequence of *Physarum* (Stalder et al. 1979) and may therefore represent a general feature of genes that are transcribed at relatively rapid rates.

B. Selective Excision of Nucleosomes Containing Transcriptionally Active Genes by Micrococcal Nuclease

During mild digestion of chromatin with micrococcal nuclease, cleavage of the DNA fiber connecting adjacent nucleosomes produces DNA molecules whose lengths are multiples of the 200-bp nucleosomal unit (see Fig. 1). A description of the fate of ovalbumin chromatin during mild digestion requires quantitation of ovalbumin gene sequences in these different-sized DNA fragments. The method outlined in Fig. 4 is not suitable for this analysis since mono- and dinucleosomal length DNA fragments bind very poorly to nitrocellulose filters (Southern 1975; Wu et al. 1979). We have therefore employed an alternative procedure for quantifying ovalbumin gene sequences in minimally digested hen oviduct DNA. Oviduct nuclei were incubated with micrococcal nuclease (75 U/ml; 1.5 min) until 5% of the DNA was converted to mononucleosomal length fragments. The DNA extracted from these nuclei was separated into four size classes by preparative agarose gel electrophoresis. Samples of the resulting DNA size classes are shown on the polyacrylamide gel in Fig. 5. The DNA in each size class was hybridized to trace amounts of ovalbumin or globin cDNAs for quantitation of ovalbumin or globin gene sequences (Fig. 6). The data in Fig. 6 are summarized in Fig. 7, which shows the relative concentration of ovalbumin or globin gene sequences in the four DNA size classes as compared with their respective concentrations in undigested DNA. The concentration of the ovalbumin-coding sequence in mononucleosomal length DNA (size class 1) was about 5-fold greater than in undigested DNA and 12-fold greater than in size class 4, which contains DNA fragments > 1300 bp (Fig. 6, top and Fig. 7). These results show that the ovalbumin gene in hen oviduct chromatin is in a conformation that is selectively cleaved by micrococcal nuclease as compared to the bulk of the oviduct DNA. This conformation is related to transcription, since the globin genes in the oviduct (Fig. 6, bottom and Fig. 7) and the ovalbumin genes in liver (Fig. 7), which are not engaged in RNA synthesis, are not preferentially attacked by this nuclease. The selective cleavage of the ovalbumin gene by micrococcal nuclease was not observed, however, when deproteinized hen oviduct DNA was used as a substrate (Fig. 7).

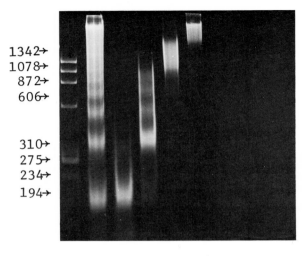

1342→
1078→
872→
606→

310→
275→
234→
194→

ϕX174 U 1 2   3 4

Fig. 5. Electrophoretic analysis of the four DNA size classes on a 3% polyacrylamide gel. Oviduct nuclei were digested with micrococcal nuclease (75 U/ml; 1.5 min) until about 5% of the DNA was converted to mononucleosomal length fragments. DNA was extracted from the digested nuclei and separated into four molecular-weight classes by electrophoresis on a preparative native agarose gel. Lane U, DNA fragments from the digested nuclei prior to separation by preparative gel electrophoresis; Lanes 1–4, the four DNA size classes in order of increasing molecular weight; Lane ϕX174, DNA fragments from an *Hae* III digest of bacteriophage ϕX174. The fragment lengths in base pairs are indicated in the figure. [See Bloom and Anderson (1978b) for additional details.]

To study further the nuclease sensitivity of the ovalbumin gene in the hen oviduct, similar studies were performed at earlier times in the digestion when less of the nuclear DNA had been converted into mononucleosome size fragments. Figure 8 shows the concentration of the ovalbumin gene in the four DNA size classes obtained from hen oviduct nuclei that were digested with micrococcal nuclease until 1, 2, 3, 4, and 5% of the DNA was monomer length. Enrichment of ovalbumin gene sequences in mononucleosomal length DNA (size class 1) was greatest when the quantity of monomeric DNA was at a minimum. When 1% of the DNA had been converted to monomer length fragments, the concentration of the ovalbumin gene in this size class was 15–20-fold greater than that in undigested DNA and 30–50-fold greater than that in the fraction containing the largest DNA fragments. As the amount of DNA in the monomer size class increased with increasing digestion, there was a progressive decrease in the concentration of ovalbumin sequences in this fraction. When 5% of the DNA was converted to mononucleosomal length fragments, this DNA was enriched only four- to five-fold in the ovalbumin sequence (Figs. 6–8). It is therefore apparent that maximal nuclease sensitivity of the ovalbumin chromatin to micro-

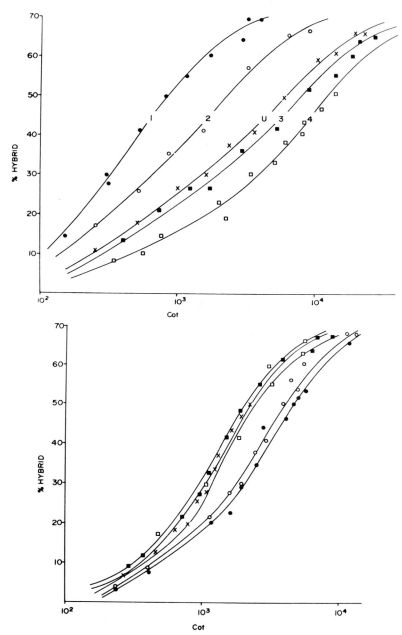

Fig. 6. Titration of ovalbumin and globin sequences in the four DNA size classes. Hen oviduct nuclei were digested with micrococcal nuclease and the fragmented DNA separated into four size classes as described in Fig. 5. The two larger-molecular-weight classes of DNA were sonicated and the DNA from all size classes was then subjected to base hydrolysis (0.3 N NaOH, 37°C for 18 h) to remove RNA. DNA (0.06–8 mg/ml) from each size class was annealed to [^{32}P]ovalbumin (top panel) or [^{32}P]globin (bottom panel) for 24 h. The size classes 1 (●——●), 2 (○——○), 3 (■——■), and 4 (□——□) contained 5, 11, 23, and 61% of the DNA, respectively. Hybridization of the cDNAs to unfractionated oviduct DNA (x——x) is shown for comparison.

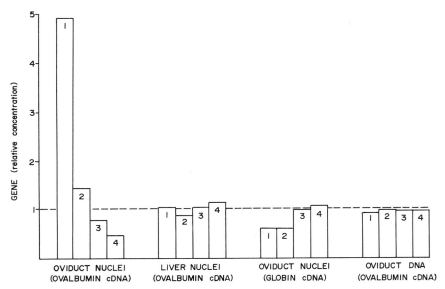

Fig. 7. Relative concentration of ovalbumin and globin genes in the four DNA size classes. DNA from the four DNA size classes prepared from hen oviduct or liver nuclei or from naked oviduct DNA was annealed to ovalbumin or globin cDNAs. In this and subsequent figures the number within each bar refers to the DNA size class under analysis. The value 1 on the ordinate corresponds to the concentration of ovalbumin or globin genes in undigested DNA. A bar that terminates above the value 1 is enriched in the gene sequence as compared to undigested DNA, whereas a bar terminating below this value is depleted in the gene under analysis. In this figure and in Figs. 8–11, 13, and 15, the relative gene concentrations were determined by DNA sequence excess analysis (see Fig. 6), moderate cDNA excess analysis (see Bloom and Anderson 1978b), and/or a standard curve analysis (see Anderson and Schimke 1976). The values shown in this figure were derived from both DNA excess (Fig. 6) and moderate cDNA excess reactions. See Bloom and Anderson (1978b) for additional details.

coccal nuclease occurred very early in the digestion when a small amount of the chromatin was converted to mononucleosomes.

The regulation of ovalbumin mRNA synthesis by steroid hormones in the immature chick oviduct represents a developmental system that is well suited for the study of the relationships between chromatin structure and specific gene activity (reviewed in O'Malley et al. 1979). Daily administration of estrogen to immature chicks results in the spectacular promotion of oviduct growth and the cytodifferentiation of tubular gland cells that synthesize ovalbumin and the other major egg white proteins. After 2–3 weeks of treatment, more than 60% of the mRNA within each gland cell codes for ovalbumin. On estrogen withdrawal, tubular gland cells persist for a time but the synthesis of ovalbumin, the level of ovalbumin mRNA, and the synthesis of ovalbumin mRNA sequences decline rapidly. Restimulation of oviducts with estrogen or progesterone (secondary stimulation) results in a rapid accumulation of ovalbumin mRNA in preexisting tubular gland cells that is

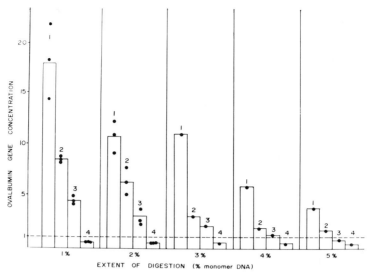

EXTENT OF DIGESTION (% monomer DNA)

Fig. 8. Quantitation of the ovalbumin gene sequences in different-sized DNA fragments with increasing micrococcal nuclease digestion. Hen oviduct nuclei were digested with micrococcal nuclease (5–50 U/ml; 1.5–10 min) until about 1, 2, 3, 4, or 5% of the DNA was converted to mononucleosomal length fragments (abscissa). These percentages were estimated by optical density scans of negatives of 1.8% agarose gels. Following digestion, the DNA was separated into four molecular-weight size classes and the concentration of the ovalbumin gene in each size class determined by standard curve analysis. The data represent the results from nine separate experiments with two to three hen oviducts/experiment. [Data from Bloom and Anderson (1981).]

coincident with the induction of ovalbumin synthesis and the synthesis of ovalbumin-specific mRNA sequences. As shown in Fig. 9, the decrease in ovalbumin gene expression following estrogen withdrawal is accompanied by a change in the ovalbumin chromatin. In these studies, chicks were injected with estradiol (1 mg/day) for 14 days and then withdrawn from the hormone for the times indicated. The concentration of ovalbumin gene sequences in the four DNA size classes generated by the nuclease digestion of oviduct nuclei was determined as described in Fig. 7. Oviduct nuclear estrogen receptors were quantified by the [^{3}H]estradiol exchange assay (Anderson et al. 1972). At 3 days after estradiol withdrawal, there was a 70% reduction in the nuclease sensitivity of ovalbumin chromatin as compared to the stimulated oviducts, and by 6 days the concentration of ovalbumin-coding sequences in the four DNA size classes was similar to total undigested DNA. The progressive decline in the ability of micrococcal nuclease to cleave selectively the ovalbumin sequences after estrogen withdrawal was paralleled by a decrease in the concentration of estrogen receptors in oviduct nuclei. The level of nuclear estrogen receptor sites decreased from about 4000 sites/nucleus in the stimulated oviducts to about 300 sites/nucleus in oviducts obtained from chicks withdrawn from the hormone for 6 days.

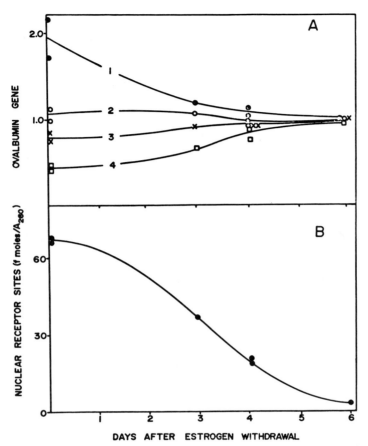

DAYS AFTER ESTROGEN WITHDRAWAL

Fig. 9. Effects of estradiol withdrawal on the ovalbumin gene in different-sized DNA fragments and nuclear estrogen receptor. Immature chicks were injected with 1 mg/day of estradiol for 14 days and then withdrawn from the hormone for the indicated days. (A) Oviduct nuclei were digested with micrococcal nuclease and the purified DNA was separated into four size classes as described in Fig. 5. Size class 1 contained 9–11% of the total digested DNA in the stimulated oviduct. The concentration of the ovalbumin gene in size classes 1–4 was determined by both moderate cDNA excess and standard curve analysis. The value 1 on the ordinate corresponds to the concentration of ovalbumin gene in undigested DNA. The data represent the results from six separate experiments with 20–30 oviducts/experiment. (B) Nuclear estrogen receptors were quantified by the [^{3}H]estradiol exchange assay (Anderson et al. 1972) in aliquots of the oviduct nuclear suspensions. The binding capacities shown were determined by Scatchard plots (not shown). [Details in Bloom and Anderson (1979).]

Additional support for the dynamic relationship between hormone-induced ovalbumin gene expression and ovalbumin chromatin structure is provided by the studies shown in Fig. 10. Oviduct nuclei obtained from DES-(diethylstilbestrol) stimulated chicks, chicks withdrawn from DES, or withdrawn chicks injected with a single dose of DES or progesterone and killed 12 h later were incubated with micrococcal nuclease. The concentration of the ovalbumin gene sequences in the four DNA size classes was then determined. The ovalbumin chromatin was preferentially cleaved by micrococcal nuclease in the estrogen-stimulated but not in the withdrawn oviducts. This effect was readily reversible, since readministration of estrogen or progesterone to withdrawn chicks induced a reorganization of the ovalbumin chromatin into a micrococcal nuclease-sensitive conformation.

The experiments summarized in Figs. 9 and 10 show that the micrococcal nuclease-sensitive conformation of the ovalbumin chromatin is dependent on steroid hormones and suggest that this conformation is dynamically related to the expression of this gene. It is possible, therefore, that selective nuclease recognition of the ovalbumin chromatin is merely a consequence of the active transcription of the ovalbumin gene. This possibility is unlikely, however, since ribonucleotide triphosphates, which are required for the synthesis of ovalbumin mRNA sequences in isolated oviduct nuclei (Swaneck et

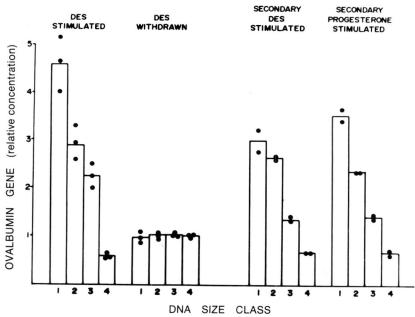

Fig. 10. Effect of steroid hormones on the distribution of the ovalbumin gene in different-sized DNA fragments. Four-day-old female chicks were injected with 2.5 mg/day of DES for 14 days (DES stimulated). Chicks withdrawn from the hormone for 4–5 days (DES withdrawn) received a single injection (4 mg) of DES (2° DES stimulated) or progesterone (2° progesterone stimulated) and were killed 12 h later. The ovalbumin mRNA represented 0.5, 0.003, 0.5, and 0.26% of the total RNA in the oviducts from these respective groups. Oviduct nuclei were incubated with micrococcal nuclease (50 U/ml; 1 min), DNA was separated into four size classes, and the concentration of ovalbumin gene determined by standard curve analysis. In all groups the size classes 1, 2, 3, and 4 contained 2–3, 2–3, 4–6, and 87–91% of the DNA, respectively. The data represent the results from 10 separate experiments with 15–19 oviducts/experiment. [From Bloom and Anderson (1981).]

al. 1979), are not present in the digestion buffer used in our studies. Digestion of hen oviduct nuclei in the presence of ribonucleotide triphosphates also did not alter the sensitivity of the ovalbumin chromatin to micrococcal nuclease (data not shown). In addition, treatment of hen oviduct nuclei with α-amanitin or actinomycin D prior to and during micrococcal nuclease digestion was without effect on the selective cleavage of the ovalbumin chromatin (Fig. 11). The concentrations of α-amanitin and actinomycin D used in these studies are greater than those required to inhibit completely the synthesis of ovalbumin mRNA sequences in isolated hen oviduct nuclei (Swaneck et al. 1979). Thus, the micrococcal nuclease-sensitive property of the ovalbumin chromatin is apparently not a direct consequence of active transcription per se. Treatment of hen oviduct nuclei with pancreatic RNase prior to nuclease digestion (Fig. 11) was also without significant effect on the selective cleavage of the ovalbumin gene in oviduct chromatin. Since the hormone-dependent conformation of the ovalbumin chromatin can be maintained in the absence of the transcriptional event, it is possible that the nuclease-sensitive conformation is required and perhaps even responsible for the transcriptional activation of the ovalbumin gene by steroid hormones.

The property of transcribing ovalbumin chromatin that is recognized preferentially by micrococcal nuclease is not yet known. Because the ovalbumin DNA within the nucleosomal core is apparently not selectively attacked by micrococcal nuclease (Figs. 2 and 3), the blurring of the higher-order oligonucleosomal bands derived from transcribing ovalbumin chromatin (Fig. 4), as well as the preferential nuclease cleavage of the internucleosomal DNA of transcribing sequences (Figs. 6–10), may result from an expansion of the nuclease cutting sites in the internucleosomal linker region upon transcriptional activation. This expansion could, in turn, result from a number of factors, including a reduction in the level of histone I in the transcribing linker regions, the binding of nonhistone proteins to these regions and/or to some higher-order conformational change that could confer onto the linker regions an elevated nuclease sensitivity. Which, if any, of these changes is responsible for the selective recognition of the ovalbumin chromatin by micrococcal nuclease is an important problem for future investigation.

C. Action of Endogenous Nuclease on Transcriptionally Active Ovalbumin Chromatin

The nuclei of eukaryotic cells contain nuclease(s) that attack the chromatin DNA during nuclear incubation in vitro (reviewed in Burgoyne and Hewish 1978). The major endogenous nuclease associated with nuclei is similar to micrococcal nuclease in that it attacks linker DNA between nucleosomes producing the typical 200-bp periodicity early in the digestion. At later incubation times, the nucleosome core is cleaved by this nuclease to produce a 10–11 base periodicity that is characteristic of a DNase I digestion. Figure 12 shows the digestion patterns produced in the DNA during the autodigestion of nuclei from a variety of avian and mammalian sources. To determine if the

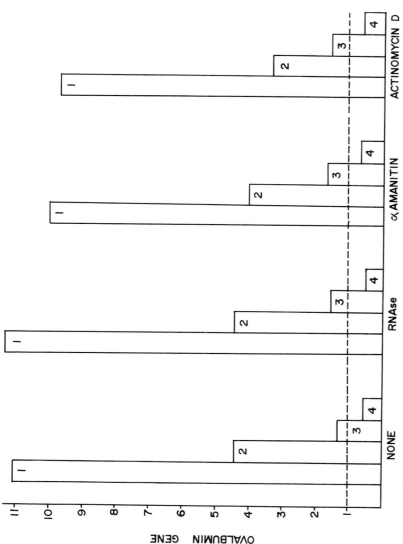

Fig. 11. Effects of α-amanitin, actinomycin D, and ribonuclease on the distribution of the ovalbumin genes in different-sized DNA fragments. Hen oviduct nuclei were incubated at 4°C for 20 min in digestion buffer alone (none) or in digestion buffer containing 5 μg/ml α-amanitin, 40 μg/ml actinomycin D, or 100 μg/ml pancreatic RNase. The nuclei were transferred to a 37°C water bath for an additional 10 min and digested by addition of micrococcal nuclease. The concentration of the ovalbumin gene in the four size classes were determined by DNA sequence excess analysis. About 20% of the total DNA was in the smallest-molecular-weight size class. [Data from Bloom and Anderson (1981).]

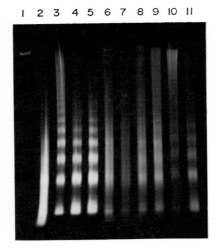

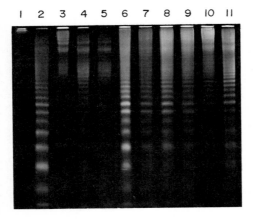

Fig. 12. Electrophoretic analysis of the DNA fragments produced by the endogenous nuclease. DNA fragments produced by the action of endogenous and exogenous nucleases are displayed on a native 1.8% agarose gel (top) and a denaturing 7 M urea–10% polyacrylamide gel (bottom). (Lane 1), undigested oviduct DNA incubated in standard digestion buffer for 6 h. (Lane 2), DNA fragments from rat liver nuclei digested with DNase I (20 μg/ml) for 5 min at 37°C. (Lane 3), DNA fragments from rat liver nuclei digested with micrococcal nuclease (75 U/ml) for 5 min at 37°C. Lanes 4–11 represent the DNA fragments from adult rat liver, 5-day-old rat liver, hen liver, hen oviduct, primary estrogen-stimulated oviduct, 4-day estrogen-withdrawn oviduct, mature erythrocyte, and immature erythrocyte nuclei incubated for 1, 1, 1, 5, 3, 3, 5, and 5 h, respectively. [Data from Vanderbilt et al. (1981).]

endogenous nuclease(s) recognizes the transcribing ovalbumin chromatin, isolated nuclei obtained from hen, estrogen-stimulated chick, and withdrawn chick oviducts were incubated for 1 h in vitro. During this time period about 3% of the nuclear DNA was converted to mononucleosome length fragments in all groups by the action of endogenous nuclease. The resulting fragmented DNA was separated into four size classes and the concentration of ovalbumin gene sequences in each size class was determined (Fig. 13). Endogenous oviduct nuclease preferentially cleaved the transcriptionally active ovalbumin chromatin in the hen- and estrogen-stimulated oviducts. The transcriptionally inert globin chromatin in the hen oviduct and the inactive ovalbumin chromatin in the withdrawn oviduct, however, were not preferentially attacked by this nuclease. A different approach, which leads to similar conclusions, is shown in Fig. 14. Nuclei from hen erythrocytes (A), primary estrogen-stimulated oviducts (B), estrogen-withdrawn oviducts (C), and secondary hormone-stimulated oviducts (D) were incubated at 37°C for the

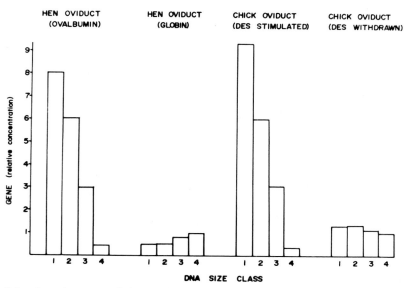

Fig. 13. Selective cleavage of the ovalbumin gene by the endogenous nuclease: solution hybridization analysis. Oviduct nuclei from hens, DES-stimulated chicks, and DES-withdrawn chicks were incubated for 1 h at 37°C. The fragmented DNA was separated into four size classes and the concentration of ovalbumin and globin genes in the hen DNA and ovalbumin genes in chick DNA was determined by moderate cDNA excess and standard curve analysis. [Data from Bloom and Anderson (1981) and Vanderbilt et al. (1981).]

times indicated. Following incubation, samples of the DNAs were electrophoresed on agarose gels (Fig. 14, top). The decrease in the molecular weight of the DNA with increasing incubation time resulted from the action of endogenous nuclease. Aliquots of the DNA samples were digested with *Hind* III and electrophoresed on agarose gels; the DNA fragments were transferred onto nitrocellulose sheets by the method of Southern (1975). The DNA fragment containing the ovalbumin gene sequence was detected by hybridization of the immobilized DNA on the sheets with ^{32}P-ovalbumin cDNA (Fig. 14, bottom). With increasing digestion time, the DNA fragment containing the ovalbumin sequence was digested by endogenous nuclease at a faster rate when the ovalbumin gene was transcribed (B and D) than when the ovalbumin gene was not transcribed (A and C).

D. Fine Structure of Ovalbumin Chromatin

Several structural and biochemical features are characteristic of the transcriptionally active regions of chromatin. Enzymatic probes for these regions, therefore, may recognize different properties associated with expressed genes. Pancreatic DNase I has been used extensively as a probe for specific genes in chromatin, since incubation of nuclei with this enzyme results in the selective destruction of transcriptionally active sequences

A B C D A B C D A B C D   A B C D A B C D A B C D
(0 MIN)   (10 MIN) (20 MIN)   (40 MIN) (60 MIN) (120 MIN)

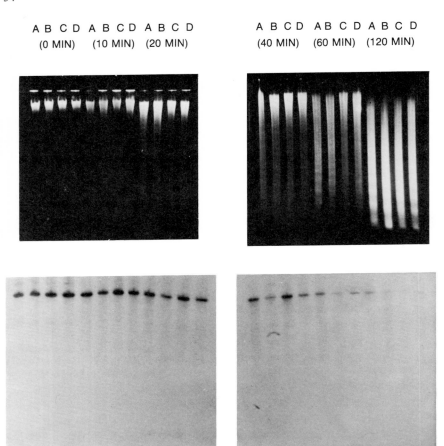

Fig. 14. Selective cleavage of the ovalbumin gene by the endogenous nuclease: blot hybridization analysis. Nuclei from (A) hen erythrocytes, (B) DES-stimulated oviducts, (C) DES-withdrawn oviducts, and (D) secondary DES-stimulated oviducts were incubated at 37°C for the times indicated. (Top) The DNA from these nuclei was deproteinized and analyzed on two agarose gels. (Bottom) Samples of this DNA were digested with *Hind* III, electrophoresed on gels, transferred to nitrocellulose sheets, and hybridized to ovalbumin cDNA. The restriction fragment shown corresponds to about 60% of the transcriptional unit of the ovalbumin gene at the 3′ end. [Data from Bloom and Anderson (1981).]

(Weintraub and Groudine 1976). Thus, the ovalbumin gene in hen or steroid-treated chick oviducts is preferentially digested by DNase I as compared to the bulk of the oviduct DNA, and as compared to the globin genes that are transcriptionally inert (Garel and Axel 1976; Bellard et al. 1978; Palmiter et al. 1978; Bloom and Anderson 1979; Lawson et al. 1981; O'Malley et al., this volume). Studies that have employed DNase I as a probe for ovalbumin chromatin, however, have revealed that a chromatin conformation recognized selectively by this enzyme is not directly related to the transcriptional activity of the ovalbumin gene. The ovalbumin gene in chick oviduct chromatin, for example, retains its sensitivity to DNase I following the withdrawal of estrogen (O'Malley et al., this volume). Since the ovalbumin chromatin in the estrogen-withdrawn oviduct is insensitive to the nucleases used in our studies (Figs. 9–14), we suggested that a chromatin conformation recognized by DNase I is different from the configuration selectively cleaved by micrococcal nuclease and the endogenous oviduct nuclease (Bloom and Anderson, 1979 and 1981).

Recent studies have revealed that the chromatin containing transcribed genes, as well as the chromatin containing DNA segments adjacent to these genes is preferentially sensitive to DNase I. For example, the ovalbumin gene and long DNA flanking segments are equally sensitive to DNase I in hen oviduct chromatin, whereas these same sequences are resistant to digestion in spleen, liver, and erythrocyte nuclei (Lawson et al. 1980; O'Malley et al., this volume). The large patch of sensitive chromatin surrounding the ovalbumin gene contains, in addition to the ovalbumin gene, the X and Y genes as well as transcriptionally inert DNA. The X and Y genes, which are closely linked to the ovalbumin gene, are also induced by steroid hormones but their rates of transcription are only about 1% and 6% of the rate of ovalbumin (Royal et al. 1979; Colbert et al. 1980). Since our studies indicated that micrococcal nuclease, unlike DNase I, recognizes a chromatin feature that is correlated with ovalbumin gene transcription (Figs. 9 and 10), we investigated the chromatin sensitivity of the ovalbumin gene and adjacent DNA sequences to micrococcal nuclease in hen oviduct chromatin (Fig. 15). These studies, as well as those described in Figs. 16–20, were performed in collaboration with G. M. Lawson, M.-J. Tsai, and B. W. O'Malley, who provided the restriction fragments used for the preparation of the hybridization probes. Hen oviduct nuclei were incubated with micrococcal nuclease until about 2% of the DNA was converted to mononucleosome length fragments. The DNA was separated into four size classes and the concentrations of coding DNA for ovalbumin (Ov–3.2), Y(Y–1.2), and X(X–4.6) genes, as well as nontranscribed DNA at the 3' side of the ovalbumin gene (OV–1.7/ 1.7) and at the 5' side of the X gene (X–2.2) were determined. The concentration of the ovalbumin-coding DNA in monomeric length fragments (size class 1) was eight-fold greater than in nondigested DNA and 20-fold greater than in the largest-molecular-weight DNA size class. This marked selectivity, however, was not observed with the X and Y genes, which are transcribed at a very low rate as compared to ovalbumin, nor with the transcriptionally inert DNA sequences. Since the nontranscribed sequences corresponding to OV(1.7/1.7) and X(2.2), as well as genes X and Y, exhibited a DNase I sensitivity indistinguishable from that observed for transcribing ovalbumin regions (Lawson et al. 1980; O'Malley et al., this volume), the results in Fig. 15 provide additional support for the proposal that micrococcal nuclease and DNase I can recognize different structural parameters of

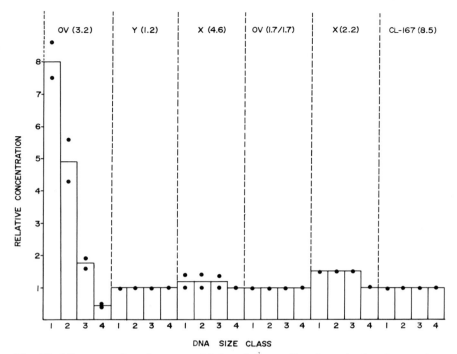

DNA SIZE CLASS

Fig. 15. Micrococcal nuclease sensitivity of the ovalbumin and related genes in hen oviduct nuclei. Hen oviduct nuclei were incubated with micrococcal nuclease and the fragmented DNA separated into four size classes as described in Fig. 5. The concentrations of coding DNA for ovalbumin (OV-3.2), Y(Y-1.2) and X(X-4.6) genes and nontranscribed DNA at the 3' side of the ovalbumin gene (OV-1.7/1.7) and at the 5' side of the X gene (X-2.2) were determined by hybridization of the DNA in each size class to the various nick-translated restriction fragments under conditions of DNA sequence excess. Restriction fragments are identified by their length in kb, and their exact locations are provided in the paper by Lawson et al. (1980). The DNA corresponding to fragment CL-167 (8.5) is outside of the DNase-I-sensitive domain. All other regions probed in the experiment are equally sensitive to DNase I digestion (Lawson et al. 1980, O'Malley et al., this volume).

the transcribing ovalbumin chromatin and that the recognition site(s) for micrococcal nuclease is more closely related to hormone-induced transcription of the ovalbumin gene.

To provide a more complete description of the nuclease-sensitive conformation of transcribing ovalbumin chromatin, we have examined the endogenous nuclease sensitivity of the regions immediately adjacent to the ovalbumin gene by blot hybridization analysis. In these studies, hen oviduct or erythrocyte nuclei were incubated for up to 1 h at 37°C. Following incubation, samples of the DNAs were electrophoresed on an agarose gel and stained (Fig. 16). Samples of the DNAs were also restricted, electrophoresed on agarose gels, blotted, and probed with various nick-translated DNA fragments (probes OV A–E) or globin cDNA (Fig. 17). The results clearly show that the bands containing the coding regions of the ovalbumin gene (corresponding to probes OV B and C) are preferentially lost in oviduct nuclei during the in vitro incubation as compared to these sequences in erythrocyte nuclei and as compared to the globin genes in oviduct nuclei that are transcriptionally silent. In contrast, sequences flanking both the 5' end (OV A) and the 3' end (OV D and OV E) of the ovalbumin transcriptional unit in oviduct chromatin appear relatively insensitive to endogenous nuclease digestion.

The results presented in Fig. 17 indicate that the entire coding region of the ovalbumin gene in oviduct chromatin is sensitive to the endogenous nuclease, whereas the flanking DNA is resistant to digestion by this enzyme. There appears, then, to be a discrete transition in chromatin structure at the ends of the transcriptional unit of the ovalbumin gene. The existence of a discrete chromatin boundary at the 3' end of the transcribing ovalbumin gene is corroborated by the studies shown in Fig. 18. As shown in the map at the top of Fig. 18, there is an internal $EcoRI$ site at 2.1 kb (kilobase pairs) from the 3' terminus and another 9.3 kb downstream. As predicted from this map, DNA obtained from nuclei that were not digested with the endogenous nuclease yielded a single band of 9.3 kb when digested with $EcoRI$ and then hybridized to the 3' flanking probe OV D (see Fig. 18, 0-time incubations). If the endogenous nuclease cleaves only the coding regions of the active ovalbumin gene and not the flanking DNA at the 3' side as the results in Fig. 17

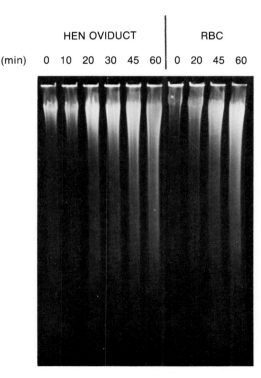

Fig. 16. Digestion of total nuclear DNA by the endogenous nuclease. Nuclei from hen oviducts or erythrocytes were incubated at 37°C for the indicated times and the DNA was analyzed on a 1.1% agarose gel.

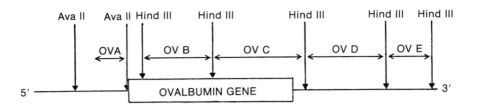

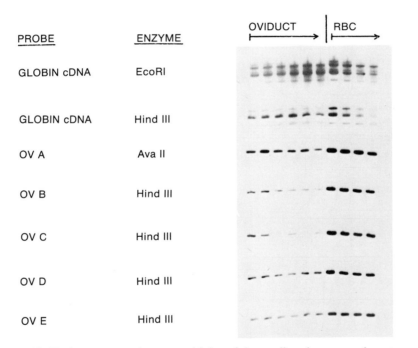

Fig. 17. Endogenous nuclease sensitivity of the ovalbumin gene and nontranscribed flanking sequences in hen oviduct and erythrocyte nuclei. The map at the top shows the location of relevant restriction sites in the ovalbumin gene and flanking sequences in chicken DNA and the restriction fragment probes (OV A–OV E) that were used in these experiments [see Lawson et al. (1980) and Woo et al. (1981) for further details]. DNA samples from the experiment described in Fig. 16 were digested to completion with *Eco*RI, *Hind* III, or *Ava* II, electrophoresed on 1.1% agarose gels, and transferred to nitrocellulose filters. The filters were hybridized to globin cDNA or to nick-translated probes OV A–OV E. Only the portion of the resulting autoradiograms that contained bands are presented. The arrows above the autoradiograms indicate the endogenous nuclease digestion times (see Fig. 16).

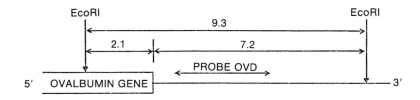

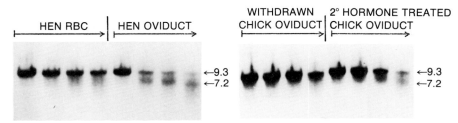

Fig. 18. Endogenous nuclease sensitivity at the 3′ end of the ovalbumin gene. The map at the top shows the location of the relevant *Eco*RI sites and probe OV D. Hen erythrocyte and oviduct nuclei were incubated at 37°C for 0, 20, 45, and 60 min. Oviduct nuclei from estrogen withdrawn and secondary hormone stimulated chicks were incubated for 0, 5, 15, and 30 min. The arrows indicate the increasing digestion times. The DNA was digested with *Eco*RI, electrophoresed on two agarose gels, blotted, and probed with OV D.

suggest, then this 9.3-kb parent band should be shortened in the digested oviduct nuclei by 2.1 kb to give rise to a 7.2-kb subband containing only 3′ flanking sequences. As shown in Fig. 18 (left), a discrete subband of about 7.2 kb was indeed produced at the expense of the 9.3-kb parent fragment in the digested hen oviduct nuclei. This subband was not observed in the digested DNA from erythrocyte nuclei (Fig. 18, left), nor was it observed when the DNA from digested hen oviduct was not restricted with *Eco*RI (data not shown). In addition, as shown at the right of Fig. 18, this 7.2-kb subband was generated in digested oviduct nuclei from secondary hormone stimulated chicks but not in oviduct nuclei obtained from chicks withdrawn from estrogen. The transition from resistant to sensitive chromatin at the 3′ end of the ovalbumin gene is therefore correlated with the gene's transcriptional activity. The discrete chromatin boundaries at the 3′ (Figs. 17 and 18) and possibly 5′ (Fig. 17) ends of the active ovalbumin gene may play an important role in delineating the transcriptional unit from its flanking sequences in oviduct chromatin.

Studies by Wu (1980) have demonstrated that specific chromatin sites upstream from the 5′ side of several heat-shock genes in *Drosophila* are exquisitely sensitive to attack by DNase I. Hypersensitive DNase I sites upstream from active chicken globin genes (Weintraub et al. 1981), rat insulin genes (Wu and Gilbert 1981), and other eukaryotic genes (summarized in Kolata 1981) have also been reported. Since the chromatin sites adjacent to the heat-shock genes are present prior to transcriptional activation of these genes by heat treatment, it was proposed that these hypersensitive regions may play an important role in determining the potential for gene expression in eukaryotes (Wu 1980). However, attempts to demonstrate specific DNase I cutting sites upstream from the ovalbumin gene in oviduct chromatin have thus far been unsuccessful (Kuo et al. 1979; Bellard et al. 1980). As shown in Fig. 19, such sites are readily detected during the autodigestion of oviduct nuclei with the endogenous nuclease. In Fig. 19, hen oviduct and erythro-

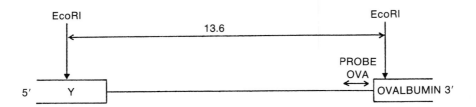

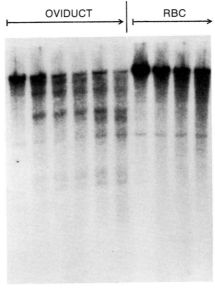

Fig. 19. Specific endogenous nuclease cutting sites upstream from the ovalbumin gene in hen oviduct nuclei. The map at the top shows the location of the relevant *Eco*RI sites and probe OV A. Hen oviduct and erythrocyte nuclei were autodigested for the times indicated in Fig. 16. The arrows indicate increasing digestion times. The DNA was restricted with *Eco*RI, electrophoresed, blotted, and probed with OV A. The numbers on the left refer to the length of the fragments in kilobases.

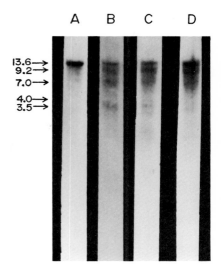

Fig. 20. Specific endogenous nuclease cutting sites upstream from the ovalbumin gene in chick oviduct nuclei. Nuclei from (A) erythrocytes, (B) hen oviducts, (C) withdrawn chick oviducts, and (D) secondary hormone stimulated chick oviducts were incubated for 30 min at 37°C. DNA from these nuclei was restricted with *Eco*RI and probed with OV A as described in Fig. 19.

cyte nuclei were incubated for up to 1 h at 37°C. Following incubation the nuclear DNA was purified, restricted with *Eco*RI, blotted, and hybridized to probe OV A. As predicted from the map at the top of the figure, DNA obtained from hen oviduct or erythrocyte nuclei that were not incubated in vitro yielded one band of 13.6 kb when digested with *Eco*RI and hybridized to probe OV A. A series of at least seven discrete subbands appear in the oviduct DNA during digestion of nuclei with the endogenous nuclease. These subbands are apparently cell-type specific, since they were not generated during the in vitro incubation of erythrocyte nuclei. Since these subfragments were not observed when the digested oviduct DNA was not restricted with *Eco*RI (data not shown), we surmise that one end of each band was produced from the *Eco*RI cleavage near the 5' end of the ovalbumin gene while the other arose from a specific endogenous nuclease cut upstream. From these experiments the endogenous nuclease cuts can be mapped to positions that are about 8.8, 6.6, 5.4, 3.6, 3.1, 1.5, and 1.3 kb upstream from the 5' end of the ovalbumin-coding DNA. These preferential endogenous nuclease cutting sites were also detected in oviduct nuclei obtained from hormone-stimulated (Fig. 20D) and withdrawn chicks (Fig. 20C). Since these discrete sites were found in oviduct chromatin from estrogen-withdrawn chicks (Fig. 20C), it is apparent that their presence can precede the transcriptional activation of the ovalbumin-coding DNA by steroid hormones.

E. A Model for the Effects of Steroid Hormones on the Organization of the Ovalbumin Gene in Oviduct Chromatin

The results presented in this paper can be considered in terms of one popular model for chromosome structure. Several features of the model are consistent with observations made on the heat-shock genes in *Drosophila* chromatin (Wu et al. 1979; Wu 1980; Levy and Noll 1981) and the globin genes in the chromatin from chicken erythroid cells (Stalder et al. 1980; Weintraub et al. 1981). In oviduct chromatin from chicks withdrawn from estrogen, the coding region of the ovalbumin gene is in a conformation that is insensitive to micrococcal nuclease (Figs. 9 and 10) and to the endogenous oviduct nuclease (Figs. 13, 14, and 18). Upstream from this region, however, is positioned a series of chromatin sites that are preferentially attacked by the endogenous nuclease (Fig. 20C). The existence of these and other (Wu 1980; Stalder et al. 1980; O'Malley et al., this volume) altered chromatin structures prior to transcriptional activation could render a specific DNA sequence within the region more accessible for interaction with regulatory proteins. For example, if the 5' flanking region of a hormone-responsive gene contains a nucleotide sequence that binds selectively to a receptor protein, this sequence may be accessible for receptor binding only in those cells in which the gene is to be activated by the hormone. Alternatively, the actual chromatin configuration near potentially active loci might play a role in the selective recognition of regulator proteins that alter the transcription of a specific subset of genes. The reader is referred to the publications by Wu (1980), and by Stalder et al. (1980), for a more thorough discussion of these potentially important concepts.

Following the administration of steroid hormones to withdrawn chicks, the hypersensitive sites flanking the ovalbumin gene are retained (Figs. 19 and 20), while the entire coding region is rendered sensitive to both micrococcal nuclease (Figs. 10 and 15) and the endogenous nuclease (Figs. 14 and 17). The chromatin boundaries of the coding region at the 3' (Figs. 17 and 18) and possibly 5' (Fig. 17) ends of the ovalbumin gene are discrete, since sequences immediately adjacent to the coding regions remain in a nuclease-insensitive state. Because the enhanced nuclease sensitivity of the ovalbumin sequences in the oviduct is not a direct consequence of active transcription at the locus per se (Fig. 11), the chromatin configuration responsible for the preferential release of active mononucleosomes by micrococcal nuclease (Figs. 6–10, 15) may represent an important feature in delineating the transcriptional unit of the ovalbumin gene. We suspect that the chromatin configuration responsible for this selective nuclease recognition is related to an expansion of the nuclease cutting sites in the linker regions because the bulk of the nucleosomal core particles associated with the ovalbumin gene in oviduct chromatin is not selectively cleaved internally by micrococcal nuclease (Figs. 2 and 3). The nature of the changes within the linker regions that might be responsible for this effect are not yet known (see Section IB).

F. Fractionation of Chromatin into Transcriptionally Active and Inactive Regions

Nuclease digestion studies such as those described in the preceding sections and references have provided some valuable information on the packaging of specific genes in chromatin. In general, however, these studies have revealed little about the nature of the chromosomal components that are responsible for the generation of potentially active or active loci. One approach that has been used to describe the general features of transcribed chromatin had involved the separation of chromatin into active and inactive regions and the subsequent characterization of the proteins in the resulting chromatin fractions. The chromatin fractionation scheme outlined in Fig. 21 is based on the ability of micrococcal nuclease to selectively excise transcribed nucleosomes from chromatin. Hen oviduct nuclei were digested with micrococcal nuclease, chilled, and centrifuged. The resulting first supernatant fraction (1SF) contained chromatin particles released from intact nuclei. The nuclei in the sediment were lysed and centrifuged again to yield a pellet (P) and a second supernatant fraction (2SF). Electrophoretic analysis of the DNA extracted from the 2SF and P fractions revealed a series of larger-molecular-weight DNA fragments, whereas only mononucleosomal DNA fragments were observed in the 1SF (Fig. 22). Hybridization studies revealed that when about 5% of the total nuclear DNA was converted to mononucleosome length fragments, the 1SF was five-fold enriched in the ovalbumin gene and the 2SF was five-fold depleted in this sequence when compared to undigested DNA (Fig. 23). As expected from the results presented in Fig. 8, a reduction in the extent of nuclease digestion resulted in a

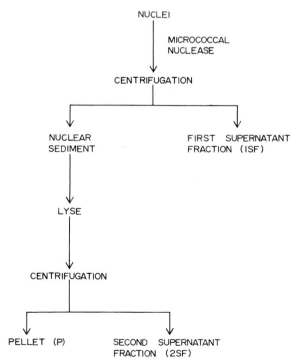

Fig. 21. Chromatin fractionation procedure. The fractionation procedure is outlined in the text. [For additional details see Bloom and Anderson (1978b).]

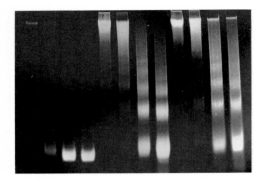

A B C D E F G H I J K L

Fig. 22. Electrophoretic analysis of DNA purified from nuclear fractions. Hen oviduct nuclei were digested with micrococcal nuclease (75 U/ml), and an aliquot of the nuclear suspension was withdrawn and separated into 1SF, 2SF, and pellet fractions at the times indicated. DNA isolated from these fractions was electrophoresed on a 3% polyacrylamide slab gel. Lanes A–D: 1SF at 0.25, 1.5, 15, and 30 min; lanes E–H: 2SF at 0.25, 1.5, 15, and 30 min; lanes I–L: pellet fraction at 0.25, 1.5, 15, and 30 min, respectively. [Data from Bloom and Anderson (1978b).]

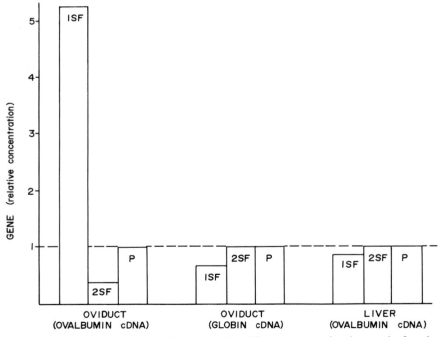

Fig. 23. Concentration of ovalbumin and globin sequences in chromatin fractions. Hen oviduct and liver chromatins were fractionated as outlined in Fig. 21 and the concentration of ovalbumin and globin genes was determined by DNA sequence excess and moderate cDNA excess hybridizations. [Data from Bloom and Anderson (1978b).]

corresponding increase in the enrichment of the ovalbumin gene in 1SF (data not shown).

Template-active nucleosomes prepared by the procedure shown in Fig. 21 are substantially enriched, as compared to bulk chromatin, in acetylated core histones (Nelson et al. 1980), "high-mobility-group" (HMG) nonhistone proteins 14 and 17 (Mathew et al. 1981), estrogen receptor proteins (Scott and Frankel 1980), and a number of other nonhistone chromosomal proteins that have not been extensively characterized (Norman and Liew 1981). The 1SF nucleosomal fraction contains the full complement of nucleosomal core histones (Nelson et al. 1980; Mathew et al. 1981) but is apparently deficient in histone H1 (Mathew et al. 1981). These variations may be a direct reflection of the composition of nucleosomes engaged in RNA transcription in vivo. However, since the active fraction prepared by this procedure is contaminated with transcriptionally inert DNA (Fig. 23), such a conclusion must remain tentative at the present time. Moreover, these studies tell us nothing about the nature of the proteins that are associated with specific active genes in chromatin. More specific probes for selected chromatin regions may therefore be required to provide a complete description of the proteins associated with individual gene loci in the chromosome.

II. Monoclonal Antibodies as Probes for Hormone-Dependent Chromatin Proteins

A. Tissue Specificity

The nonhistone proteins in chromatin are thought to play an important role in the control of gene expression in eukaryotes (reviewed in Stein et al. 1978). Two-dimensional electrophoretic analysis has revealed over 450 different nonhistone polypeptides in chromatin from a variety of sources (Peterson and McConkey 1976; Nikodem et al. 1981). A small fraction of these polypeptides are tissue specific and these proteins may be involved in the control of genes that are expressed in a tissue-specific manner. However, to our knowledge, no tissue-specific chromosomal proteins have been isolated and very little is known about their biochemical properties. Indeed, only a few nonhistone chromosomal proteins have been prepared from eukaryotic sources and the limited availability of these purified proteins and their specific antibodies has represented a major problem in understanding the role of these proteins in gene control mechanisms. A conventional immunologic approach for the preparation of such antibodies is severely limited because of the difficulties involved in isolating specific chromosomal proteins. The application of monoclonal antibody technology (reviewed in Yelton and Scharff 1981) to chromosomal proteins (Saumweber et al. 1980; Howard et al. 1981) circumvents this limitation, since specific antibodies directed against single immunologic determinants are readily produced by immunization with an impure antigen.

We have prepared a library of 25 monoclonal antibodies by first immunizing mice with total chromatin from the hen oviduct. The mouse spleen cells were fused to myeloma cells (SP2/0-Ag14) and the resulting hybridomas subcloned on soft agar by standard procedures. Since oviduct-specific chromosomal proteins might be involved in the control of genes that are expressed only in the oviduct, antibodies from each clone in the library were tested for their ability to bind chromatins from different tissues of the chicken. In Figs. 24 and 25, increasing amounts of solubilized chromatin from hen oviduct, liver, lung, and erythrocyte were adsorbed to the wells of a polystyrene microtiter plate. The plate was washed and media containing antibody 3 (Fig. 24) or antibody 11 (Fig. 25) was incubated in the wells. After washing, the wells were incubated with ^{125}I-rabbit anti-mouse-IgG developing antibody and washed again; the bound radioactivity was then determined. The antigen recognized by antibody 3 is tissue specific, since binding was observed with oviduct chromatin but not with chromatins prepared from

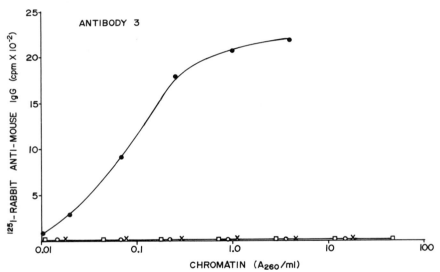

Fig. 24. Binding of antibody 3 to chromatins from four tissues of the hen. Chromatin was solubilized by brief sonication. The indicated amounts of soluble chromatin from hen oviduct (●), liver (○), lung (x), or erythrocyte (□) were incubated in the wells of microtiter plates. The plates were washed, blocked with BSA, and incubated with a ¼ dilution of the spent supernatant from the hybridoma culture producing antibody 3. The plates were then washed, incubated with ^{125}I-rabbit anti-mouse IgG, and washed again; the bound radioactivity was then determined. Nonspecific binding was determined by the incubation of a ¼ dilution of supernatant from an IgG-secreting myeloma culture in place of the hybridoma supernatant in the assay system. The levels of nonspecific binding, which represented no more than 3% of the specific binding observed with oviduct chromatin, were subtracted from all data points in the figure. The relative amounts of specific antigen recognized by antibody 3 as determined from these data were 100, <3, <3, and <3 for oviduct, liver, lung, and RBC chromatins, respectively.

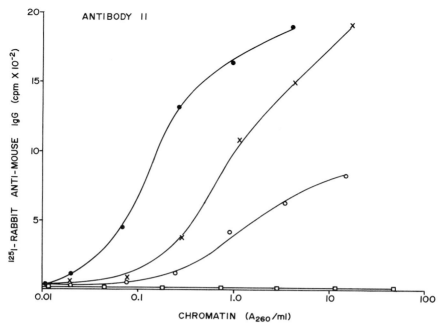

Fig. 25. Binding of antibody 11 to chromatins from four tissues of the hen. The procedure and the symbols designating the tissue sources of the chromatins are the same as those in Fig. 24. The relative amounts of the specific antigen recognized by antibody 11 were 100, 6, 29, and <3 for oviduct, liver, lung, and RBC chromatins, respectively.

liver, lung, or erythrocyte. Antibody 11, in contrast, recognizes a determinant that is present in lung and liver as well as in the oviduct. The results summarized in Table 1 (columns 1–4) reveal that 20 out of 25 of the monoclonal antibodies recognize determinants that are found in hen oviduct but not in liver, lung, or erythrocyte. Only five of the antibodies tested (11, 13, 17, 25, and 39) reacted with chromatin from nonoviduct sources. The marked tissue specificity of the antibody library was retained when proteins extracted from oviduct and liver chromatins were used in the assay system (Table I, columns 5 and 6). Thus, all the antibodies are directed against protein components of the chromatin. Contamination of oviduct chromatin with the major egg white proteins does not account for the presence of the large number of oviduct-specific antibodies in the library because none of the oviduct-specific antibodies reacted with a mixture of ovalbumin, conalbumin, ovomucoid, and lysozyme (Table 1, column 7).

To further characterize the proteins recognized by the antibodies in the library, the proteins in hen oviduct chromatin were fractionated according to their chromatin binding by the hydroxyapatite dissociation method (Bloom and Anderson 1978a). The electrophoretic patterns of the chromatin proteins presented in Fig. 26 illustrate the utility of this method for fractionation of chromosomal proteins. Sheared hen oviduct chromatin was applied to a

Table 1. Monoclonal Antibodies Against Oviduct Chromatin: Tissue Specificity[a]

Antibody no.	Oviduct chromatin	Liver chromatin	Lung chromatin	RBC chromatin	Oviduct chromatin proteins	Liver chromatin proteins	Ovalbumin conalbumin ovomucoid lysozyme
2	100	<3	<3	<3	100	<3	<3
3	100	<3	<3	<3	100	<3	<3
4	100	<3	<3	<3	100	<3	<3
6	100	<3	<3	<3	100	<3	<3
8	100	<3	<3	<3	100	<3	<3
10	100	<3	<3	<3	100	<3	<3
12	100	<3	<3	<3	100	<3	<3
14	100	<3	<3	<3	100	<3	<3
15	100	<3	<3	<3	100	<3	<3
16	100	<3	<3	<3	100	<3	<3
18	100	<3	<3	<3	100	<3	<3
19	100	<3	<3	<3	100	<3	<3
26	100	<3	<3	<3	100	<3	<3
27	100	<3	<3	<3	100	<3	<3
28	100	<3	<3	<3	100	<3	<3
29	100	<3	<3	<3	100	<3	<3
30	100	<3	<3	<3	100	<3	<3
35	100	<3	<3	<3	100	<3	<3
36	100	<3	<3	<3	100	<3	<3
40	100	<3	<3	<3	100	<3	<3
11	100	6	29	<3	100	5	<3
13	100	12	49	<3	100	27	<3
17[b]	100	100	100	100	100	80	90
25	100	81	53	<3	100	97	<3
39	100	<3	19	4	100	<3	<3

[a] The table shows the relative concentration of the indicated antigens in soluble chromatins (columns 1–4), in total chromatin protein preparations (columns 5 and 6), and in a mixture of the major egg white proteins (column 7). Antigen concentrations were determined by the solid-phase radioimmune assay described in Fig. 24. Total chromatin proteins were prepared by hydroxyapatite chromatography as described by Bloom and Anderson (1978a).

[b] Antibody 17 reacted with all preparations tested, including egg white proteins. Since this antibody recognized a large number of individual proteins as determined by protein blotting analysis, we presume that its binding characteristics are relatively nonspecific.

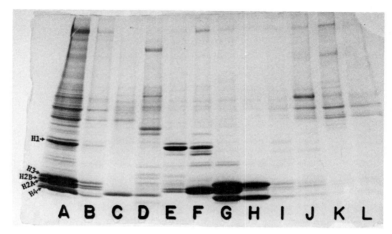

Fig. 26. Fractionation of hen oviduct chromosomal proteins by the hydroxyapatite dissociation method. Unfractionated chromatin (U) proteins are shown in lane A. Twenty A_{260} units of chromatin at 2 A_{260} U/ml of NaP (10 mM, pH 7.0) were applied to a hydroxyapatite column. Chromatin that failed to bind to the matrix is shown in lane B. The column was washed with 80 mM NaP (lane C) and the proteins selectively dissociated from the immobilized chromatin by NaCl (D–H) followed by NaCl and urea (I–K). Solutions were 80 mM in NaP (pH 7.0) until lane L, where phosphate molarity was made 500 mM to elute nucleic acid. Twenty milliliters of each solution were applied to the column. The concentrations of NaP, NaCl, and urea are shown below.

A	B	C	D	E	F	G	H	I	J	K	L	
U	10	80	80	80	80	80	80	80	80	80	500	NaP (mM)
U	—	—	0.25	0.5	0.75	1.25	2	2	2	2	2	NaCl (M)
U	—	—	—	—	—	—	—	2	4	8	5	urea (M)

[Data from Bloom and Anderson (1978a).]

hydroxyapatite column in a solution that does not dissociate chromosomal proteins from DNA (10 mM phosphate). Chromatin proteins were subsequently dissociated from the immobilized DNA or other chromatin components still bound to the DNA by stepwise elution with NaCl (lanes C–H) followed by stepwise elution with urea in the presence of 2 M NaCl (lanes I–K). Table 2 shows the relative concentrations of the proteins recognized by the various antibodies in the loosely bound, moderately tightly bound, and very tightly bound fractions of oviduct chromatin. There was no apparent relationship between chromatin-binding properties and tissue specificity of the chromatin proteins recognized by the antibodies. Oviduct-specific proteins appeared in the loosely bound (e.g., antibody 36), moderately tightly bound (e.g., antibody 8), and tightly bound (e.g., antibody 40) fractions of chromatin. These results suggest that there are a number of specific proteins unique to the hen oviduct that differ in their chromatin binding properties.

Table 2. Fractionation of Hen Oviduct Chromatin Antigens by the Hydroxyapatite Dissociation Method[a]

	Total activity in hydroxyapatite fractions (%)		
	I	II	III
Hen oviduct specific antibody no.			
2	94	<3	5
3	<3	38	61
4	4	24	61
6	<3	43	57
8	<3	71	28
10	97	<3	3
12	30	30	40
14	36	28	36
15	27	27	45
16	20	40	43
18	<3	47	53
19	<3	56	44
26	<3	64	34
27	<3	25	75
28	<3	67	31
29	<3	34	64
30	<3	39	61
35	94	<3	6
36	100	<3	<3
40	10	10	80
Nontissue specific antibody no.			
11	<3	<3	100
17	<3	96	3
25	8	8	83
39	49	49	<3

[a] Hen oviduct chromatin proteins were separated into loosely bound (fraction I), moderately tightly bound (fraction II), and tightly bound (fraction III) fractions as described in Fig. 26. These three fractions correspond respectively to lanes C to D, E to H, and I to K on the gel presented in Fig. 26. The relative amounts of total protein in these three fractions were 21, 42, and 37% respectively. Relative antigen concentrations were determined as described in Fig. 24.

The factors responsible for the presence of the large number of oviduct-specific antibodies in the library are not completely understood. One-dimensional electrophoretic analysis of chromatin proteins from oviduct, liver, lung, and erythrocyte revealed that only a small fraction of the detectable bands were oviduct specific (not shown). However, sera from six different mice inoculated with hen oviduct chromatin reacted preferentially with oviduct chromatin as compared to chromatins from liver, lung, or spleen (not shown). It is generally assumed that many of the major nonhistone proteins play a structural or enzymatic role and thus, like histones, may be highly conserved in eukaryotes. Chicken oviduct specific chromatin proteins, in contrast, are expected to be found only in avian species and therefore should be highly antigenic in the mouse. These speculations aside, the results presented in this section show that a library of antibodies against tissue-specific chromatin proteins is readily attainable.

B. Hormone-Dependent Changes in the Protein Composition of Oviduct Chromatin

Electrophoretic analysis has revealed that steroid and thyroid hormones affect the levels of a large number of nuclear proteins in a variety of target cells (see Spelsberg et al. 1973; Stein et al. 1978; Nikodem et al. 1981; and the references cited therein). For example, 102 out of 500 hepatic nuclear proteins detected by two-dimensional gel electrophoresis either disappeared or were markedly reduced after thyroidectomy. By 24 h after administration of thyroid hormone, 80 of these protein subunits reappeared in the hepatic nuclear fraction (Nikodem et al. 1981). The results shown in Figs. 27 and 28 and those summarized in Table 3 show that a large number of the proteins recognized by the clonal antibodies were also hormone dependent in the chick oviduct system. As shown in Fig. 27, the protein recognized by antibody 3, which is oviduct specific in the hen (Fig. 25, Table 1), is also hormone dependent in the chick oviduct, since it declines to near nondetectable

Table 3. Effects of Hormone Treatment on Chromatin Antigens in the Chick Oviduct

	DES stimulated	WD-2 day	WD-6 day	2° DES stimulated	2° P stimulated
Hen oviduct specific antibody no.					
2	100	111	12	55	78
3	100	74	<10	36	46
4	100	75	<10	38	42
6	100	66	<10	34	36
8	100	69	<10	37	46
12	100	84	<10	35	42
14	100	73	<10	27	38
15	100	89	<10	36	53
18	100	83	11	48	65
26	100	74	<10	30	25
27	100	80	<10	28	32
28	100	89	<10	52	61
29	100	83	<10	70	31
30	100	71	<10	20	23
Nontissue specific antibody no.					
11	100	120	100	100	100
13	100	110	140	92	81
17	100	100	100	100	100
25	100	126	104	112	115

[a] Chicks were treated as described in Fig. 27. Relative antigen concentrations were determined as described in Fig. 27.

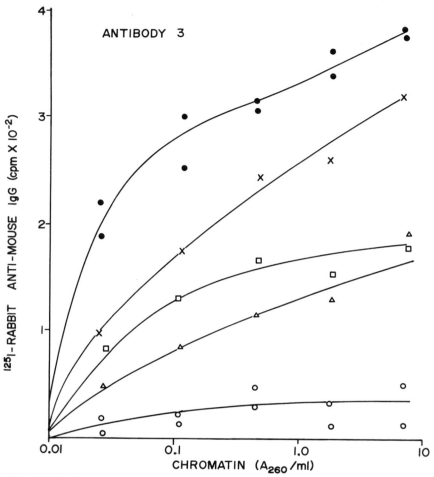

Fig. 27. Binding of antibody 3 to chick oviduct chromatins. Estrogen-stimulated chicks (○) were withdrawn from hormone for 2(x) or 6(○) days. The 6-day-withdrawn animals received a single injection of DES (△) or progesterone (□) and were killed 12 h later. The procedure is the same as that in Fig. 24.

levels following estrogen withdrawal and reappears in oviduct chromatin after secondary stimulation with estrogen or progesterone. In contrast, the determinant recognized by antibody 11, which is present in hen lung and liver, as well as in oviduct (Fig. 26, Table 1), is hormone independent in the chick, since its level is essentially equivalent in hormone-treated and withdrawn oviduct chromatins (Fig. 28). In addition, as summarized in Table 3, all hen oviduct-specific antibodies tested recognized chromatin proteins that were hormone dependent in the chick oviduct. In contrast, the antibodies that reacted with chromatin from nonoviduct sources in the hen (Table 1) reacted with proteins in chick oviduct chromatin that were not dependent on

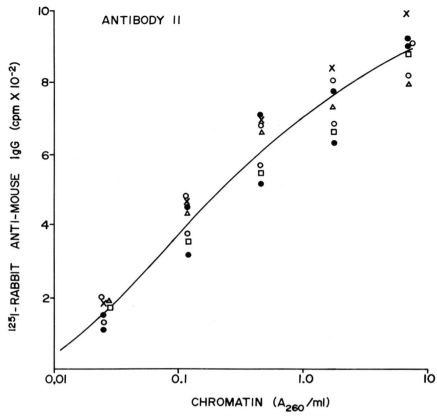

Fig. 28. Binding of antibody 11 to chick oviduct chromatins. The procedure and the symbols designating the oviduct sources of the chromatin are the same as those in Figs. 24 and 27, respectively.

steroid hormones. Another approach illustrating this interesting relationship is shown in Fig. 29. In these studies, solubilized chromatin from four tissues of the hen or from oviducts obtained from chicks in various hormonal states were incubated with squares of nitrocellulose filter paper. Following incubation, the filters were washed, incubated with the indicated antibodies, washed again, and incubated with ^{125}I-rabbit anti-mouse IgG. Inspection of the resulting autoradiogram confirms the results shown in Tables 1 and 3 in that the hen oviduct-specific chromatin proteins recognized by antibodies 2, 3, 4, 8, 12, 15, 27, 28, and 29 were induced by steroid hormones, whereas the non-oviduct-specific proteins recognized by antibodies 11, 13, and 17 were not.

The determination of the apparent molecular weights of the proteins recognized by two of the antibodies in the library is shown in Fig. 30. Total chromosomal proteins from estrogen-stimulated and withdrawn oviducts were separated on SDS gels and the antigens recognized by antibody 3 and

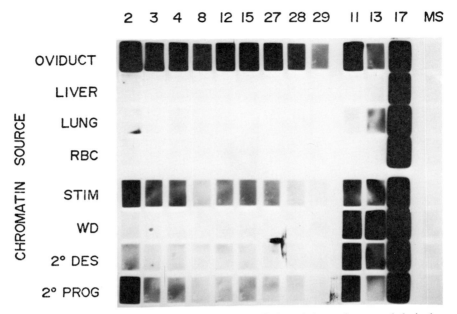

Fig. 29. Relationship between the tissue specificity of the antigens and their dependence on steroid hormones. Solubilized chromatins from four tissues of the hen (oviduct, liver, lung, and RBC) and from oviducts obtained from chicks in various hormonal states (Stim, WD, 2° DES, 2°P) were incubated with nitrocellulose filters (1 × 1.5 cm). Filters were blocked with BSA and incubated with the indicated hybridoma supernatants containing the antibodies or with myeloma supernatant (MS). The filters were then washed, incubated with ^{125}I-rabbit anti-mouse IgG, washed again, and visualized by autoradiography.

antibody 11 were identified by protein blotting analysis. A single protein species of $\sim$180,000 MW was recognized by antibody 3 in the protein mixture from stimulated oviduct chromatin. This polypeptide was not detected in the withdrawn oviduct chromatin (Fig. 30), nor was it detected in chromatins from liver, lung, or erythrocyte (not shown). These observations are in agreement with the results presented in Tables 1 and 3 and Figs. 27 and 29. The antigen recognized by antibody 11, which is hormone independent as determined by the assays described in Figs. 28 and 29, has a molecular weight of $\sim$50,000 and is detected in both stimulated and withdrawn oviduct preparations. Because we cannot as yet assign the two specific bands on the autoradiograms shown in Fig. 30 to corresponding bands on the stained gels of chromatin proteins, the precise concentration of the proteins recognized by antibodies 3 and 11 cannot be determined. However, we estimate that the oviduct-specific protein species recognized by antibody 3 represent less than 1% by weight of the total chromosomal proteins in the hen or hormone-stimulated chick oviduct.

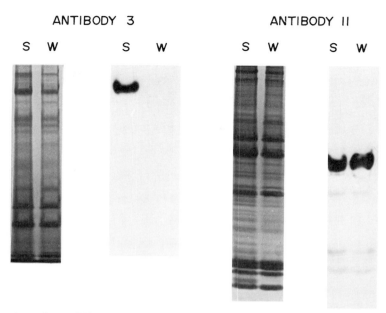

Fig. 30. Detection of two different chromatin antigens. Oviduct chromatin proteins from stimulated (S) or withdrawn (W) chicks were separated on 7.5% (for antibody 3) or 15% (for antibody 11) SDS–polyacrylamide gels. The proteins in the gels were stained (left portion of each panel) or were transferred electrophoretically onto nitrocellulose filters. Detection of the proteins recognized by the two antibodies was performed as described by Burnette (1981), and the resulting autoradiograms are shown at the right of each panel.

The results presented in this section indicate that steroid hormones exert some very specific effects on the protein composition of chick oviduct chromatin. Whether the appearance of these proteins in the oviduct chromatin preparations following hormone treatment results from a change in protein synthetic or degradative rates or from a change in the distribution of these proteins between cytoplasm and nucleus is not yet known. In addition, it is possible that some (but probably not all) of the antibodies used in these studies are directed against oviduct cytoplasmic proteins that might contaminate the chromatin preparations. Immunohistochemical techniques are currently being used with the hope of resolving these questions.

III. A Concluding Remark

Steroid hormones elicit rapid alterations in the general structural organization of chromatin in a variety of vertebrate target cells. These structural changes, which include nuclear swelling and alterations in chromatin thermal denaturation patterns, circular dichroism spectrum, and polylysine, antinomycin, and RNA polymerase binding (see Mainwaring and Jones

1975; Schwartz et al. 1975; Tsai et al. 1975; Kalimi et al. 1976; Loor et al. 1977; Snow et al. 1978; Clark and Peck 1979, and the references cited therein) may be a reflection of alterations in the structure and composition of specific chromosomal loci. The results presented in this paper show that steroid hormones do induce some very specific changes in the chromatin environment of the ovalbumin gene and in the protein composition of the chromatin in the oviduct. How these changes are related to each other and, more importantly, how they relate to specific gene expression and hormone-induced differentiation and growth represent important problems for future investigations.

Acknowledgments. We would like to thank Connie Philbrook for typing the manuscript. K. B. and J. V. were supported by predoctoral traineeships from the NIH. This work was supported by PHS NCI Grant #R01 CA25799.

References

Anderson JN, Clark JH, Peck EJ (1972) Biochem J 126: 561
Anderson JN, Schimke RT (1976) Cell 7: 331
Ashburner M, Bonner JJ (1979) Cell 17: 241
Bellard M, Gannon F, Chambon P (1978) Cold Spring Harbor Symp Quant Biol 42: 779
Bellard M, Kuo MT, Dretzen G, Chambon P (1980) Nucl Acids Res 8: 2737
Bloom KS, Anderson JN (1978a) J Biol Chem 253: 4446
Bloom KS, Anderson JN (1978b) Cell 15: 141
Bloom KS, Anderson JN (1979) J Biol Chem 254: 10532
Bloom KS, Anderson JN (1983) (submitted to J Biol Chem)
Bonner J, Huang RC (1963) J Mol Biol 6: 169
Burgoyne LA, Hewish DR (1978) In: Busch H (ed) The Cell Nucleus IV. Academic Press, New York. p 48
Burnette WN (1981) Anal Biochem 112: 195
Chambon P (1978) Cold Spring Harbor Symp Quant Biol 42: 1209
Clark JH, Peck EJ Jr. (1979) Female Sex Steroids. Springer-Verlag, New York
Clever U (1966) Am Zool 6: 33
Colbert DA, Knoll BJ, Woo SLC, Mace ML, Tsai M-J, O'Malley BW (1980) Biochemistry 19: 5586
Comings DE (1966) J Cell Biol 28: 437
Foe VE (1978) Cold Spring Harbor Symp Quant Biol 42: 723
Garel A, Axel R (1976) Proc Natl Acad Sci USA 73: 3966
Howard GC, Abmayr SM, Shinefeld LA, Sato VL, Elgin, SCR (1981) J Cell Biol 88: 219
Kalimi M, Tsai SY, Tsai, M-J, Clark JH, O'Malley BW (1976). J Biol Chem 251: 516
Kolata GB (1981) Science 214: 775
Kornberg RD (1977) Ann Rev Biochem 40: 931
Kuo MT, Mandel JL, Chambon P (1979) Nucl Acids Res 7: 2105
Lawson GM, Tsai M-J, O'Malley BW (1980) Biochemistry 19: 4403
Levy A, Noll M (1981) Nature 289: 198
Littau VC, Allfrey VG, Frenster JH, Mirsky AE (1964) Proc Natl Acad Sci USA 52: 93
Loor RM, Hu A-L, Wang TY (1977) Biochim Biophys Acta 477: 312
McGhee JD, Felsenfeld G (1980) Ann Rev Biochem 49: 1115
McKnight SL, Bustin M, Miller OL Jr (1978) Cold Spring Harbor Symp Quant Biol 42: 741

Mainwaring WIP, Jones DM (1975) J Steroid Biochem 6: 475
Mathew CGP, Goodwin GH, Wright CA, Johns EW (1981) Cell Biol Int Rep 5, No 1: 37
Nelson D, Covault J, Chalkley R (1980) Nucl Acids Res 8: 1745
Nikodem VM, Trus BL, Rall JE (1981) Proc Natl Acad Sci 78: 4411
Norman GL, Liew CC (1981) J Cell Biol 21: 66a
O'Malley BW, Roop DR, Lai EC, Nordstrom JL, Catterall JF, Swaneck GE, Colbert DA, Tsai M-J, Dugaiczyk A, Woo SLC (1979) Recent Prog Horm Res 35: 1
Palmiter RD (1975) Cell 4: 189
Palmiter RD, Mulvihill ER, McKnight GS, Senear AW (1978) Cold Spring Harbor Symp Quant Biol 42: 639
Pelling C (1964) Chromosoma 15: 71
Peterson JL, McConkey EH (1976) J Biol Chem 251: 548
Pruitt SC, Grainger RM (1981) Cell 23: 711
Royal A, Garapin A, Cami B, Perrin F, Mandel JL, LeMeur M, Bregegegre F, Gammon F, LePennec JP, Chambon P, Kourilsky P (1979) Nature 279: 125
Saumweber H, Symmons P, Kabisch R, Will H, Bonhoeffer (1980) Chromosoma 80: 253
Schimke RT, McKnight GS, Shapiro DJ, Sullivan D, Palacios R (1975) Recent Prog Horm Res 31: 175
Schwartz RJ, Tsai M-J, Tsai SY, O'Malley BW (1975) J Biol Chem 250: 5175
Scott RW, Frankel FR (1980) Proc Natl Acad Sci USA 77: 1291
Snow LD, Eriksson H, Hardin JW, Chan L, Jackson RL, Clark JH, Means AR (1978) J Steroid Biochem 9: 1017
Southern EM (1975) J Mol Biol 98: 503
Spelsberg TC, Mitchell WM, Chytil F, Wilson EM, O'Malley BW (1973) Biochim Biophys Acta 312: 765
Stalder J, Seebeck T, Braun R (1979) BBA 561: 452
Stalder J, Larsen A, Engel JD, Dolan M, Groudine M, Weintraub H (1980) Cell 20: 451
Stein GS, Stein JL, Thomson JA (1978) Cancer Res 18: 1181
Swaneck GE, Kreuzaler F, Tsai M-J, O'Malley BW (1979) Biochem Biophys Res Commun 88: 1412
Tsai SY, Tsai M-J, Schwartz RJ, Kalimi M, Clark JH, O'Malley BW (1975) Proc Natl Acad Sci 72: 4228
Vanderbilt JN, Bloom KS, Anderson JN (1981) J Biol Chem (submitted)
Weintraub H, Groudine M (1976) Science 193: 848
Weintraub H, Larsen A, Groudine M (1981) Cell 24: 333
Woo SLC, Beattie WG, Catterall JF, Dugaiczyk A, Staden R, Brownlee GG, O'Malley BW (1981) Biochemistry 20: 6437
Wu C (1980) Nature 286: 854
Wu C, Gilbert W (1981) Proc Natl Acad Sci USA 78: 1577
Wu C, Wong Y-C, Elgin SCR (1979) Cell 16: 807
Yelton DE, Scharff MD (1981) Ann Rev Biochem 50: 657

Discussion of the Paper Presented by J. Anderson

ROY: How do you know the antibodies that you used are directed against the chromosomal proteins and not against cytoplasmic contaminants?

ANDERSON: We do not. We have found relatively small amounts of all antigens in a crude cytosol preparation from the hen oviduct. The relative concentrations of the antigens in cytosol ranged from 5% to 50% of their concentrations in chromatin. At this time we are uncertain of the interpretation of this observation. It is possible that some of the antibodies are directed against oviduct cytoplasmic proteins that have contaminated the chromatin during preparation. Alternatively, the antigens are chromosomal proteins that are also found in the cytoplasm. We do know that about half of the antibodies recognize proteins that are tightly bound to the chromatin. In addition, the proteins recognized by the antibodies are associated with soluble chromatin fragments prepared by mild digestion of hen oviduct nuclei with micrococcal nuclease. Although these results provide some very indirect evidence that at least some of these proteins are authentic components of chromatin, we feel that immunofluorescent studies are required for the verification of the nuclear localization of the antigens.

SCHRADER: If you look at the pattern of chromosomal proteins on an SDS gel following hormone administration, you don't see many evident changes. Have you tried doing protein blotting with your antibody to see whether they are major chromosomal proteins that you have antibodies to or whether the antibodies are recognizing minor proteins that are hormone inducible?

ANDERSON: One-dimensional electrophoretic analysis of the proteins from estrogen-stimulated and withdrawn oviduct chromatins does indeed reveal a marked similarity in the intensities of the major bands. The proteins recognized by two of our antibodies, however, do not correspond to any of these major polypeptides and are therefore minor components in the chromatin mixture.

SCHRADER: My second question is an alternate interpretation of something that you presented. I believe you showed that the Y gene in the ovalbumin domain was not digestible by micrococcal nuclease and didn't show the size distribution in the various size classes. Is that correct?

ANDERSON: As much as we can determine.

SCHRADER: That gene is hormone inducible. Even though it is a minor amount compared to ovalbumin, one has to remember that ovalbumin is about 60% total cell protein. Therefore, the Y-gene product is a few percent, 3–6% of the total cell protein. That is a fair amount of the message being made. Doesn't the Y-gene then represent an example of a gene that is hormone inducible but is not micrococcal nuclease sensitive?

ANDERSON: Yes, except we feel it is a matter of detection. With ovalbumin, the first size class was enriched about 10-fold as compared to unfractionated DNA. Since the transcriptional rate of Y is only about 6% of the rate of ovalbumin, one would expect the first size class to be enriched about 1.5-fold in the Y sequences if sensitivity is directly proportional to transcriptional activity. Our current techniques are not sufficiently sensitive to detect a 1.5-fold enrichment.

SCHRADER: It could be just a sensitivity problem in the assay?

ANDERSON: Yes.

SCHRADER: If it is a sensitivity question, then doesn't that call into question your controls? In other words, maybe it is the sensitivity of the assay, as far as the globin gene is concerned and that that is not as good a control as it might initially have seemed.

ANDERSON: The best control is the bulk of the DNA and not globin. You are comparing a specific sequence to the bulk of the DNA.

BAXTER: I was a bit confused about your correlations with the transcriptional activity. You presented experiments that showed no change after you gave actinomycin D. Please clarify.

ANDERSON: The experiment was to isolate nuclei from hen oviduct and incubate the nuclei in actinomycin D.

BAXTER: So you did not block transcription in vivo.

ANDERSON: No, we have wanted to do those studies, but we have not yet done them.

O'MALLEY: I'd like to reemphasize a point derived from your presentation that is very important to people working in hormone action. Many investigators wish to obtain effects of hormones on nuclei, specifically on initiation of RNA synthesis. In fact, people cannot measure significant initiation of RNA synthesis; they can see chain elongation, but not initiation. You showed data on this very active endogenous nuclease that preferentially degrades active genes. As you mentioned, this could of course affect transcription of nuclear chromatin in vitro. The extension of this point that I want to make is that since the entire gene is destroyed within minutes of breaking the cell open, consider what must happen at the hypersensitive sites for DNase, which are at the 5' flanking region of the gene and which are in the region of promoters. It is possible that unless one could block this nuclease during homogenization and cell fractionation, one might never be able to get initiation of transcription of a cell-specific gene in nuclei in vitro, because you have this DNase destroying the promoter region where polymerase must bind and initiate.

Discussants: A. K. ROY, J. N. ANDERSON, W. SCHRADER, J. D. BAXTER AND B. W. O'MALLEY

Chapter 3

Estrogen Receptor Regulation of Vitellogenin Gene Transcription and Chromatin Structure

DAVID J. SHAPIRO, MARTIN L. BROCK, AND MARSHALL A. HAYWARD

I. Introduction

Despite remarkable advances in our understanding of the organization of eukaryotic genes, the mechanisms by which eukaryotic gene expression is regulated remain one of the major unresolved questions in biochemistry and molecular biology. A major approach to this problem has been through analysis of the mechanisms of steroid hormone action at the gene level. For the last several years we have been investigating the estrogen induction of vitellogenin gene transcription in liver cells of male *Xenopus laevis* and using this as a relatively simple model system for regulated gene expression in eukaryotes.

In egg-laying or oviparous vertebrates, the egg yolk precursor proteins, the vitellogenins, are synthesized in the liver as high-molecular-weight precursors, secreted into the serum, and taken up and processed by the ovary (Bergink and Wallace, 1974). *Xenopus* vitellogenins are phospholipoglycoproteins. This set of related proteins each exhibits a monomer molecular weight of approximately 200,000 (Wiley and Wallace 1978). Vitellogenins are encoded by a family of four related mRNAs, each approximately 6.5 kb in length (Shapiro and Baker 1977, 1979; Wahli et al. 1976, 1981). The vitellogenin mRNAs are transcribed from a small gene family composed of four related genes, each of which is approximately 18–20 kb in length and contains at least 33 intervening sequences (Wahli et al. 1980, 1981). Early in our studies it became apparent that the induction of vitellogenin synthesis is accompanied by a corresponding rise in the level of vitellogenin messenger RNA, indicating that it is the amount of messenger RNA in cells that constitutes the primary level of control over the rate of vitellogenin synthesis (Shapiro et al. 1976; Shapiro and Baker 1979). We therefore concentrated our attention on the factors regulating vitellogenin mRNA levels.

The rationale for many of our current studies is based on observations we made several years ago in which we examined the kinetics of vitellogenin mRNA accumulation during primary and secondary estrogen stimulation. (We define primary stimulation as the administration of estrogen to liver cells that have never expressed the vitellogenin genes before, and secondary stimulation as the administration of estrogen to liver cells that are inactive in vitellogenin messenger RNA synthesis at the time of restimulation but have previously expressed these genes.) Three major conclusions emerged from these studies, which are summarized in Fig. 1 (Baker and Shapiro 1977, 1978; Shapiro and Baker 1979; Shapiro 1981). (a) In both unstimulated liver cells and in liver cells withdrawn from estrogen for an extended period of time, there is no detectable vitellogenin mRNA down to a level of less than one molecule per cell. (b) In the primary response there is a lag of several hours before vitellogenin messenger RNA begins to accumulate in cells, whereas in the secondary estrogenic response, there is essentially no lag and messenger RNA is detectable in cells in as little as 1 h. (c) A second major difference between the primary and secondary estrogenic responses is that throughout primary and secondary estrogen stimulation, the rate of vitello-

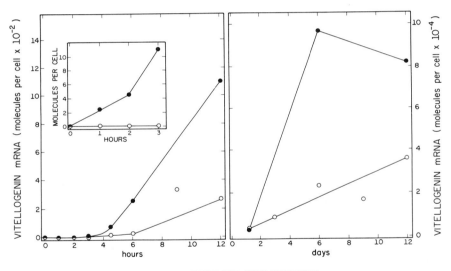

TIME AFTER ESTROGEN ADMINISTRATION

Fig. 1. Comparison of vitellogenin mRNA sequence accumulation during primary and secondary estrogen stimulation. Male *Xenopus laevis* were either unstimulated (primary stimulation; [○——○]) or injected one time with estrogen and withdrawn for 60–65 days (secondary stimulation; [●——●]) prior to administration of estrogen at 0 h. Liver RNA was isolated at the indicated times and hybridized in RNA excess to vitellogenin cDNA. The measured value of $C_rt_{1/2}$, the yield of RNA per gram of liver, and the size of vitellogenin mRNA were used to calculate the number of molecules of vitellogenin mRNA per cell, as described by Baker and Shapiro (1977, 1978). The inset in the left panel illustrates the rapid onset of vitellogenin mRNA accumulation during the first few hours of secondary (but not primary) stimulation. The data demonstrate that the accumulation of vitellogenin mRNA is several times more rapid during both early (left panel) and late (right panel) secondary estrogen stimulation than it is in primary estrogen stimulation. (From Baker and Shapiro 1978.)

genin messenger RNA accumulation is approximately four times greater in secondary estrogen stimulation than in primary estrogen stimulation.

A good deal of our recent work represents an effort to examine two major features of the vitellogenin system, both for their intrinsic interest as potential mediators and regulators of gene expression and also as possible explanations for this kind of cellular memory effect.

II. Estrogen Regulation of Vitellogenin Gene Transcription

In order to localize more precisely the site at which estrogen regulates vitellogenin mRNA it was necessary to move beyond the type of mRNA accumulation study illustrated in Fig. 1. We therefore made direct measurements of the rate of vitellogenin gene transcription. In these experiments we pulse-label the cells with large amounts of tritiated uridine, isolate the labeled nuclear RNA, and hybridize it to vitellogenin cDNA clones immobilized on nitrocellulose filters. The data (Table 1) demonstrate that there is no significant transcription of the vitellogenin genes in either unstimulated or withdrawn *Xenopus* liver cells. The rate of vitellogenin gene transcription during secondary estrogen stimulation is approximately four times greater than the rate during primary estrogen stimulation (Table 1), indicating that the observed four-fold difference in the rates of vitellogenin mRNA accumulation in primary and secondary estrogen stimulation (Fig. 1) is due to differences in the rates of vitellogenin gene transcription. The maximum relative

Table 1. Vitellogenin Gene Transcription Rates During Primary and Secondary Estrogen Stimulation[a]

Time after estrogen (h)	Primary stimulation (ppm)	Secondary stimulation (ppm)
0	0[b]	0[b]
1	—	166
2	—	135
4	75	1390
6	510	2720
8	470	2230
12	670	2240
24	610	2390

[a] Male *X. laevis* received estrogen in vivo and were pulse-labeled for 45 min in organ cultures using 5 mCi of [^{3}H]uridine. Nuclear RNA was isolated by phenol chloroform extraction and hybridized in DNA excess to vitellogenin cDNA clones immobilized on nitrocellulose filters (McKnight et al. 1980). Nonspecifically bound RNA was eliminated by stringent washes and exhaustive digestion with ribonucleases. These data are the relative proportion (in parts per million) of nuclear transcription devoted to vitellogenin RNA synthesis. The pulse-labeling time was chosen with due regard for the fact that approximately 20 min is required to transcribe a 20-kb gene in *Xenopus* (Anderson and Smith 1978). cDNA clones are used as hybridization probes because genomic clones contain numerous repeated DNA sequences within their introns. These repeat sequences can, and do, hybridize to a variety of nuclear RNAs.
[b] Less than 2 cpm/filter. (Primary 0 time was pulse-labeled for 60 min.)

rate of vitellogenin gene transcription in primary estrogen stimulation is
674 ± 157 (S.D.) parts per million of nuclear RNA synthesis. Once this rate is
attained, approximately 12 h after primary estrogen stimulation, it is maintained, but not exceeded, for many days. Following withdrawal of estrogen,
transcription of the vitellogenin genes ceases, indicating that leaky low-level
transcription of the vitellogenin genes into nuclear RNA does not continue in
the absence of estrogen. Restimulation of these withdrawn cells with estrogen produces significant transcription of the vitellogenin genes in as little as
1 h (Table 1) and results in maximal transcription $(2451 \pm 171$ ppm) in
approximately 6 h.

These data indicate that the expression of the vitellogenin genes is under
precise transcriptional control. These genes are not transcribed in the absence of estrogen either in *Xenopus* liver cells or in a heterologous cell
[cultured *Xenopus* kidney cells (Brock and Shapiro 1982)]. Our data are
consistent with the view that estrogen-mediated changes in the rate of vitellogenin gene transcription represent the primary locus of control in this
system.

III. The Estrogen Receptor and Vitellogenin Gene Expression

In the general model for steroid hormone effects on gene expression the
cytoplasm of target cells contains high levels of protein(s) capable of stereo-
specific high-affinity hormone binding. The binding of hormone to the receptor results in "transformation" of the receptor to a form with a high affinity
for nuclear components. The receptor then translocates into the nucleus,
binds to chromatin, and exerts its nuclear effects (Yamamoto and Alberts
1976; Jensen 1979). The estrogen receptor systems involved in the induction
of vitellogenin mRNA deviate from this model in several important respects.
Cytoplasm of unstimulated male *Xenopus* liver cells contains little, if any
estrogen receptor (Hayward et al. 1980; Fig. 2). Nuclei of unstimulated
Xenopus liver cells contain approximately 550 high-affinity estrogen binding
sites (Fig. 2).

Administration of estradiol-17β results in the induction of nuclear receptor
that reaches a level of approximately 2000 sites per cell (Fig. 2). Withdrawal
of estrogen results in a decline in the nuclear receptor level that ultimately
returns to the level prevailing in unstimulated cells (approximately 500 sites/
nucleus; Fig. 2). In striking contrast, estrogen receptor, which is virtually
absent in cytosol of unstimulated and estrogen-stimulated cells increases to a
level of approximately 1000 sites/cell in withdrawn cells (Hayward et al.
1980; Fig. 2). This cytoplasmic receptor is able to translocate rapidly into the
nucleus on restimulation with estrogen and may play a role in the shorter lag
period required for the initiation of vitellogenin gene transcription early in
secondary estrogen stimulation.

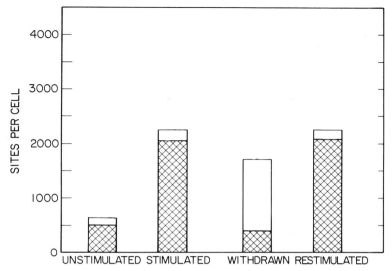

Fig. 2. Estrogen receptor levels in unstimulated, stimulated, withdrawn, and restimulated *Xenopus* liver cells. The level of specific high-affinity estrogen binding to cytoplasmic (open bars) and nuclear (cross-hatched bars) estrogen receptors was determined in unstimulated adult male *Xenopus* liver; 2–3 days after primary estrogen stimulation (stimulated) in animals withdrawn from estrogen for 60–70 days (withdrawn); and 2 days after restimulation of withdrawn *Xenopus* (restimulated). (From Hayward et al. 1980 and unpublished data.)

One explanation for the more rapid transcription of the vitellogenin genes on restimulation with estrogen that we can exclude is that there is more nuclear estrogen receptor in those restimulated cells and that the rate of transcription is in some direct way related to the level of nuclear receptor. As indicated in Fig. 2, the level of nuclear estrogen receptor is essentially constant (approximately 2000 sites/cell) in primary and secondary estrogen stimulation. The maximum rate of vitellogenin gene transcription in primary estrogen stimulation is therefore not limited or controlled by the level of nuclear estrogen receptor.

Early in primary and secondary estrogen stimulation there is a reasonable correlation between the increase in nuclear estrogen receptor level and the onset of vitellogenin messenger RNA accumulation. The insert in Fig. 3 shows that the level of nuclear estrogen receptor changes very little until about 3 h or so after estrogen stimulation, when it starts to rise very sharply. This coincides reasonably well with the time at which vitellogenin mRNA first appears in these cells. A similar correlation between the induction of increased levels of nuclear estrogen receptor and the appearance of vitellogenin mRNA is observed early in secondary estrogen stimulation. A careful examination of the kinetics of estrogen receptor induction and vitellogenin mRNA accumulation suggests that the induction of nuclear estrogen receptor may precede the appearance of vitellogenin messenger RNA (Fig. 3). We therefore decided to examine the question of whether or not the induction of additional nuclear estrogen receptor is a prerequisite for the activation of vitellogenin gene transcription. This question relates to the broader issue of whether or not the estrogen induction of vitellogenin gene transcription represents a direct or early effect of the hormone or requires the prior induction of other regulatory proteins.

We approached this question by quantitatively (>99%) inhibiting protein synthesis using cycloheximide and assaying the levels of nuclear estrogen receptor and of cellular vitellogenin mRNA. The data (Table 2) demonstrate that inhibition of protein synthesis effectively blocks the estrogen induction of nuclear estrogen receptor. However, quantitative inhibition of protein synthesis and nuclear estrogen receptor accumulation does not abolish the induction of vitellogenin mRNA in either primary or secondary estrogen stimulation (Table 2). These data and earlier kinetic data (Shapiro and Baker 1977, 1979; Baker and Shapiro 1977, 1978; Shapiro 1982) are all consistent with the view that the induction of vitellogenin gene transcription is a direct early effect of estrogen in this system. The fact that the level of vitellogenin mRNA accumulation is markedly reduced following inhibition of protein synthesis suggests that the induction of estrogen receptor is required to

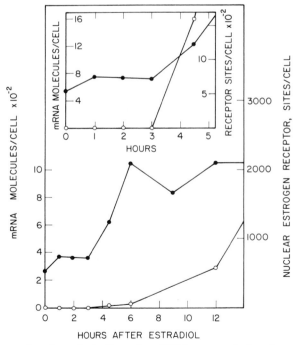

Fig. 3. Nuclear estrogen receptor levels and vitellogenin mRNA accumulation during primary estrogen stimulation. The level of nuclear estrogen receptor (●——●) was determined at various times after estrogen stimulation. The kinetics of vitellogenin mRNA accumulation (○——○) was determined in a separate series of experiments (Baker and Shapiro 1978).

Table 2. Absence of an Absolute Requirement for Protein Synthesis in Estrogen Induction of Vitellogenin mRNA[a]

	Nuclear estrogen receptor (sites/cell)	Vitellogenin mRNA (molecules/cell)
Primary stimulation 6 h after estrogen	2000	25
Cycloheximide-treated, primary stimulation 6 h after estrogen	500	5
Secondary stimulation 6 h after estrogen	1200	256
Cycloheximide treated, secondary stimulation 6 h after estrogen	600	42

[a] Protein synthesis was quantitatively inhibited by injection of cycloheximide 60 min before administration of estradiol-17β. At this dose, protein synthesis is quantitatively inhibited for at least 12 h. Six hours after estrogen administration the livers were excised and aliquots were assayed for nuclear estrogen receptor and vitellogenin mRNA content.

allow vitellogenin gene transcription to reach its maximum rate. Once the level of nuclear estrogen receptor reaches a maximum (approximately 2000 sites/cell), other factors appear to control the rate of vitellogenin gene transcription, as that level of nuclear estrogen receptor is capable of supporting the very different vitellogenin gene transcription rates observed in primary and secondary estrogen stimulation (Table 1).

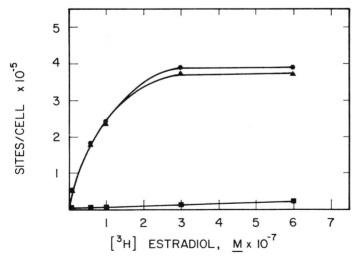

Fig. 4. Estrogen binding as a function of estradiol-17β concentration in a cytoplasmic extract from unstimulated male *Xenopus* liver. Cytosol was clarified by centrifugation at 100,000 g_{av}. Supernatants were incubated with increasing concentrations of [^{3}H]estradiol alone to determine total binding (●) or with [^{3}H]estradiol plus a 100-fold excess of unlabeled estradiol-17β to determine nonsaturable binding (■). Specific binding (▲) is derived by subtracting nonsaturable from total binding at each estradiol concentration. (From Hayward and Shapiro, 1981.)

Table 3. Specificity of Estradiol-17β Binding to the Estrogen Receptor and Estrogen-Binding Protein[a]

Cold competitor 100× excess	% Inhibition of binding	
	Receptor	Binding protein
Estradiol-17β	100	100
Estrone	70	94
Diethylstilbesterol	87	69
Estradiol-17α	N.D.	88
Progesterone	5	17
Testosterone	9	4
Dexamethasone	5	3

[a] Incubations for competition experiments involving estrogen receptor contained $5 \times 10^{-9}\ M$ [^{3}H]estradiol-17β and $5 \times 10^{-7}\ M$ unlabeled competitor. Incubations for competition experiments involving the estrogen-binding protein contained $5 \times 10^{-7}\ M$ [^{3}H]estradiol-17β and $5 \times 10^{-5}\ M$ unlabeled competitor. Data are from Hayward et al. (1980) and Hayward and Shapiro (1981).

IV. A Middle Affinity Estrogen-Binding Protein In *Xenopus* Liver Cytosol

The apparent absence of significant levels of estrogen receptor in *Xenopus* liver cytosol led us to examine cytosol extracts for other estrogen-binding entities. *Xenopus* liver cytosol contains high levels of the middle-affinity estrogen-binding protein whose binding isotherm is shown in Fig. 4. In contrast to the estrogen receptor, which exhibits a K_d of $4 \cdot 10^{-10}\ M$ (Westley and Knowland 1978; Hayward et al. 1980), the estrogen-binding protein, or EBP exhibits a K_d for estradiol-17β of $4 \times 10^{-8}\ M$ (Hayward and Shapiro 1981). The EBP is not a nonspecific steroid-binding protein, as it exhibits the same highly specific estrogen binding as does the estrogen receptor (Table 3). The EBP is unlikely to play a direct role in the induction of vitellogenin gene expression, as it does not translocate into the nucleus following estrogen administration (Hayward and Shapiro 1981) and exhibits a K_d for estradiol-17β that is inconsistent with the dose–response curve for the induction of vitellogenin in organ cultures. The EBP, which is present at several hundred thousand molecules/cell (Fig. 4), may function as a kind of "shuttle" protein that transfers estrogen from the cell membrane to the nuclear estrogen receptor, or as an agent that protects estradiol from rapid cytoplasmic metabolism by sequestering it.

V. Methylation of the Vitellogenin Genes

A major goal of our research has been to identify unique properties or structural features of the vitellogenin genes that correlate with their expression. We were also interested in those properties that might allow us to distinguish between a gene in an unstimulated cell, which has never been expressed, and one in a withdrawn cell, which seems to have an exceptional potential for efficient expression and more rapid transcription. In many vertebrate systems the demethylation or undermethylation of DNA within and around a particular gene appears to be positively correlated with the expression of that gene. (For reviews see Razin and Riggs 1980; Burdon and Adams 1980; Ehrlich and Wang 1981.) We therefore examined the methylation state of one of the four vitellogenin genes, designated the A1 gene (Wahli et al. 1979, 1980, 1981). We identified and mapped nine methylation sites in the 3' ⅔ of the gene (approximately 13 kb of DNA starting from the 3' end of the gene). The restriction map of the A1 gene methylation sites was obtained by digesting cloned DNA with the appropriate enzymes (Fig. 5). Since this genomic *Xenopus* DNA clone is being synthesized as λ-DNA within *E. coli*, it is unmethylated and all the potential methylation sites are cut by methylation-sensitive restriction endonucleases. The methylation state of these sites in *Xenopus* liver and in a control nonexpressing tissue, red blood cells, was determined by isolating the cellular DNA, digesting it with Eco RI or with Eco R1 plus the appropriate methylation-sensitive restriction endonuclease, fractionating the digested DNA by agarose gel electrophoresis, transferring it to nitrocellulose by Southern blotting, and hybridizing with appropriate nick-translated cloned probes (Folger-Bruce, 1981).

Two striking conclusions emerged from this work: (a) There is no difference between the methylation pattern of the vitellogenin genes in nonexpressing tissues (red blood cells, uninduced liver, and withdrawn liver) and estrogen stimulated liver in which the vitellogenin genes are being transcribed at a rapid rate (Fig. 5). (b) Even during periods of rapid transcription the vitellogenin genes remain extensively methylated. Six of the nine methylation sites we identified remain fully methylated in estrogen stimulated liver, two sites are partially methylated, and one site is unmethylated in all cases (Fig. 5).

Xenopus liver is a mixture of at least four different cell types. Hepatocytes that are capable of inducing vitellogenin gene transcription comprise only about 50% of the cells in *Xenopus* liver. One possible objection to our work would be to argue that the reason we don't see any change in methylation patterns in induced liver is that only a small fraction of the cells whose DNA we have looked at actually transcribe the vitellogenin genes and exhibit a change in methylation. This type of minor change could perhaps escape detection on our blots. To examine this possibility we were able to take advantage of a unique feature of the vitellogenin system. The induction of vitellogenin synthesis and secretion is so massive that in order to secrete the very large amounts of vitellogenin produced, cells that are actively expressing the vitellogenin genes must elaborate and synthesize a large and complex secretory apparatus. These cells contain so much lipid that they exhibit a physical change in cell density relative to both unstimulated hepatocytes and to all the other cell types in *Xenopus* liver. These hepatocytes, by virtue of this density shift, can be separated from all the other cells in *Xenopus* liver by fractionation on metrizamide gradients. This allows us to isolate a pure population of cells that are actively expressing the vitellogenin genes. When

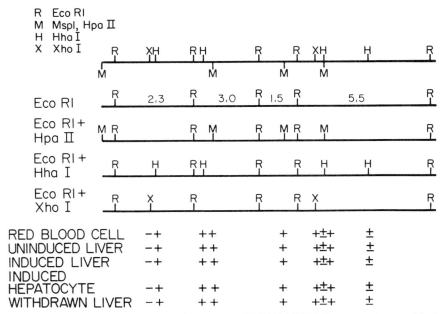

Fig. 5. Methylation of the vitellogenin A1 gene. X1V106 DNA (generously provided by W. Wahli) was digested with all the available methylation-sensitive restriction enzymes. The vitellogenin clone X1V106 is devoid of sites for some enzymes, while others (such as Ava 1) are less specific than the enzymes used and revealed no additional sites. The restriction map for all the enzymes used is shown on top with the maps for all the individual pairs of enzymes used in the digests shown below. Sites that are fully methylated ($+$), methylated in some but not all cells ($\pm$), and unmethylated ($-$) are shown below. The partially methylated and unmethylated sites are in introns.

we isolate these cells, which we call induced hepatocytes, we find that there is no change in their methylation pattern relative to nonexpressing cell types (Fig. 5). We therefore conclude that the expression of the vitellogenin genes is not correlated with DNA demethylation (at least at internal sites) and that rapid transcription of the vitellogenin genes occurs even when they are extensively methylated.

VI. DNase Sensitivity of the Vitellogenin Genes in Chromatin

The absence of any changes in internal methylation sites of the vitellogenin A1 gene leads to the question of whether transcriptionally active vitellogenin genes exhibit the properties typical of other transcriptionally active genes. It has been known for several years that genes in chromatin that are undergoing rapid transcription exhibit enhanced susceptibility to digestion by DNase 1 (Weintraub and Groudine 1976) and by micrococcal nuclease

(Bloom and Anderson 1978) and preferential solubility at low magnesium concentrations (Bonner et al. 1975). We examined the conformation of the vitellogenin genes in chromatin by investigating their sensitivity to DNase 1 digestion.

In these experiments, nuclei are isolated, the intact nuclei are digested with DNase 1, nuclear DNA is isolated and fractionated by agarose gel electrophoresis, and the DNA corresponding to the various size classes is isolated from the gels. The four size fractions we use are illustrated in Fig. 6, which shows an agarose gel of the digested size fractionated DNA. Genes undergoing active transcription will be cut more frequently (and into smaller fragments) by DNase 1 and will be enriched in the smallest of these size classes (fraction 1 DNA). The relative abundance of vitellogenin sequences in each of the four size classes of DNA was determined by moderate probe excess hybridization to vitellogenin DNA (Bloom and Anderson 1978). In unstimulated liver cells that do not transcribe the vitellogenin genes (Table 1) and do not contain detectable vitellogenin mRNA (Baker and Shapiro 1978; Fig. 1) the vitellogenin genes are depleted in the transcriptionally active fraction 1 DNA and enriched in the bulk DNA (fraction 4), which is transcriptionally inactive (Fig. 7, upper panel). Following the estrogen induction of vitellogenin gene transcription, the vitellogenin genes become DNase sensitive (Folger et al 1980; Felber et al. 1981; Fig. 7, middle panel). These data suggested a pattern of DNase sensitivity that is typical of that observed for many transcriptionally active genes. The question of whether DNase 1 sensitivity parallels transcription or is an irreversible property that the vitel-

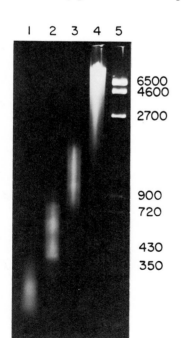

Fig. 6. Agarose gel electrophoresis of the four DNA size classes obtained from DNase 1 digested DNA. *Xenopus* liver nuclei were digested with DNase 1, the DNA was extracted, and size fractionated by electrophoresis on a preparative 1.5% of agarose gel. The regions corresponding to the four DNA molecular weight size classes were excised from the gel, and the DNA was isolated. These fractions were run on another agarose gel. Lanes 1–4 are the four DNA size classes used in these studies. Lane 5 is a Hind II + Hind III digest of the 10-1A plasmid. The fragment lengths in base pairs are on the right.

Fig. 7. Distribution of the vitellogenin genes in different size classes of DNA obtained after DNase 1 digestion. Nuclei from unstimulated (upper panel), estrogen-stimulated (middle panel), and withdrawn (lower panel) liver cells of male *Xenopus laevis* were digested with DNase 1. The digested DNA was separated into four molecular-weight size classes by agarose gel electrophoresis. The DNA in each of the four size classes (see Fig. 6) was hybridized in moderate cDNA excess (approximately 10-fold excess over the amount of vitellogenin cDNA in unfractionated DNA) to [^{3}H] vitellogenin cDNA and the number of counts in hybrids was determined. The DNA size classes are 1 (●) the smallest DNA (Fig. 6), 2 (▲), 3 (□), 4 (△), and unfractionated (○). The fraction 2 and 3 curves are omitted from the center and lower panels for simplicity. (Redrawn from data of Folger et al. 1980.)

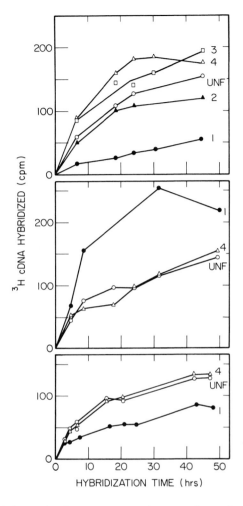

logenin genes retain even after withdrawal of estrogen and cessation of transcription was a critical one. Nuclease sensitivity studies of the globin (Stalder et al. 1980a,b) and ovalbumin (Palmiter et al. 1977) genes suggest that the globin genes become DNase sensitive prior to transcription and that both sets of genes remain nuclease sensitive even after transcription ceases. Estrogen administration does not produce a permanent change in the conformation of the vitellogenin genes in chromatin. The vitellogenin genes in withdrawn cells have returned to the transcriptionally inactive DNase-insensitive conformation found in unstimulated cells (Fig. 7, bottom panel). The data indicate that the shorter lag period observed during secondary stimulation is not the result of permanent conversion of the vitellogenin genes to a more "open" conformation—at least with respect to those structural features that are susceptible to investigation using DNase 1. These data also suggest that the vitellogenin genes may become DNase sensitive without a requirement for the cycle of DNA replication observed in many other systems, since induction of vitellogenin does not require, and may not involve, DNA replication (Green and Tata 1976). In this system in which vitellogenin gene transcription occurs for relatively brief periods over the life span of a differentiated liver cell, DNase 1 sensitivity appears to be closely correlated with transcription.

VII. Methylation of Xenopus Chromatin Fractions

Our inability to demonstrate a role for changes in DNA methylation in the expression of the vitellogenin genes led us to approach the broader question of whether transcriptionally active DNA fractions are undermethylated in *Xenopus laevis*. Our ability physically to isolate DNA enriched in transcriptionally active sequences (Fig. 7) allowed us to approach this question. We developed a method for separation of the various DNA bases by HPLC using methods similar to those described by Kuo et al. (1980). The separation of 5-methyl cytosine from the other bases and from the small amount of ribo G that contaminates all DNA preparations we have examined was excellent (Fig. 8). Integration of the various peaks yields data similar to litera-

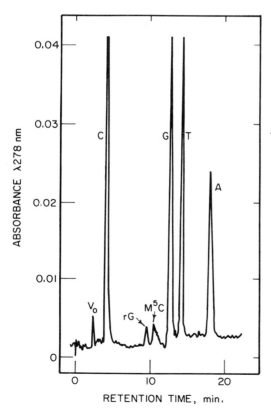

Fig. 8. Fractionation of the constituents of DNA by HPLC. Extracted, deproteinized DNA was exhaustively digested with DNase 1, Nuclease P1, and bacterial alkaline phosphatase (Kuo et al. 1980). The nucleosides were separated (at 20°C) on an Altex Ultrasphere ODS column in a buffer containing 10 mM $NH_4H_2PO_4$, pH 5.3, and methanol (6–15%). Nucleosides were quantitated by monitoring absorption at 278 nm (relative to standards) and integrating the area under the peak with a Hewlitt-Packard integrator. Data are for a 10-ng sample of salmon sperm DNA.

Table 4. 5-methyl Cytosine Content of *Xenopus* Liver Chromatin Fractions[a]

Source of DNA	Source of data	m^5C	Mole % $m^5C/C + m^5C$
Salmon	Authors	1.5	6.8
sperm	Literature[b]	1.6	7.2
Calf	Authors	1.3	5.8
thymus	Literature[b]	1.4	6.2
Xenopus	Authors	1.8	8.5
red blood cells	Literature[c]	1.4	6.9
Xenopus liver,	Authors	1.5	7.4
unstimulated	Literature[d]	1.7	—
Xenopus liver, estrogen stimulated			
Unfractionated		1.7	7.9
Fraction 1[e]		1.6	7.5
Fraction 4		1.7	7.9

[a] The determination of m^5C content is as described in the legend to Fig. 8.
[b] Data of Kuo et al. (1980) and 5 to 15 others.
[c] Data of Dawid (1965) done by TLC.
[d] Data of Bauer and Kroger (1976), m^5C content only.
[e] The fraction 1 and 4 DNA is that shown in the middle panel of Fig. 7.

ture values for such well-studied DNAs as calf thymus and salmon sperm (Table 4). The 5-methyl cytosine content of fraction 1 DNA (which is enriched in transcriptionally active sequences [Fig. 7]) was identical to that of unfractionated DNA and fraction 4 DNA (Table 4). These preliminary studies have therefore failed to reveal a role for DNA methylation changes in the overall pattern of *Xenopus* gene expression. Of course, the possibility that there exists a small number of critical methylation sites, perhaps in the 5' flanking region, is not excluded by any of our investigations.

IX. Conclusions

Although the detailed molecular mechanisms regulating vitellogenin gene expression and mRNA levels remain obscure, several key features of these processes have been revealed in recent studies. Unstimulated liver cells preserve the potential to respond to estrogen by transcription of the vitellogenin genes, but they are in no sense primed to do so. In the liver cells of unstimulated male *Xenopus* the vitellogenin genes are insensitive to DNase and are not transcribed at an appreciable rate (Table 1; Fig. 7; Folger et al. 1980; Felber et al. 1981). The cytoplasm of these cells contains little if any estrogen receptor. Estrogen entering the cells presumably binds to the cytoplasmic estrogen binding protein and eventually reaches the nucleus. In the nucleus it binds to preexisting nuclear estrogen receptor and elicits at least two responses. The estrogen–receptor complex both activates vitellogenin gene transcription and autoinduces the synthesis of the additional receptor that is required for maximal transcription of the vitellogenin genes. The induction of vitellogenin gene transcription, which is accompanied by a shift of the genes to a transcriptionally active DNase 1 sensitive conformation, does not involve changes in the methylation pattern of internal sites in the vitellogenin genes.

This type of overview of events involved in the induction of vitellogenin gene transcription allows us to begin to ask questions on the mechanism of estrogen action at a mechanistic rather than a phenomenologic level.

References

Anderson DM, Smith LD (1978) Devel Biol 67: 274–285

Baker HJ, Shapiro DJ (1977) J Biol Chem 252: 8428–8434

Baker HJ, Shapiro DJ (1978) J Biol Chem 253: 4521–4524

Baur R, Kroger H (1976) Hoppe-Seylers Z Physiol Chem 357: 308

Bergink EW, Wallace, RA (1974) J Biol Chem 249: 2897–2903

Bloom KS, Anderson JN (1978) Cell 15: 141–150

Bonner J, Gottesfeld J, Garrard W, Billing R, Uphouse L (1975) Meth Enzymol 40B: 97–102

Brock ML, Shapiro DJ (1982) Submitted for publication

Burdon RH, Adams, RLP (1980) Trends Biochem Sci 11: 294–297

Dawid I (1965) J Mol Biol 12: 581–597

Ehrlich M, Wang RY-H (1981) Science 212: 1350–1357

Felber K, Ryffel U, Weber R (1978) Molec Cellular Endocrinol 12: 151–166

Felber B, Gerber-Huber S, Weber R, Ryffel GU (1981) Nuc Acids Res 15: 2455–2474

Folger KR, Anderson JN, Shapiro DJ (1980) Fed Proc 39: 1886

Folger-Bruce K (1981) Ph.D. Thesis University of Illinois, Urbana IL

Green CD, Tata JR (1976) Cell 7: 131–139

Hayward MA, Mitchell TA, Shapiro DJ (1980) J Biol Chem 255: 11308–11312

Hayward MA, Shapiro DJ (1981) Develop Biol 88: 333–340

Jensen EV (1979) Pharmacol Rev 30: 477–491

Kuo KC, McCune RA, Gehrke CW, Midgett R, and Ehrlich M (1980) Nuc Acids Res 8: 4763–4776

McKnight GS, Lee DC, Palmiter RD (1980) J Biol Chem 255: 148–153

Palmiter RD, Mulvihill ER, McKnight GS, Senear AW (1977) Cold Spring Harbor Symp Quant Biol XLII: 639–648

Razin A, Riggs AD (1980) Science 210: 604–609

Shapiro DJ (1982) CRC Crit Rev Biochem, 12: 187–204

Shapiro DJ, Baker HJ, Stitt DT (1976) J Biol Chem 251: 3105–3111

Shapiro DJ, Baker HJ (1977) J Biol Chem 252: 5244–5250

Shapiro DJ, Baker HJ (1979) Ontogeny of Receptors and Reproductive Hormone Action, edited by TH Hamilton, JH Clark, WA Sadler, Raven Press, New York 1979, pp. 309–330

Stalder J, Groudine M, Dodgson JB, Engel JD, Weintraub H (1980a) Cell 29: 973–980

Stalder J, Larsen A, Engel JD, Dolan M, Groudine M, Weintraub H (1980b) Cell 20: 451–460

Wahli W, Wyler T, Weber R, Ryffel GU (1976) Eur J Biochem 66: 457–465

Wahli W, Dawid IB, Wyler T, Jaggi RB, Weber R, Ryffel GU (1979) Cell 16: 535–549

Wahli W, Dawid IB, Wyler T, Weber R, Ryffel GU (1980) Cell 20: 107–117

Wahli W, Dawid IB, Ryffel GU, Weber R (1981) Science 212: 298–304

Weintraub H, Groudine M (1976) Science 193: 848–856

Westley B, Knowland J (1978) Cell 15: 367–375

Wiley WH, Wallace RA (1978) Biochem Biophys Res Commun 85: 153–159

Yamamoto KR, Alberts BM (1976) Ann Rev Biochem 45: 722–746

Discussion of the Paper Presented by D. J. Shapiro

O'MALLEY: Do you have a methylation site at the 5' end of the gene that you could monitor. As I'm sure you are aware, demethylation at a specific promotor sequence could still occur.

SHAPIRO: This is a possibility that's very difficult for us to exclude in this system. The methylation sites we've looked at are not all the way out of the 5' end of this gene, both because of lack of some of the appropriate probes and also because the DNA around the 5' end of the gene is very heavily middle-repetitive DNA and it is very difficult to probe. It's worth pointing out that a little bit of data from Barbara Felber and Suzanne Gerber-Huber in the laboratories of Gerhart Ryffel and Rudolph Weber does suggest that there is no particular DNase-sensitive site at the 5' end of the vitellogenin genes. However, the possibility that there is a single methylation site somewhere near the 5' end of these genes that does change on expression can't be excluded, and I emphasize that the data that we have presented here just indicate that undermethylation at internal sites isn't positively correlated with expression of the vitellogenin genes.

O'MALLEY: I have a question on another statement you made. Unless I misunderstood the DNase digestion experiments, I did not see anything that is incompatible with what goes on in oviduct tubular gland cells. As I understand it, you're using the Anderson-type assay including size fractionation. That's really not looking at DNase-1 sensitivity as we're looking at it. You are really looking at a hypersensitive region. In other words, you're looking for regions that are preferentially clipped and which produce smaller pieces of DNA which you detect in your fraction-1. This is a similar assay to the micrococcal nuclease assay that John (Anderson) mentioned. In fact, I would think those were equivalent assays even though they use different enzymes. I don't think the enzymes are as important as the method of preparation of DNA, the time course of digestion, and the assay protocol.

SHAPIRO: Let me make a few points first: I don't say that these are incompatible; I think there are some real differences between the systems that we can get at. If you do DNase-1 sensitivity studies by much the same methodology as this, in oviduct (and this has been done in Dick Palmiter's lab, and also by John), at least over the withdrawal times that people look at, the genes do appear to be DNase-1 sensitive. We consider that when we look at DNase-1 sensitivity, by a hybridization method as opposed to a blotting method, what we are looking at, in a sense, is the average sensitivity across a long stretch of DNA, because we are looking at a situation in which a large number of particular cuts have to be made, to take a 20-kb DNA down to a few hundred nucleotides. So we are not looking for one or a small number of hypersensitive sites. It is also true, however, that if you do a blotting experiment (and this has been done by the Ryffel and Weber labs, with discrete vitellogenin DNA fragments), one obtains essentially the identical result to the DNase sensitivity data that I have presented here, which is that the gene is not sensitive in unstimulated cells. So I think there really is some difference between the two systems. There are at least two possible explanations that I can see for this: One is that in our system, which employs liver cells that do not require estrogen for viability, it is possible to withdraw the cells for a very long period of time. At the end of this withdrawal period, it is possible to show that vitellogenin genes are not being transcribed at any rate at all. A number of DNase sensitivity studies starting with the original Weintraub and Groudine studies and ranging through some studies on MMTV have indicated that in some cases even a very low level of expression is enough to make a gene DNase 1 sensitive. So it is at least possible that in the short-term withdrawal experiments that are feasible in the oviduct there is enough residual transcription to allow the ovalbumin gene to attain DNase-1 sensitivity. I think the other possible explanation is that a liver cell is in fact not a terminally differentiated specialized cell; it's basically a multipotential differentiated cell. An oviduct cell is one whose differentiation has been dedicated to the production of egg white protein, and in the absence of its hormone it won't continue to exist. A *Xenopus* liver cell, of course, in half the animals, may maintain the potential to express that gene, but would never expect to express it. So in this kind of multipotential differentiated state, I think it's likely that the hormone creates the situation in which that gene can become responsive, but in the absence of the hormone, the liver cell reverts to its more or less generalized differentiated functions. So I think it's perhaps a little bit different, in terms of what kind of differentiating system it may be, from one that is even more specified, such as the oviduct system.

O'MALLEY: I don't agree with what you are saying. The DNase I assays that Axel and Palmiter used on oviduct tissue are quite different from yours. They were looking at digestion of 15–20% of the DNA with overall depletion of specific sequences. The anderson assay is a much more sensitive assay. It can detect a region of greater sensitivity with a stretch of moderately sensitive DNA. As I see it, our results could be similar in both cell types. Maybe John (Anderson) wants to make a comment on this point.

ANDERSON: The correlation between transcription and DNase I sensitivity in the vitellogenin system and the lack of such a relationship with ovalbumin could result from differences in experimental protocols such as digestion extent, or from an actual conformational difference between the vitellogenin and ovalbumin genes in the hormone-withdrawn chromatins. Minimal and extensive DNase I digestions should be performed in both systems for clarification of this point.

SCHRADER: You made the point about the methylation pattern being essentially unchanged. I think it should be pointed out that when you use that restriction nuclease method, it depends on the luck of the draw of having a critical methylation at a site in the DNA for which you have the pair of isoschizomers that will detect it. You have mapped five or six sites or so, out of all the C residues in that domain that are potentially methylatable. So there's nothing to exclude the possibility using that assay that there are other sites that are methylated in the domain that are very acutely regulated. I think it might be a little bit too general to say that the methylation pattern does not change. I just like to point out one that should say for those five or six sites for which you have the probe that it doesn't change.

SHAPIRO: What we have done is all that can be done at this time. We have employed all known methylation-sensitive enzymes, not just the HpaII–MspI isoschizomers. It appears in the DNA sequences that have been looked at, these restriction enzymes do identify most of the known methylation sites. So it's quite likely that the nine sites we have looked at actually represent a reasonable subset of methylation sites and probably include most of the internal methylation sites in this system.

SCHRADER: My second question is with respect to the weak estrogen binder and the strong estrogen receptor in the nucleus. Could you review for us the evidence that the strong estrogen binding sites in the nucleus are in fact the estrogen receptor that is biologically active? For example, the simplest experiment would be to show that if you give labeled hormone to a *Xenopus* that the label ends up on the strong sites in the nuclei of the liver. Has that experiment been done?

SHAPIRO: I'll do just what you said. I'll review the evidence that that binding entity that we call the estrogen receptor is involved in the induction. Since there are only a very small number of sites, this receptor system is not as extensively characterized as some others. The evidence is basically this: the dose response curve for induction is the same as the binding curve for the estrogen receptor. Using in vitro induction systems, you can get induction beginning at concentrations at which you begin to load that receptor. The second major point is that the binding affinity of the receptor for various steroid hormones correlates very well with the ability of these hormones to induce vitellogenin gene expression in organ cultures. The hormone specificity is quite different from that in the oviduct system. The oviduct is secondary estrogen stimulation becomes quite a promiscuous system in which many substances, such as kepone, and several hormones can turn on the ovalbumin genes. In the vitellogenin system the same rigid and precise estrogen specificity is maintained in secondary stimulation. Finally, there are not other detectable estrogen binding entities in these cells. The specific experiment you mentioned is difficult to do, given the very small number of estrogen receptor sites in these cells.

SCHRADER: Could you clarify the last point you made? You said that it had been a difficult experiment to do. If one injects labeled hormone, do you mean that you just don't have enough counts to do the experiment or that, for example, there are too many counts in the nucleus?

SHAPIRO: If you inject labeled hormone into animals the combination of rapid metabolism of the injected labeled estradiol-17β and the small numbers of receptor sites results in essentially no detectable label in the nucleus. This type of experiment is usually done using the in vitro "liver cube" induction system. Even in this system very few counts are found in the nucleus. The level of nuclear estrogen receptor is only a few hundred sites per cell in unstimulated cells, and this is 10 to 100 times lower than the level prevailing in most other systems that are responsive to estrogen.

LIAO: I remember you use the word *potentiation* of mRNA synthesis or production by steroid receptor complex. I just wonder whether you have any specific view on that and also whether you include something like protection of RNA.

SHAPIRO: Well, we have some data which I haven't talked about today which suggest that estrogen does exert a pleotropic effect in this system. In addition to inducing the transcription of the vitellogenin genes it also induces a selective stabilization of vitellogenin mRNA against degradation in the cytoplasm and therefore that the induction of massive amounts of vitellogenin mRNA is a consequence of both the initiation of rapid transcription of the vitellogenin genes and the fact that vitellogenin mRNA is extremely stable during induction. As far as the actual molecular mechanisms for either of those two processes is concerned, I'm afraid that we don't have any more compelling mechanistic data than do any of the other individuals who have spoken.

LIAO: Maybe I should ask one more question. Can you release the steroid receptor complex from nuclei by RNase?

SHAPIRO: We have not actually attempted to release the complex with ribonuclease, so I really don't know what the results will be.

Discussants: B.W. O'MALLEY, D.J. SHAPIRO, J.N. ANDERSON, W. SCHRADER, and S. LIAO

Chapter 4

Ecdysterone, Ecdysterone Receptor, and Chromosome Puffs

B. Dworniczak, S. Kobus, K. Schaltmann-Eiteljörge, and O. Pongs

I. Introduction

Steroid hormones modulate the rates of transcription of specific genes (Ashburner 1972b; Becker 1959; Becker 1962; Berendes 1967; Clever and Karlson 1960; Clever 1961; Zhimulev 1974). The paradigm of hormone action is that the hormone interacts with a specific cytoplasmic receptor. These hormone–receptor complexes then translocate rapidly into the nucleus. Thereupon, they elicit their regulation of transcription by mechanisms that are not known.

Recent studies on steroid hormone regulation of ovalbumin and conalbumin gene transcription have demonstrated a direct relationship between the level of transcription of these genes and the concentration of estrogen receptor in the nucleus. Studies on the interaction of glucocorticoid receptor with cloned MTV–DNA sequences have shown that the receptor indeed binds with high affinity to specific DNA sequences and exhibits the kind of selectivity that is known for prokaryotic transcriptional regulatory proteins.

The steroid hormone ecdysterone is a key hormone in the development of *Drosophila melanogaster*. There are many target tissues for the hormone, among them the salivary glands of *Drosophila* larvae. The salivary glands contain highly polytenized chromosomes. They represent an important experimental system to study the regulation of gene expression, potentially combining the wealth of *Drosophila* genetic information with suitable cytological as well as biochemical analysis. We have recently shown that endogenous ecdysterone can be photoactivated in salivary glands of *D. melanogaster* larvae (Gronemeyer and Pongs 1980). Thereupon, the hormone is covalently bound to component(s) of the polytene chromosomes. The bonded hormone was visualized by indirect immunofluorescence microscopy. This technique allowed us for the first time to demonstrate the presence of ecdysterone on chromosomal loci, whose transcription is hormonally regulated. These data showed that ecdysterone binds directly and specifically to the genes that are under hormonal control.

The development of larval salivary glands is governed by ecdysterone. It is manifest in the concomitant temporal control of the puffing pattern on the polytene chromosomes. A detailed cytological analysis of the effect of ecdysterone in vivo as well as in vitro on the puffing pattern of polytene chromosomes has been undertaken by Ashburner (Ashburner 1972a; 1972b). Since these puffing patterns are characteristic of a particular developmental stage, they are referred to as puff stages. The sequence of events has been divided into 21 puff stages, each characterized by a unique set of puff sites. All in all, 194–341 such sites have been described based on cytological examinations (Ashburner 1972b; Zhimulev 1974).

In general, there are three overall effects of ecdysterone with respect to puffing. The hormone induces a regression of puff size in some sites [e.g., puffs at 3C on chromosome X, 25B (previously 25AC) on the left arm of chromosome 2, and 68C on the left arm of chromosome 3]. The second and third effects are that the hormone induces puffing immediately at some sites (early puffs) and with a time lag of some hours at other sites (late puffs). Early puffs are induced within minutes and reach maximal size within 2–4 hr; concomitant increases in the concentrations of puff site-specific RNAs have been demonstrated by in situ hybridization (Bonner and Pardue 1977; Bonner et al. 1977). Protein products that are translated from early mRNAs are apparently important for the formation of late puffs. Induction of most of these late puffs is inhibited by cycloheximide, which suggests that protein synthesis is important for induction of late puffs (Ashburner 1974). As photoactivation of ecdysterone induces specific covalent binding of hormone to defined chromosomal sites, we could investigate whether late puffs are directly induced by ecdysterone. Moreover, it was important to find out how long the hormone was actually needed for the activation of transcription at a particular locus and how the hormonal regulation of a particular gene was terminated. Also, it was intriguing to identify the nature of the molecule(s), which reacted with ecdysterone upon irradiation.

II. Results and Discussions

Figure 1 illustrates the experimental procedure using salivary glands of third instar *D. melanogaster* larvae. Salivary glands were dissected, incubated in vitro in the presence of 5×10^{-6} M ecdysterone and then were irradiated for 5 min with a high-pressure mercury lamp. Ultraviolet light below 320 nm was filtered by using a cutoff filter (Gronemeyer and Pongs 1980). Since we did not use radioactively labeled hormone during the incubation, we had to devise an indirect immunofluorescence assay for detecting bound hormone on polytene chromosomes. After fixation of the salivary glands they were squashed. The chromosomes were postfixed in ethanol and afterwards incubated with primary antibodies against ecdysterone, which were raised in rabbits injected with ecdysterone–thyroglobulin adducts. This was followed by an incubation with secondary fluorescein-labeled goat anti-rabbit antibodies.

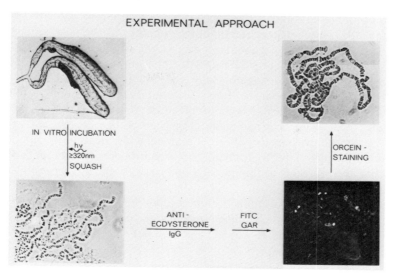

Fig. 1. After dissection, salivary glands of *D. melanogaster* were incubated in vitro in Grace's medium (Grace 1962) in the presence of $5 \times 10^{-6} M$ ecdysterone. Squash preparations of irradiated glands were made in 45% acetic acid–formaldehyde. Preparations were then incubated with rabbit antiserum directed against ecdysterone and subsequently with FITC labeled goat anti-abbit antibodies. After the fluorescence pattern was photographed, chromosomes were stained with lactoacetic acid–orcein for phase-contrast microscopy.

Unfortunately, this indirect assay for endogenous, bonded steroid hormone does not allow any straightforward controls for demonstrating the specificity of the photoreaction taking place inside the salivary glands. There are, however, some obvious and important experiments to demonstrate the selectivity of the photoreaction as well as of the rabbit antibodies employed. First of all, when chromosomes were squashed without irradiation and analyzed by immunofluorescence microscopy, no specific fluorescence could be detected (Gronemeyer and Pongs 1980). Similarly, the fluorescence at a particular locus disappeared when the gland was transferred into hormone-free medium prior to analysis (see below). Heat shock, which blocks transcription of almost all the genes except for the heat-shock genes and maybe a few others, completely abolishes fluorescence on polytene chromosomes (Gronemeyer and Pongs 1980). This incidentally also indicates that not just any puff site lights up in our immunofluorescence assay by virtue of being a puff.

It was important to establish that the photoreaction, which alters the structure of ecdysterone (Dworniczak, Schaltmann, and Pongs, unpublished results), does not alter the antigenicity of the hormone. When ecdysterone was irradiated in aqueous solutions for 5–10 min, it was completely altered into a new compound of an as yet unknown structure as judged by thin-layer chromatography. We assume that the photoactivation, which induces in the hormone itself an apparent intramolecular rearrangement, is the same that gives rise to covalent reaction of the hormone with polytene chromosomes.

Therefore, radioimmunoassays were carried out using either ecdysterone or irradiated ecdysterone to measure the binding capacities of the used antiserum. The results of these experiments showed that irradiation of ecdysterone does not destroy the antigenic sites for the antiserum, which was employed in our immunofluorescence assay and which was raised against nonirradiated ecdysterone, because the radioimmunoassays demonstrated that 80% of irradiated ecdysterone was bound to the antibodies.

More recently, we have obtained monoclonal antibodies against ecdysterone. The immunofluorescence pattern, obtained with these monoclonal antibodies and fluorescein labeled goat anti-mouse antibodies, were identical to the ones that are reported in this paper (Dworniczak and Pongs, manuscript in preparation).

In summary, these data show that the antibodies recognized specifically bonded hormone on the polytene chromosomes. Fluorescence caused by unspecific stickiness of antibodies was negligible in our experiments (Gronemeyer and Pongs 1980).

Salivary glands of puff stage 1 were dissected from third instar larvae of *D. melanogaster*. They were incubated in the presence of $5 \times 10^{-6} M$ ecdysterone for various times, as indicated in Fig. 2. Then the glands were irradiated, squashed, and analyzed by indirect immunofluorescence microscopy for bound hormone. Figure 2 illustrates the data for the left arm of chromosome 3, which were obtained from a series of such experiments. The incubations spanned puff stages 1–10, i.e., the temporal appearance of early puffs, such as at 74EF/75B, and the onset of late puffs, such as those at 63E and 78D. An examination of Fig. 2 readily demonstrates that appearance of fluorescence due to bound hormone and induction of puffing at a particular site coincide. For example, the puffs at 74EF and 75B increase dramatically in size in going from puff stage 1 to 6. Similarly, the fluorescence at these sites increases and subsequently decreases when the puffs regress. On the other hand, there is no detectable fluorescence at locus 78D in very early puff stages; only when puffing is induced at this locus can fluorescence be detected. In general, the cytological classification into early and late puffs is mirrored by the fluorescence pattern obtained at a particular puff stage. This holds for all the polytene chromosomes (Dworniczak and Pongs, manuscript in preparation).

These data show that ecdysterone is accessible for binding antibodies in early puff stages at early puffs only and, vice versa, in late puff stages at late puffs only. This observation might be simply due to the fact that in loci that are not puffed, bound hormone is inaccessible to our antibodies. Based on the following arguments, however, we believe that the absence of fluorescence at a particular site indeed demonstrates the absence of bound hormone. Squashed chromosomes are denatured structures that allow binding of antibodies to any chromosomal site, whether band, interband, or puff, as has been amply demonstrated with a variety of antibodies (Christensen et al. 1981; Jamrich et al. 1977; Saumweber et al. 1980).

Figure 3 shows a blowup of the left tip of chromosome 3 at puff stage 7. At this puff stage the early puff at 63F regresses and the late puff at 63E

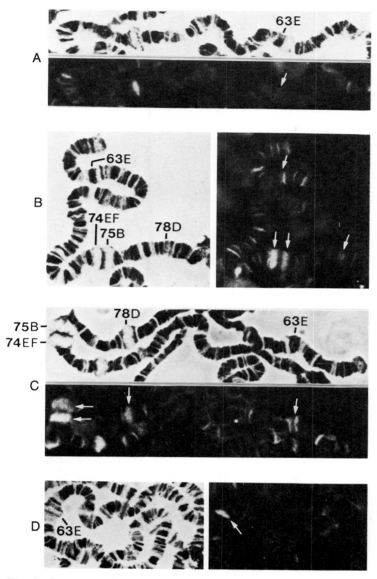

Fig. 2. Sequence of puffing and of fluorescence of the left arm of chromosome 3. Salivary glands of puff stage 1 were dissected and incubated for various times in the presence of 5×10^{-6} M ecdysterone. (A) 0 h incubation, puff stage 1; (B) 2 h incubation, puff stage 4/5; (C) 4 h incubation, puff stage 7/8; (D) 6 h incubation, puff stage 9/10). After incubation glands were analyzed as described in Fig. 1.

Fig. 3. Blowup of the left tip of chromosome 3 at puff stage 7/8. Chromosomes were prepared as described in Figs. 1 and 2.

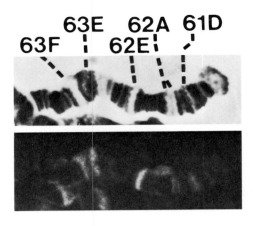

commences, whereas the puff at 62E has reached its peak (Ashburner 1972b). If we compare the puff sizes with the corresponding fluorescence intensities in Fig. 3, it is quite clear that they do not correlate. The largest puff, i.e., at 63F, exhibits the weakest fluorescence intensity. Moreover, the intensity at 63E, although this locus is hardly puffed at all, is stronger than that at 63F. The fluorescence intensity at 63E is an example, which shows that the accessibility of bound ecdysterone for antiecdysterone antibodies is not limited to largely puffed loci. Similar observations were made with polytene chromosomes of *Chironomus tentans* larvae. Then the data of Fig. 2 demonstrate that ecdysterone binding, puff induction, and transcriptional activity are closely interwoven. As long as a locus is puffed (i.e., actively transcribed), bound ecdysterone is present at this site. The active state of a locus requires the presence of hormone. These results immediately raise several questions:

1. Is the active state induced by the hormone?
2. What terminates the active state and causes dissociation of bound hormone?
3. What makes the hormone recognize early sites versus late sites?

There is good evidence that the concentration of nuclear receptors and the rate of transcription of a hormonally regulated gene are interdependent (Palmiter et al. 1981). The recent finding that purified glucocorticoid receptor selectively binds within 500 bp upstream of the TATAT box of cloned MTV-proviral DNA strongly suggests that the hormone–receptor complex is directly regulating the rate of transcription of a gene by acting on the transcriptional initiation process. Puffing, however, also implies a transition from inactive to active chromatin, which can visually be seen as a transition from highly condensed to unfolded chromatin. Figure 2 shows that the fluorescence intensities gradually increase with an increase of the puffed site. This does not tell us, however, at what stage during the induction of puffing the hormone comes into action. When we compared the fluorescence intensities of various puff stages at loci 63E and 78D, which are both late puffs, it occurred to us that a significant fluorescence intensity was apparent before the loci appeared cytologically as real puffs. Compare, for instance, puff stages 2–4 with puff stage 6 for locus 78D and puff stages 7 and 10 for locus 63E. These observations tentatively suggest that hormone binding triggers puff induction, which would also explain why more and more hormone becomes bound as puff induction progresses.

How is puff regression regulated? In the case of ecdysterone, this problem is intimately related to switching bound hormone from early to late puff sites. At the time that ecdysterone is dissociated from early puff sites to allow their regression, the hormone starts to recognize loci, where late puffs are induced. This notion implies that the temporal migration of ecdysterone from early to late puff sites is regulated. That there is such a regulatory mechanism has been hinted by the finding that inhibition of protein synthesis at early puff stages prohibits the induction of late puffs (Ashburner 1974). We were interested to show that ecdysterone migrates from early to late puff sites. To demonstrate this we choose the following experimental approach. The salivary gland consists of a pair of nearly equal-sized lobes joined by a common duct. After irradiation we cut the common duct and one lobe was fixed immediately and analyzed by immunofluorescence microscopy to determine the puff stage at the time of irradiation. The sister lobe was incubated in vitro in the presence of ecdysterone for 2 h and then fixed and analyzed too. The results, shown in Fig. 4, demonstrate that most of the nuclei within an irradiated gland developed in the normal way during the incubation and reached the same puff stage as a nonirradiated gland during a 2-h incubation. It did not make any difference for the puffing as well as fluorescence pattern whether the gland was irradiated before or after the 2-h incubation.

We assume that salivary gland nuclei do not contain a large pool of free nuclear hormone–receptor complexes. Under this assumption, our data indicate that ecdysterone migrates from early to late puff sites. It implies a temporal control of the recognition of early puff sites. As the recognition and the following induction of late puffs require proteins, it occurred to us that termination of hormone binding to early puff site may be much more complex. The intermolt puffs 3C, 25B, and 68C obviously differ in their type of regulation and at least 68C requires proteins for regression. This could be demonstrated in the following sequence of experiments. The two lobes of a salivary gland of puff stage 1 were separated. One lobe was incubated in the presence of $7 \times 10^{-4}\ M$ cycloheximide and $5 \times 10^{-6}\ M$ ecdysterone; the other lobe was incubated solely in the presence of $5 \times 10^{-6}\ M$ ecdysterone. After 1 h of incubation the lobes were transferred to hormone-free medium and incubated for another 2 h (the first in the presence of $7 \times 10^{-4}\ M$ cycloheximide). Afterward, hormone binding to polytene chromosomes was analyzed.

Figure 5 shows the results. In the absence of cycloheximide, 68C is completely regressed and does not show any fluorescence, whereas the site remained to be puffed in the presence of cycloheximide and the fluorescence

Fig. 4. Left arm of chromosome 3. Salivary glands were dissected and irradiated immediately. After irradiation one lobe was incubated for 2 h in the presence of $5 \times 10^{-6} M$ ecdysterone; afterward it was fixed and analyzed by immunofluorescence microscopy (B) and phase contrast (A).

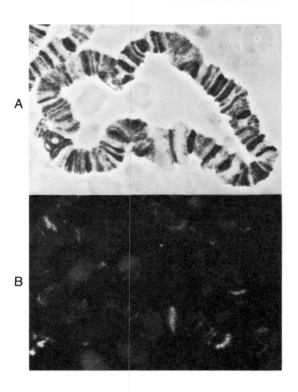

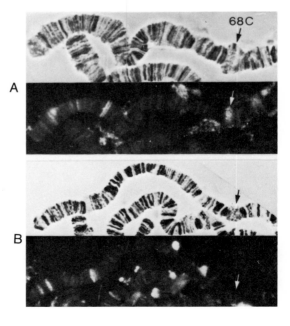

Fig. 5. Left arm of chromosome 3. Salivary glands of puff stage 1 were cultured in the presence of ecdysterone $(5 \times 10^{-6} M)$ for 1 h and then incubated in hormone-free medium for another 2. (A) $7 \times 10^{-4} M$ cycloheximide was continually present; (B) control experiment with no cycloheximide. Arrows indicate the chromosomal locus 68C. (Pictures are from different salivary glands in the same series of experiments).

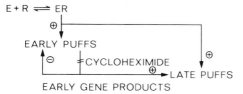

Fig. 6. Model of ecdysterone action. The hormone combines with receptor in the cytoplasm; the hormone–receptor complex then translocates into the nucleus and induces early puffs. Transcription in early puffs gives rise to gene products, which in turn inhibit binding of the hormone–receptor complex at early puff sites and promote binding to late puff sites. The switch in affinity of the hormone–receptor complex for late puff sites is inhibited by cycloheximide.

intensity is comparable to that at puff stage 1. This demonstrates that active protein synthesis is necessary to remove ecdysterone from puffed loci and to make it available for binding to new sites.

Thus, we propose a modification of a previous model on ecdysterone action (Ashburner et al. 1973). The hormone activates transcription at early loci, such as 63F, 74EF, 75B. This activation is temporally regulated by proteins, which are synthesized during the same time and which are probably products of some of the transcribed and hormonally regulated genes. These proteins exert both forward and backward control (i.e., they inhibit the binding of hormone to early sites, thereby inducing regression of puffs, and they promote the binding of hormone to late sites, thereby inducing late puffs). Figure 6 summarizes this proposal about ecdysterone action. The model, shown in Fig. 6, predicts that the switch from early to late puff induction in effect depends on the ratio of nuclear ecdysterone–receptor complex to the gene products necessary for late puff induction. Thus, it should be possible to manipulate this switching not only by inhibiting protein synthesis but also by lowering the concentration of nuclear receptor.

Many futile attempts have been made to characterize and to isolate the ecdysterone receptor of D. melanogaster. For many years there were even doubts that invertebrates might in fact contain any hormone receptor molecules with properties similar to mammalian steroid receptors. It was demonstrated recently that Drosophila tissue culture cells contain a receptor for ecdysteroids (O'Connor et al. 1980). The receptor has not yet been purified. As photoactivation of ecdysterone induced specific covalent binding of hormone to defined chromosomal sites, it was intriguing to identify the nature of the molecule(s) that reacted with ecdysterone on irradiation and to see whether ecdysterone was bound to its nuclear receptor on irradiation.

Salivary glands were, therefore, irradiated on a preparative scale. Proteins as well as DNA were analyzed for bonded hormone. It turned out that irradiation linked ecdysterone to a single protein, which has a molecular weight of 130,000 and an isoelectric point of 5.8. Ecdysterone becomes covalently linked to the same molecule in tissue culture cells in terms of charge and molecular weight (Schaltmann-Eiteljörge and Pongs 1982). The compartmentalization of this protein inside the cell, where it resides in the cytoplasm and then is translocated into the nucleus, as well as sedimentation characteristics and its binding properties for ecdysteroids demonstrated that the irradiation induced a covalent bond between ecdysterone and its receptor (Schaltmann-Eiteljörge and Pongs 1982). The translocation of the ecdysterone–receptor complex into the nucleus leads to its association with chromatin. This is demonstrated in Fig. 7. Kc tissue culture cells were incubated with the ecdysteroid [³H]ponasterone A. This ecdysteroid becomes cross-linked on irradiation to the same protein as ecdysterone (Schaltmann-Eiteljörge and Pongs 1982). It is, however, available with a specific radioactivity higher than ecdysterone. Therefore, it was employed in the following experiment. Nuclei were prepared, irradiated, and incubated with micrococcal nuclease, until less than 5% of total DNA was digested. Nucleosomes were then separated on isokinetic sucrose gradients and analyzed for bound hormone-receptor complex (Fig. 7). It should be mentioned that irradiation was necessary for this analysis, since without irradiation no bound hormone–receptor complex was detected. The results, of Fig. 7, show that the ponasterone A–receptor complex associated with nucleosomes on sucrose gradi-

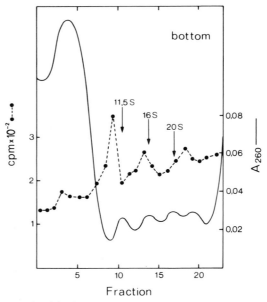

Fig. 7. Association of [³H]ponasterone A with chromosomal fragments. Kc cells at 2.5×10^8 cells/ml were incubated with [³H]ponasterone A (10^{-7} M) in D22 medium (2 h, 25°C). Nuclei were isolated as described by Worcel et al. (1978) and Wu et al. (1979), irradiated, and incubated with micrococcal nuclease until less than 5% of total DNA was digested. After extraction chromosomal fragments were separated on isokinetic sucrose gradients containing 1 mM Na EDTA, PH7, ($c_t = 5.1\%$, $c_r = 31.4\%$, $V_m = 9.6$ ml, McCarty et al. 1974). Centrifugation was in a Beckmann rotor SW41 for 15 h at 35,000 rpm at 2°C. Gradients were fractionated and the absorbance at 260 nm (——) and the bound [³H]ponasterone A (-----) (TCA) precipitation were measured.

ents migrates ahead of the bulk nucleosomal fractions. This might be due to a protection of the linker region against nuclease digestion. This interpretation, however, has to be substantiated by further experiments.

In our irradiation experiments ecdysteroid was selectively bound to its nuclear receptor in chromatin. This result has important consequences for the immunofluorescence data reported in Figs. 2–5. It shows that the fluorescence, which we have detected in puffs on the polytene chromosomes, is due to hormone–receptor comlex present at the particular chromosomal locus. This observation strongly suggests that steroid hormone–receptor complexes modulate transcriptional activity directly at the level of the gene. They do so by binding to a cytoplasmic receptor, which on hormone binding translocates to the nucleus. The hormone–receptor complex then binds directly to the genes, which are hormonally regulated. It should be emphasized again that other proteins modulate the affinity of the hormone–receptor complex to particular chromosomal sites as well as the stability of this interaction, as described in Fig. 6.

References

Ashburner M (1972a) Chromosoma 38: 255–281
Ashburner M (1972b) In: Beerman W (ed) Results and Problems in Cell Differentiation. (Springer-Verlag, Berlin, Vol. 4, pp. 102–151
Ashburner M, Chihara C, Meltzer P, Richards G (1973) Cold Spring Harbor Quant Biol 38: 655–662
Ashburner M (1974) Dev Biol 39: 141–157
Becker HJ (1959) Chromosoma 10: 645–678
Becker HJ (1962) Chromosoma 13: 341–384
Berendes HD (1967) Chromosoma 22: 274–293
Bonner J, Berninger M, Pardue ML (1977) Cold Spring Harbor Symp Quant Biol 42: 803–814
Bonner J, Pardue ML (1977) Cell 12: 219–225
Christensen ME, LeStourgeon WM, Jamrich M, Howard GC, Serunian LA, Silver LM, Elgin SCR (1981) J Cell Biol 90: 18–24
Clever U, Karlson P (1960) Exp Cell Res 20: 623–626
Clever U (1961) Chromosoma 12: 607–675
Grace TDC (1962) Nature 195: 788–789
Gronemeyer H, Pongs O (1980) Proc Natl Acad Sci USA 77: 2108–2112
Jamrich M, Greenleaf A, Bautz EKF (1977) Proc Natl Acad Sci USA 74: 2079–2083
McCarty KS, Vollmer RT, McCarty K (1974) Anal Biochem 61: 165–183
O'Connor JD, Maroy P, Beckers C, Dennis R, Alvarez CM, Sage BA (1980) In: Roy AK, Clark JH (eds) Gene Regulation by Steroid Hormones. (Springer-Verlag, Berlin, pp 263–277
Palmiter RD, Mulvihill ER, Shepherd JH, McKnight GS (1981) J Biol Chem 256: 7910–1916
Saumweber H, Symmons P, Kabisch R, Will H, Bonhoeffer F (1980) Chromosoma 80: 253–275
Schaltmann-Eiteljörge K, Pongs O (1982) Proc Natl Acad Sci USA 79, 6–10
Worcel A, Han S, Wong ML (1978) Cell 15: 968–977
Wu C, Bingham PM, Livak KJ, Holmgren R, Elgin SCR (1979) Cell 16: 797–806
Zhimulev IF (1974) Chromosoma 46: 59–76

Discussion of the Paper Presented by O. Pongs

O'MALLEY: What is the relative affinity of your antibody for native and photolized ligand and what part of the molecule is it against.

PONGS: I do not know which part of the hormone is recognized by the antibody. If one does a competition between cold and radioactive irradiated ecdysterone in a radioimmunoassay, the result is that the antibody does not differentiate between irradiated and nonirradiated ecdysterone.

O'MALLEY: The final question is how ecdysone might be concentrated in puff areas yet not necessarily be involved in an important biologic reaction. For instance, ecdysone creates puffs; in puffs, the DNA is less condensed and less tightly bound to protein. Perhaps free steroid is simply bound nonspecifically to DNA in this region, as steroids have the capacity to do. Have you ever just added ecdysone to DNA and then added antibody? Is it conceivable that a new puff could pick up steroid and the antibody could see the free steroid even though the steroid was not bound to receptor?

PONGS: Well, first of all if you had simply binding to the DNA it wouldn't be necessary to do cross-linking.

O'MALLEY: Well, no, it is a reversible reaction, so that you may still need cross-linking.

PONGS: In the heat-shock experiment, of course, the hormone is around.

O'MALLEY: What if you did heat-shock induction in the presence of ecdysone?

PONGS: That is done in the animal. It is an in vivo experiment. We do a heat shock of the larvae, take out the gland, and look at fluorescence, and we don't see any.

O'CONNOR: What is the sensitivity of your antibody to ecdysone?

PONGS: Compared to what?

O'CONNOR: How much ecdysone do you have to have in order for your antibody to see it?

PONGS: I'll have to look.

CLARK: My question has to do with Bert's (O'Malley) because I was wondering about the nonspecific reactivity of the compound with the DNA. Have you ever used other compounds—for instance, R5020—to see whether or not irradiation would cause it to link convalently, as a control?

PONGS: Not in that system, but we used it in the glucocorticoid system. In the glucocorticoid system if you used receptor minus cells (tissue culture cells), you would not get any cross-linking whatsoever.

SCHRADER: I have a couple of question; first of all, are there developmental times in *Drosophila* in which the ecdysone titer is zero?

PONGS: It is below detection.

SCHRADER: If there is developmental stage at which the ecdysone level is below detection and you do such an irradiation at that time, it would seem to me to be a very good control with respect to the specificity of the reaction. Have you done such a control?

PONGS: No, but we have tried to preabsorb our serum—for instance, with ecdysone–protein conjugates, in which case you can saturate it.

SCHRADER: My second question is with respect to the efficiency of coupling of the tritiated ponasterone A to the protein. Do you have any estimates of the efficiency of that coupling reaction >4? In our hands with the progesterone receptor, the efficiency was fairly low.

PONGS: With the glucocorticoid receptors with dexamethasone, it is in our hands between 30 and 40%, with ecdysterone and Kc tissue culture cells, it is in the order of 20–25%. With salivary glands yields are certainly higher, because it is an ideal system. We just have 140 cells in a bilayer. So you do not have much that can interfere with your light that irradiates the tissue.

SCHRADER: At what wavelength do you irradiate?

PONGS: In the case of dexamethasone, at 310 μm; in the case of ecdysone, 325 μm. It depends on the steroid hormone that you use.

ROY: Almost all of the puffs that you showed looked like they are lighting up with your antibody. Did you try to induce the puffs with some other hormone, such as juvenile hormone, to see whether they do bind the antibody or not? That is one question; the second question is, "Did you try to treat your squash with DNase, wash it out, and then put the antibody and see whether they light up?"

PONGS: The answer to the first question is that heat shock puffs are an example where we do not see fluorescence. On the slides that I showed, many puffs did not fluoresce—for example, 25AC. A very prominent puff is at this locus, where we don't see any fluorescence. Another one would be, for instance, the puff at locus 2B on the X chromosome. In general, there are many puffs without fluorescence. On the tip of the chromosome 3, which I showed on one slide, there were large puffs (I forgot to point that out) that don't light up. We did not do any DNase-I digestion, no.

COFFEY: I would have anticipated that only a small area of your puff might have been fluorescently lit up with the hormone. How much DNA is in that fluorescent area? You wouldn't expect all of that DNA to be binding hormone, would you? How many base pairs of DNA are located within one of those fluorescent bandwidths?

PONGS: First of all, we are looking at more than 1000 strands of DNA. Second, we are using an indirect immunofluorescence approach. So the fluorescence is attenuated. That is probably the reason for the blur. We actually have obtained pictures of *Chironomus* chromosomes where the fluorescence has distinct features and appears in confined areas. I have no idea how the DNA looks in a puff. We don't know whether all 1000 strands are decondensed or just 10; that we can't tell.

PECK: Would you clarify a point? Am I right, when you were studying the late puff system you could irradiate early or late and see this covalent complex moving from one puff to another. Is that your interpretation?

PONGS: That is our interpretation. The hard fact is: If we take the early puff stage and irradiate the gland at this early puff stage and then carry on the in vitro incubation to reach a later puff stage, the fluorescence is no longer on early puffs, only on late ones, without any further irradiation. There are two possible interpretations for this observation. All nuclear receptors are bound to chromosomes. Then this would only mean that the same receptor molecules that were on early puffs, migrate to late puffs. However, if there is a sizable pool of nonbound nuclear receptors, then that doesn't hold. Then you could infer that molecules leave early sites, and the receptor molecules, which bind to late puffs, come from the pool. What we have to know, and that we can only measure with receptor antibodies, is how much free receptor is there around in the nucleus, but I bet it is none.

Discussants: B. W. O'MALLEY, O. PONGS, J. D. O'CONNOR, J. L. CLARK, W. T. SCHRADER, A. K. ROY, E. J. PECK AND D. S. COFFEY

Chapter 5

Ecdysteroid Effects on the Cell Cycle of *Drosophila melanogaster* Cells

JOHN D. O'CONNOR AND BRYN STEVENS

Ecdysteroids are polyhydroxylated steroids possessing a 5α hydrogen and a conjugated 7-ene-6-one function in the B ring. Their most impressive biological effects are seen in insects, in which they regulate the timing of the larval, pupal, and adult molts. For the past several years a number of laboratories (Courgeon 1972; Savakis et al. 1981; Stevens et al. 1981; Berger 1978) have been examining the mode of action of two ecdysteroids, 20-OH-ecdysone and its more potent analog, ponasterone A, on the Kc cell line established from embryos of *Drosophila melanogaster* by Echalier and Ohanessian (1970). The results of these studies reveal ecdysteroids to be very similar in their mode of action to vetebrate steroid hormones. Thus shortly after the administration of 20-OH-ecdysone to cultured cells there is apparently a marked increase in the synthesis of several specific proteins (Savakis et al. 1980). Following longer exposure to hormone the Kc cells respond with increased enzyme activity (Cherbas et al. 1977; Best-Bel pomme et al. 1978), alteration of both cell shape (Cherbas et al. 1980; Cougeon 1972; O'Connor and Chang 1981) and cellular agglutinability (Rosset 1978), and modifications of cell surfaces (Metakovski et al. 1975). The existence of high-affinity binding sites for ecdysteroids in cultured cells (Maroy et al. 1978) is consistent with the preceding events being receptor mediated, as suggested by the current steroid paradigm (Yamamoto and Alberts 1976).

A condition common to the preceding responses of cultured cells to ecdysteroids is the arrest of the cell cycle in the postsynthetic G_2 phase (Stevens et al. 1980). The present chapter will delineate the kinetics of this response and examine the alteration of intracellular receptor concentration as a function of the exposure to hormone.

I. Arrest of the Cell Cycle

The cycle of cultured *Drosophila* cells is different from that of exponentially growing mammalian cells. In most mammalian cell lines G1 occupies the major portion of the cell cycle (Pardee et al. 1978). In contrast, Dolfini (1971) suggested that the G_2 phase was quite long and occupied the greatest portion of the cell cycle in cultured *Drosophila* cells. The recent advances in flourescent flow cytometry have permitted alterations in the cell cycle of cultured cells to be examined in detail. Figure 1 illustrates the normal cell cycle of exponentially growing Kc cells. Because this cycle distribution was quite different from that normally seen in mammalian cells, it was useful to confirm the two peaks of fluorescence as G_1 and G_2M, respectively. Thus, Fig. 1b illustrates the effect of hydroxyurea on the profile of fluorescence. The interruption of DNA synthesis causes a buildup of cells in G_1 resulting in the profile observed in Fig. 1b. In contrast, treatment of cells with vinblastine prevents mitosis and results in the accumulation of cells in G_2M (Fig. 1c). Analysis of hydroxyurea- and vinblastine-treated populations of cells and comparisons of their fluorescence makes it virtually certain that the first peak in the fluorescent profile of Kc cells is G_1 and the second peak is G_2. Exposure of Kc cells to physiological concentrations of 20-OH-ecdysone results in a fluorescence profile nearly identical to that seen following vinblastine treatment. Microscopic examination of ecdysteroid-treated cells revealed very few in metaphase, thus establishing the arrest to be in G_2. The time course of G_2 arrest is rapid. Twelve hours following the administration of 20-OH-ecdysone to exponentially growing cells, there are none detectable in G_1 (Stevens et al, 1980). Furthermore, the maintenance of the G_2 arrest is dependent on the continued exposure to hormone.

It is interesting to note that the hormonally induced accumulation of cells in G_2 occurs at the same rate as colchicine-induced arrest (Fig. 2). The similarity of the curves suggests that the cells are not being accelerated to a specific block point in G_2 but rather are accumulating in G_2 after they complete a normal G_1S phase. It is also interesting to note that the arrest may not

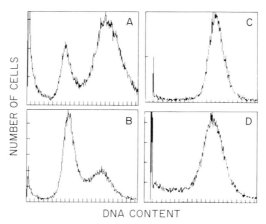

Fig. 1. Fluorescent flow cytometry of Kc cells. The analyses were performed as indicated in Stevens et al. (1980). Panel A is the histogram of fluorescence from exponentially growing untreated Kc cells. Panels B–D represent the effect of different exogenous agents on the histogram of fluorescence in Kc cells. Panel B is obtained following 18 h exposure to hydroxyurea at a concentration of 200 ug/ml. Panel C is obtained following 12 h exposure to vinblastine at a concentration of 10^{-6} M. Panel D is obtained following 12 h exposure to 2×10^{-7} M 20-OH-ecdysone.

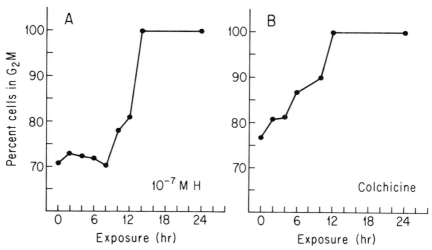

Fig. 2. Time course of colchicine-induced mitotic arrest. Cells of the clone 7C4 were treated with 10^{-7} M 20-OH-ecdysone (A) or 10^{-7} M colchicine (B), harvested, fixed, and analyzed for fluorescence as previously indicated (Stevens et al. 1980).

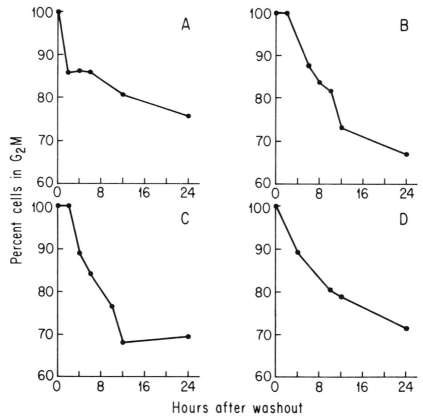

Fig. 3. Time course of reentry into the cell cycle after varying periods of homonal exposure. Cells of the clone 7E10 were treated with 10^{-7} M 20-OH-ecdysone for (A) 12 h, (B) 24 h, (C) 72 h, (D) 96 h; then they were washed free of hormone and analyzed for fluorescence.

occur at a specific point in G_2. This was suggested by the observations illustrated in Fig. 3. The kinetics of reentry into the cell cycle of ecdysteroid-arrested cells demonstrate a marked lack of synchrony; consequently, it is likely that arrest occurs at varying points in the long G_2 period.

II. Acquisition of Resistance to Ecdysteroids

After prolonged exposure to ecdysteroids Kc cells reenter the cell cycle and begin dividing again (Stevens et al. 1980). This resumption of division is not due to metabolism of the hormone, since the addition of excess 20-OH-ecdysone fails to prevent the escape from arrest. The acquisition of resistance is illustrated in Fig. 4. After a four- to five-day exposure to 20-OH-ecdysone the cycle in treated cells is virtually indistinguishable from that of

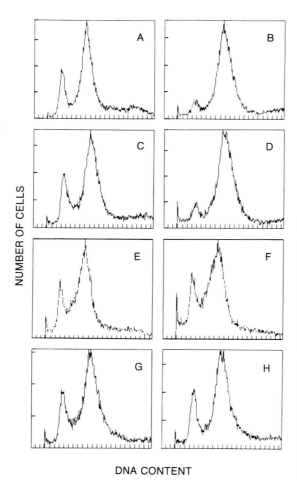

Fig. 4. Acquisition of resistance to ecdysteroid-induced G_2 arrest. Cells of the Kc-derived clone 7C4 were treated for 1(A), 3(C), 4(E), or 5(G) days with 10^{-7} M 20-OH-ecdysone, washed free of hormone, and then assayed for the ability to respond to a second exposure to 10^{-7} M 20-OH-ecdysone. The right column of panels represents the cell cycle distribution following the second exposure to hormone, which was for 24 h. The left column of panels represents a control population of cells unexposed to hormone for a second 24 h period but previously exposed for 1–5 days.

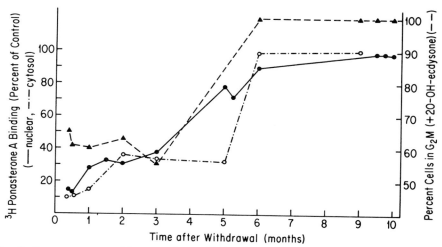

Fig. 5. Reappearance of hormonal sensitivity in Kc cells following removal from prolonged exposure to 20-OH-ecdysone. Cells were exposed to 2×10^{-7} M 20-OH-ecdysone for 7 days, after which they were placed in ecdysteroid-free medium. At the times indicated they were challenged by a 24-h exposure to 20-OH-ecdysone. The percent of the 7C4 cells in G_2M after 24 h in the presence of 20-OH-ecdysone (▲) was determined by fluorescent flow cytometry. Nuclear (●) and cytosolic (○) ecdysteroid binding were measured as previously described (Stevens et al. 1980).

untreated cells. Moreover, the treated cells are refractive to the inductive action of ecdysteroids. This acquires ecdysteroid insensitivity has been correlated with a marked decrease in intracellular ecdysteroid receptors (Stevens et al. 1980). Figure 5 illustrates the relative ecdysteroid-binding capacities of both the nuclear and cytosolic compartments of Kc cells at various times after withdrawal of 20-OH-ecdysone. As indicated, the cells were exposed to 20-OH-ecdysone for 24 h and the percent arrested in G2 was determined by fluorescent flow cytometry. In addition, estimations of both nuclear- and cytosolic-binding capacity were made at these times. Both the nuclear and cytosolic receptor concentrations are markedly reduced for the first month after withdrawal and then begin a slow return to near normal levels, which are achieved nearly 6 months after hormone withdrawal. Coincident with the elevation of receptor concentrations is a return to a hormonally sensitive state as manifested by the G_2 arrest in the presence of 20-OH-ecdysone. Such a "downregulation" of ecdysteroid receptors has not been demonstrated in whole organisms. Although downregulation is apparent in Kc cells only after a 4- or 5-day exposure to hormone, it may be that cells exposed to the cyclic ecdysteroid fluctuation characteristic of *Drosophila* (Kraminsky et al. 1980) can downregulate more rapidly. This possibility can be explored once Kc cells are adapted to chemostat conditions and when receptors are measured in whole organisms at various times in development. It should be underscored that exposure to hormone is inducing the downregulation of receptors as opposed to selecting for a receptor-deficient variant

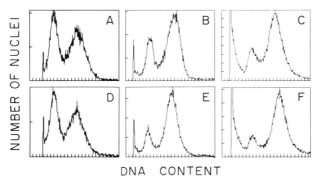

NUMBER OF NUCLEI

DNA CONTENT

Fig. 6. Cell cycle distribution of wing (A–C) and leg (D–F) imaginal discs. Animals were staged and dissected and nuclei obtained as described elsewhere (Fain and Stevens 1982). Histograms of fluorescence were obtained from discs of 94 h (A and C), 115 h (band D), and 120 h (C and F).

already present in the population. The latter possibility has been carefully studied, and it has been determined that there are no detectable cycling cells during the G_2 arrest period. Considering that there are no cycling cells through 100 h of hormone treatment, that the cell cycle distribution at 160 h is identical to that of control populations, and that cell viability remains high throughout the period of hormonal exposure, it is unlikely that a selective process is responsible for the acquisition of resistance to ecdysteroids.

III. Cell Cycle in Imaginal Discs

Although the downregulation of ecdysteroids receptors has not been demonstrated in vivo, the induced G_2 arrest can be demonstrated in imaginal discs. Figure 6 illustrates the DNA content of both wing disc nuclei and leg disc nuclei obtained from *Drosophila* at various times of development. It is obvious that in nuclei from both types of disc there is a progression from a G_1 DNA content to a G_2 DNA content. This shift is demonstrable 115 h after egg deposition—a time at which the ecdysteroid titers are increasing very rapidly (Kraminsky et al. 1980). A more detailed examination of the cell cycle in imaginal discs has been presented elsewhere (Fain and Stevens 1982). Suffice it to say that the pattern illustrated in Fig. 6 is observed in every imaginal disc that has been examined, except for the eye portion of the eye antennal disc. In this case the cells accumulate in the G_1 phase of the cycle.

IV. Conclusions

Although it is not known if the in vitro responses of the Kc cell line and derived clonal sublines to ecdysteroids are identical to in vivo differentiative responses, these cells provide an excellent model for investigating the mode of action of ecdysteroids. Many of the responses of the Kc cells can be observed in the intact organism (O'Connor et al. 1980; Cherbas et al. 1980), and the similarity of the ecdysteroid receptors in Kc cells and imaginal discs (Yund et al. 1978) underscores the utility of the cell line as a model for ecdysteroid action. Furthermore, several phenomena observed in these cells, such as the induced downregulation, suggest obvious experiments to be performed with tissue from the developing organism.

References

Berger E, Ringler R, Alahiotis S, Frank M (1978) Dev Biol 62: 498–511

Best-Belpomme M, Courgeon AM, Rambach A (1978) Proc Nat Acad Sci 75: 6102–6106

Cherbas P, Cherbas L, Williams C (1977) Science 197: 275–277

Cherbas P, Cherbas L, Demetri G, Manteuffel-Cynborowska M, Savakis C, Yonge CD, Williams CM (1980) In: Roy A, Clark J (eds) Gene Regulation by Steroid Hormones. Springer-Verlag, New York, p 278

Courgeon A (1972) Exp Cell Res 74: 327–336

Dolfini S, Courgeon AM, Trepolo L (1971) Experientia 269: 1020–1021

Echalier J, Ohanessian A (1970) In vitro 5: 162–172

Fain and Stevens (1982) Dev Biol 92: 247–258

Kraminsky GP, Clark WC, Estelle MA, Gietz RO, Sage BA, O'Connor JD, Hodgetts RB (1980) PNAS 71: 4175–4179

Maroy P, Dennis R, Beckers C, Sage B, O'Connor JD (1978) PNAS 75(12): 6035–6038

Metakovskii EV, Cherdantseva EM, Gvozdev VA (1977) Mol Biol 11: 124–133

O'Connor JD, Maroy P, Dennis R, Alvarez CM, Sage BA (1980) In: Roy AK, Clark JA (eds) Gene regulation by steroid hormones, Springer-Verlag, New York, p 263–277

O'Connor JD, Chang ES (1981) In: Gilbert LI, Frieden E (eds) Metamorphosis. Plenum, New York, p 241–261

Pardee AB, Dubrow R, Hamlin J, Kleitzen RF (1978) Ann Rev Biochem 47: 715–750

Rosset R (1978) Exp Cell Research, 111: 34–36

Savakis C, Demetri G, Cherbas P (1980) Cell 22: 665–674

Stevens B, Alvarez CM, Bohman R, O'Connor JD (1980) Cell 22: 675–682

Yamamoto KR, Alberts BM (1976) Ann Rev Biochem 45: 721

Yund MA, King DS, Fristom JW (1978) Ecdysteroid receptors in imaginal discs of Drosophila melanogaster. PNAS 75: 6039–6043

Chapter 6

Structure and Function of Growth Hormone—A Target for Glucocorticoid Action

JACK L. KOSTYO

The GHs comprise a rather unique family of small globulins produced by the anterior pituitary glands of vertebrates. Structural analysis of the GHs of a number of species including man have indicated that the hormone is a single-chain polypeptide of about 22,000 daltons, with two intrachain disulfide bridges, but there are some variations from one species to the next in primary structure. One of the chief physiological functions of GH is to stimulate somatic growth, which it does by regulating the synthesis of proteins and nucleic acids in the young organism. However, one of the features that makes the GHs unique is the fact that they exhibit a variety of biological effects in addition to their growth-promoting action. These effects are difficult to reconcile by a common mechanism of action. In addition to being anabolic, GH can be diabetogenic (i.e., it can produce insulin resistance and hyperinsulinemia when administered chronically or when secreted in excess by a tumor, and it can cause permanent fasting hyperglycemia and glucosuria in susceptible animals, like the dog and cat). Paradoxically, the hormone can also be insulinlike. This effect can usually only be demonstrated in GH-deficient subjects or with tissues of hypophysectomized animals and consists of a transient stimulation of glucose transport and metabolism, which in the whole animal is observed as a transient period of hypoglycemia. The GHs of primates are also lactogenic, and this reflects their close structural relatedness to the prolactins and the placental lactogens. Lastly, the GHs exhibit species specificity, so that the GHs of the infraprimate species are not biologically active in primates, including man, although primate GHs are active in species that are lower on the phylogenetic tree. It appears that the GH receptors of primates are quite discriminating and that the differences in primary structure, and hence conformation, between primate and infraprimate GHs render the latter incapable of interacting with the GH receptors of the primate.

A major consequence of this property of species specificity has been that children with growth defects have had to be treated with hGH prepared from pituitaries obtained at autopsy, a fact that more than 20 years ago contributed to the formation of the National Pituitary Agency, an organization that has sponsored the worldwide collection of human pituitaries and the preparation of hGH, among others. However, this source of hGH has never been adequate to treat the many children with GH-responsive growth defects. Further, the scarcity of hGH precluded any full-scale exploration of the therapeutic usefulness of its anabolic property in other clinical situations, such as the treatment of severe burns or the hastening of woundhealing. The scarcity of hGH prompted some investigators, such as myself, to devote a substantial number of years to an analysis of the structure–function relationships of GH, in the hope that only a small portion of the molecule would be required to produce growth and that this small portion might be produced synthetically. As it turned out, it appears that a substantial portion of the molecule is required to produce growth, dashing any hopes for direct synthesis of an active agent. However, as all are now aware, work on the GH gene and successful efforts to have it expressed in bacteria, have provided a solution to the problem of the supply of hGH.

However, I should hope that this will only be a first step in the application of the study of GH genes and the regulation of their expression. I say this because this approach may be useful in solving many of the questions that still exist concerning the chemistry and physiology of GH and diseases related to it.

One of the successful outcomes of the GH structure–function studies is that the various biological activities exhibited by GH are probably not produced by precisely the same region of the GH molecule; that is, there are probably multiple active sites in the molecule responsible for its several activities. Hence there are probably a family of receptors responsible for its anabolic, diabetogenic, and insulinlike effects. Thus, it might be possible, once the appropriate active regions of the molecule are identified, to produce modified GH molecules that exhibit only one property predominantly. Indeed, if the therapeutic usefulness of hGH is ever to be explored adequately (i.e., its pharmacology), it will be necessary to administer it in doses that will undoubtedly produce symptoms of diabetes, and possibly islet cell damage. Thus one hope might be to eliminate the diabetogenic property and retain the anabolic activity. In any event, producing modified GHs possessing only one of their several properties would make these substances highly useful as probes to study the receptors and the molecular mechanisms involved in the expression of the individual properties of the hormone.

Another potential application might be to gain insight into the molecular nature of various GH-related growth defects. It has been well known from the work of Lewis, Chrambach, and others that even very highly purified native hGH preparations exhibit molecular heterogeneity, and it seems clear that some of this heterogeneity cannot be related to alterations of the native hormone during the isolation and purification procedure. For example, Lewis has identified a hGH variant that lacks 15–20 amino acid residues in the N-terminal end of the molecule. This so-called 20K variant has altered biological properties. It is conceivable that other variants of the GH molecule exist and that they may be responsible for some of the puzzling growth defects observed in humans. It has been established from GH structure–function work that the GH molecule may be modified to virtually eliminate its growth-promoting activity; yet such modified molecules show a high degree of cross-reactivity in the RIAs used routinely in the clinic. It is conceivable, for example, that individuals with short stature, who have normal levels of circulating GH by RIA but who respond dramatically with accelerated growth to the administration of exogenous GH, might be producing and secreting a GH variant, with low growth-promoting activity but high cross-reactivity in the RIA, as has been suggested recently by Rudman and his co-workers. Another group of individuals comes to mind, those with so-called cerebral gigantism, that grow dramatically like children with GH-secreting tumors yet have normal circulating levels of GH by RIA. Although the cause of this disease has not been established, could it be possible that these individuals produce a supraactive GH variant with normal cross-reactivity? Lewis and Chrambach have both reported that they were able to modify hGH enzymatically and in so doing enhance its growth-promoting potency several-fold. In view of this the existence of a supraactive GH variant may not be so farfetched.

Chapter 7

Mechanism of Extinction of Growth Hormone and Prolactin Genes in Somatic Cell Hybrids

E. BRAD THOMPSON AND JEANNINE S. STROBL

I. Introduction

Two major unanswered questions exist regarding the nature of steroid regulation of genes. First, how is it that steroid hormones act on the genome of differentiated cells programmed to respond to them in a particular way? A common end result of this effect is the accumulation of specific mRNA sequences. Second, what are the controls that dictate whether a tissue responds to a particular hormone at all and, if so, which of many tissue-specific effects the hormone elicits? Earlier, it was believed that the presence or absence of specific steroid receptors in various cells might be the answer, but virtually every cell type in the body contains glucocorticoid receptors. It may be, of course, that there are subtle differences in the glucocorticoid receptors in different cell types, or as is more popularly believed, there may be control molecules that organize the DNA in different differentiated cells so as to prevent the expression or induction of unwanted genes.

For some years we have been employing somatic cell hybrid systems to explore this second problem of differentiated control over glucocorticoid-stimulated genes. We believe that this approach is of value for several reasons: (1) Pure cell populations can be prepared in quantity, thus avoiding some of the major difficulties inherent in studying naturally differentiating tissues that frequently are available only during fetal life, are difficult to obtain free of contaminating cell types, and often are present in very small numbers. (2) In hybrid-cell tissue culture systems one can also hope to manipulate the process of differentiation. For example, when one fuses a cell programmed to respond to a hormone with a cell not so programmed, it is often the case that the hormonal responses in the hybrid cell are shut off (Gehring and Thompson 1979; Ringertz and Savage 1976a; Ringertz and Savage 1976b). On some occasions it has been observed that as these hybrid cells grow and segregate chromosomes, the differentiated response returns. Such systems allow one to study how the hormonal responsiveness of particular genes is controlled.

The types of control over gene expression demonstrated in somatic hybrids to date are varied. Frequently, however, there appears to be specific control over certain genes, although the extent of control is somewhat variable for different cell crosses and for the specific genes under study. The original generalization, that one would find all "luxury functions" under dominant negative control in hybrids, does not seem to be the case (Ringertz and Savage 1976b). Examples of dominant positive control and of co-dominance as well have since been discovered (Ringertz and Savage 1976b; Lipsich et al. 1979; Pfahl and Bourgeois 1980; Orly and Schramm 1976). Nevertheless, for a given gene product studied in the appropriate cell types one can demonstrate that the controls seen appear to be relatively specific. For example, in an instance of negative dominance, the phenotype of the dominant cells does not predominate for all proteins or genes in the hybrid, but only for a very limited number. Thus, one must conclude that whatever reorganization of structure–function is brought about in the hybrid cells, it results in controls that appear to be acting at specific organizational points in the genome.

One class of experiments in which negative dominance over specific gene expression has been the rule is that of proteins induced by steroid hormones. Most of these studies have been carried out on genes controlled by glucocorticoids; however, in one case an estrogen-inducible product was shown to be extinguished (Thompson et al. 1980). Since we have previously reviewed and tabulated these results (Gehring and Thompson 1979), we will not reexamine these issues now, but summarize by simply stating that cells of several tissue types and species have been shown to follow this general rule. One striking exception to negative dominance is that of mouse mammary tumor virus. One study only has been carried out in cell hybrids examining expression of this virus, which in certain infected cells is inducible by glucocorticoids; in that case no negative dominance was seen (Parks et al. 1976).

However, the general result, that of negative dominance, raised the question that if such controls are indeed specific, exactly what are they and how are their effects exerted? Some of the possible control mechanisms that might explain the data include gene loss, gene alteration, repression of transcription, abnormal processing of transcripts, failure to translate mRNA, and rapid induced peptide decay.

There may be preferential loss of an entire chromosome or of a large piece of a chromosome that carries the gene of interest early after the cell hybrid is formed. Thus, by the time one grows enough hybrid cells to test for the gene product in question, expression is absent because the chromosome carrying the gene has been lost from the hybrid cells. This possibility can be tested if one knows the chromosome assignment of the gene being studied and if the cell lines employed are sufficiently euploid to allow karyotypic analysis. If not, it is often impossible to say whether or not the chromosome or chromosome piece carrying the gene is present. On a more micro scale there may be structural loss of the gene because of a deletion of all or part of the gene as a result of mutation or structural damage that in some way is the result of the hybridization or indeed of spontaneous events. In addition, there could be loss of expression of the gene even though its structural sequence is intact, if key control sequences required for its expression are deleted or mutated. These may include those directly involved in the normal transcription of the gene or in its processing. The gene sequence may be also intact but unexpressed if DNA methylation occurs and is for the gene in question a critical factor in the control of its expression. Finally, it is possible that following cell hybridization, translocation of the gene to a region of the genome in which it cannot be expressed occurs. Controls of all of these types are known to occur in eukaryotic cells. The case for methylation is well known. Deletions and rearrangements have been shown to account to lack of expression of immunoglobulin genes and certain viruses (Hieter et al. 1981; Coleclough et al. 1981; Yamamoto et al. 1980). All the preceding kinds of controls concern loss of gene expression via structural or functional changes in the gene at the DNA level.

Another level at which the control obviously could occur would be that of nuclear control factors. Such factors in the nucleus might interact with the genome so as to prevent the expression of an otherwise normal gene. "Position effects" or nucleosome phasing may be relevant to this model, as may location of genes in the so-called euchromatin or heterochromatin regions of the genome (Yamamoto et al. 1980; Spofford 1981). If these sorts of transdominant negative controls do occur in hybrids they might exert their effects on a broad or relatively limited scale (i.e., their influence might extend to one or a few genes or to large regions of the genome). A few limited experiments with cell heterokaryons suggest that these types of transdominant negative control do occur in somatic cell hybrids. In these experiments specific cell products were assayed, shortly after the cells were fused, but before the two nuclei had fused (Ringertz and Savage 1976a; Thompson and Gelehrter 1971). The failure to express a specific product at such a time precludes chromosome loss and makes unlikely DNA structural modification. One is thus obliged to use a model in which some element from the negative dominant parent either destroys that product (or its essential precursors or cofactors) before it can be analyzed or prevents its transcription. Relevant to these latter models are "cybrid" experiments in which enucleated cytoplasts from the negative dominant cell are fused to whole expressing cells. In some studies of this sort lasting negative dominant control has been seen (Gopalakrishnan et al. 1977). In this case it seems unlikely that the cytoplasts could be contributing enzymes responsible for the rapid decay of the product studied over many cell generations. Therefore the control factors contributed by the cytoplast may well be acting at a nuclear level to cause permanent self-regulated repression.

To return to the list of general types of control that may result in a phenotype of negative dominance, one must include the possibilities that such dominance is due to interference with or altered processing of primary gene transcripts or that the dominant factor is causing instability of a specific mature mRNA. In the former instance one would imagine that the primary transcript of the gene studied might be produced but then destroyed so rapidly that no messenger was able to be spliced out of it and therefore no protein synthesized. And finally, of course, there might be inhibition of translation of an otherwise normal mRNA or posttranslational destruction of the product itself. These major kinds of models are those that one must address in order to determine the actual types of controls operative in cell hybrids. Any insight into such controls, we believe, would pertain also to the general question of the nature of differentiation and control of gene expression.

II. Choice of Experimental System

To address these general questions in a modern molecular biological construct one needs first, a cell line that expresses a hormonally regulated gene and also possesses a marker that allows for selection of hybrid cells. In addition, the hormone-induced products should lend themselves to study by molecular techniques. One such system is that of growth hormone and prolactin induction in the GH_3 cell line. These cells, originally isolated from a rat pituitary adenoma by Tashjian et al. (1968), respond to glucocorticoids and triiodothyronine with increased synthesis of growth hormone and to estrogens and thyrolibrin with increased synthesis of prolactin. GH_3 cells possess receptors for all four of these hormone inducers. Of interest also is the fact that there appears to be reciprocal control of the two genes; i.e., at the same time that growth hormone is induced by its inducers, prolactin is repressed, and vice versa. Furthermore, GH_3 cells possess a marker used in the classic "HAT" somatic cell selection system (Lyons and Thompson 1977), in that they lack hypoxanthine–guanine–phosphoribosyl transferase.

Growth hormone and prolactin are related peptides with molecular weights of slightly greater than 20,000 daltons. For each of them cDNAs have been prepared and cloned. We have isolated and restriction endonuclease mapped clones of the genomic genes for both prolactin and the growth hormone from rat liver (Chien and Thompson 1980). From independently isolated genomic clones, the sequence of not only the cDNA but the genomic DNA for rat growth hormone has been published (Page et al. 1981; Barta et al. 1981; Seeburg et al. 1977). The sequence for rat prolactin genomic DNA is under active study and should be published soon. Thus, these cells and these genes in them represent a system that appears to offer many advantages for both the cell biology and the molecular biology required for a thorough analysis of the general phenomenon of negative control in somatic cell hybrids. As to the choice of the negative dominant parent, we elected for two reasons to utilize a mouse L cell strain which lacks thymidine kinase. First, the reversion rate of this marker is very low in these cells, allowing a stringent selection. Second, we and others had already observed that L cells consistently exert a negative dominant effect over several steroid-inducible genes.

III. Prolactin And Growth Hormone In Somatic Cell Hybrids

The basic experiment was to fuse L cells (TK^-) with GH_3 cells ($HPRT^-$), to isolate independent clones of the hybrids, and to examine these for growth hormone and prolactin production. Clones were obtained from two independent hybridization experiments carried out several years apart. As we have reported previously, both growth hormone and prolactin production and their induction were shut off in the early hybrids (Thompson et al. 1980). In the offspring of two hybrid clones we have observed expression of either growth hormone or prolactin. In each case, these apparent segregants have lost all L cell chromosomes identifiable by light or fluorescence microscopy. We observed that the occurrence of these cells as the hybrid cells were propagated correlated with morphological changes. The original clone of "flat" hybrid cells became gradually intermingled with "rounded" segregants. Recloning showed the "rounded" cells to be the prolactin–growth hormone producers. Thus it appeared that the GH_3 cell genome can move from the state in which it has basal expression of the genes to one in a hybrid in which the genes were neither expressed nor induced and back to one in which they are normally expressed and induced. However, since to date we have only observed this return of expression in extreme segregants, we cannot relate the control to any specific L cell chromosome or set of chromosomes. In addition, the majority of hybrids we have followed reach that state seen in many hybrids, in which there appears to be a relative balance of both parental chromosome groups and in which little further segregation takes place. In these hybrids reexpression has not occurred. Since this is so, it is formally impossible, without further experiments, to say whether the actual gene for either growth hormone or prolactin of rat origin is present in the nonexpressing hybrids.

To address this question we have examined hybrid clones of this type in further detail, with hormonal and molecular probes. Despite exposure to full inducing amounts of thyrolibrin, triiodothionine, sodium butyrate, cyclic AMP analogs, dexamethasone, dimethyl sulfoxide, 17β-estradiol, bromodeoxyuridine, or azacytidine, we observed no expression of either growth hormone or prolactin. In addition, we prepared either total cellular RNA or an RNA fraction enriched for polyadenylated sequences and translated it in wheat germ or reticulocyte cell-free systems. In no case did the RNA preparations from the hybrids show evidence for production of immunoreactive growth hormone or prolactin using antisera specific to each. Figure 1 shows the results of such an experiment in which polyadenylated RNA was translated from the two parental cell lines, from two nonexpressing hybrids (GL12 and GL14), and from an expressing hybrid (GL16). Furthermore, in the case of growth hormone, RNA analysis was carried out by the "northern blot" technique for the presence of structural GH messages. In these experiments our data indicated that the hybrid cells with or without hormonal stimulation failed to produce any detectable structural GH mRNA. Empirical titration experiments indicated that we would have been able to recog-

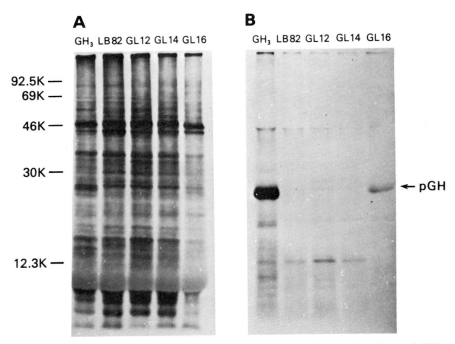

Fig. 1. Translation of polyadenylated RNA from GH_3 cells, LB82 cells, and $GH_3 \times$ LB82 hybrid clones GL12, GL14, and GL16. RNA was extracted from each of the cell cultures as described, the poly(A)-enriched fraction obtained by chromatography on oligo-dT cellulose, and equal amounts of the poly(A)–RNA translated in a reticulocyte cell-free system. The products of translation were either (A) electrophoresed on polyacrylamide gels directly or (B) first allowed to react with antigrowth hormone antiserum, collected by reaction with staph A protein, and then electrophoresed. The positions of protein markers of known molecular weights are shown on the left, with K signifying 1000. The position of authentic pregrowth hormone is shown on the right. Details of procedure are given in Strobl et al. (1982).

nize the presence of ~10 messages per cell, probably less (Strobl et al. 1982). We conclude from these experiments that the hydrids' failure to produce the products of these particular genes is not due to the absence of translatable or structural mRNA from the genes in question. Thus, the negative dominant control causes either a failure of production of transcripts or an altered stability of transcripts, such that very shortly after transcription the nuclear RNA is hydrolyzed.

The complete absence of expression again raises the question as to whether the genes simply have been deleted, or if present, in what state they do exist. In order to determine the presence of the rat growth hormone and prolactin genes in the hybrid cells, we performed cDNA hybridization experiments in the following manner. Cellular DNA was prepared and digested with a variety of specific restriction endonucleases, the digested DNA electrophoresed in agarose gels, transferred by "Southern blotting" to filters and fixed there. A labeled cDNA probe specific for the structural information contained in either the rat prolactin or rat growth hormone gene was then allowed to hybridize with the DNA on the filters. Thus, the presence of the gene in question can be demonstrated by subsequent radioautography of the hybridized DNA sequences. As probes we used cDNA to the complete or nearly complete sequence for the messages in question. Each of these cDNAs has been sequenced and shown to be correct (Seeburg et al. 1977; Cooke et al. 1980; Gubbins et al. 1979; Gubbins et al. 1980). By knowing the entire rat growth hormone genomic sequence (Barta et al. 1981), we were able to choose restriction endonucleases specifically in portions of the structural gene as well as in portions of the surrounding nonstructural sequences. We found that our probes also reacted with mouse L cell growth hormone and prolactin genes (no doubt because of the close similarity in sequence of these genes with those in the rat genome). In many cases, the rat and mouse genes electrophoresed in positions that allowed their unambiguous identification in the hybrids (Fig. 2). Our data for the growth hormone gene are much further advanced; by analysis with eight restriction endonucleases the nonexpressing hybrid cells contain the growth hormone gene in a state indistinguishable from that in the rat parent. They also show that the hybrids contain the mouse growth hormone gene as it exists in its parent. More preliminary data with the prolactin gene indicate clearly that both rat and mouse prolactin genes are present in the hybrid cells, although unexpressed, and are unaltered in any large region of their sequence. These results (Table 1) rule out the following possibilities: (1) that the gene is deleted and

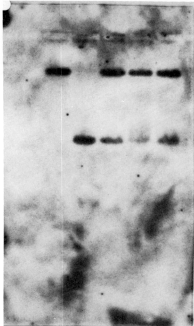

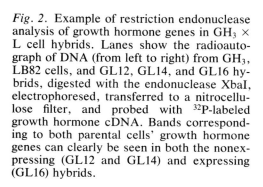

Fig. 2. Example of restriction endonuclease analysis of growth hormone genes in GH_3 × L cell hybrids. Lanes show the radioautograph of DNA (from left to right) from GH_3, LB82 cells, and GL12, GL14, and GL16 hybrids, digested with the endonuclease XbaI, electrophoresed, transferred to a nitrocellulose filter, and probed with ^{32}P-labeled growth hormone cDNA. Bands corresponding to both parental cells' growth hormone genes can clearly be seen in both the nonexpressing (GL12 and GL14) and expressing (GL16) hybrids.

Table 1. Restriction Enzyme Analysis of Growth Hormone and Prolactin Genes in Parent and Hybrid Cell Lines

| | Size of DNA fragments hybridizing with probe[a] | | |
| | Parent cells | | Hybrid cells |
Enzyme	GH_3	LB82	(3 cloned lines)[c]
I. Growth Hormone[b]			
*Bam*HI	6.4 kb		6.4 kb
		5.9 kb	5.9
*Eco*RI	9.5		9.5
		7.0	7.0
	6.6		6.6
	4.5		4.5
*Hind*III	5.9		5.9
		4.5	4.5
*Kpn*I		4.4	4.4
	2.0		2.0
	1.8		1.8
*Msp*I	4.1		4.1
		0.9	0.9
	0.6	0.6	0.6
*Pst*I		7.2	7.2
	5.6		5.6
	1.3		1.3
	0.9		0.9
*Xba*I	≥23		≥23
		5.4	5.4
II. Prolactin[d]			
*Bgl*II		12.5	12.5
	6–7		6–7
	4.5–5.5		4.5–5.5
		3	3
		2	2
*Hind*III	6	?	6
	5.3		5.3

[a] Probes were full-length cDNAs for rat prolactin and growth hormone, kindly supplied by Drs. Richard Maurer (Gubbins et al. 1980) and John Baxter (Seeburg et al. 1977), respectively.
[b] From full data in Strobl et al. 1982.
[c] Two nonexpressing hybrids, GL12 and GL14, plus one expressing cell, GL16.
[d] Preliminary data, from Thompson and Strobl, unpublished results.

(2) that the gene has been translocated to a region of the genome in which it is unable to be expressed. The latter conclusion, of course, must be limited to the gene as defined within the size limits of the widest endonuclease cut made. Thus if a very large portion of the genome were translocated to another part of the genome so that all our endonuclease studies produced cuts inside that region, we would not see it. Finally, our current data show that within the growth hormone gene specifically, there have been no gross alterations in gene size or structure.

The limits of resolution of these techniques are such that we cannot rule out small changes in gene structure that would not be detected by altered size in the agarose gel electrophoresis experiments. Also, the experiments do not speak to the question that the control region that is responsible for the transcription and induction of these two genes may be remote from the genes themselves, and that it may have been damaged in the hybrids. There is no direct evidence at the moment for such a control region, but in fact neither is there direct evidence that rules it out. Finally, these results do not speak to the question of gene methylation and gene expression.

Methylation of cytosine residues in DNA has been correlated with failure of gene expression for a number of genes (Razin and Riggs 1980). In other cases, however, no such correlation has been found (Wolf and Migeon 1982; Chen and Nienhuis 1981; van der Ploeg and Flavell 1980). In the GL hybrid cells we have undertaken analysis of the growth hormone gene for its methylation state by use of the isoschizomer pair of restriction endonucleases, Mspl and Hpall. Mspl and Hpall cut DNA in the same specific sequence, CCGG, and both enzymes cut this sequence between the internal cytosine and guanine residues. However, Mspl will cleave regardless of the methylation state of the internal cytosine, whereas Hpall will only do so if the cytosine is not methylated. Thus, a comparison of the "Southern blots" following the kind of analysis described earlier for other restriction endonucleases, after DNA digestion with either Mspl or Hpall, allows one to determine the extent of methylation of the growth hormone gene. We have compared the methylation state of the growth hormone gene in the GH_3 parent cell with that in several hybrid clones (Strobl and Thompson, unpublished). Two of these hybrids, GL12 and GL14, were clones that contained a relatively stable number of rat and mouse chromosomes and that were used in the restriction endonuclease analysis described above. These cells have never expressed growth hormone but contain both rat and mouse growth hormone sequences intact. We also studied a third hybrid, GL16, and subclones from it. This hybrid was one that initially failed to express growth hormone but quickly began to do so as it was grown out from the original clone (Thompson et al. 1980; Stroble et al. 1982). We found that subclones of hybrid GL16 were either expressors or nonexpressors. The nonexpressors resembled the previous two clones. The expressors showed both basal and inducible levels of growth hormone and had lost all discernible mouse chromosomes (see also preceding). When Mspl–Hpall analysis was carried out on all these cells, our data showed no correlation between the extent of methylation of the growth hormone gene and its expression or its inducibility (Table 2). Indeed, some of the growth hormone expressing hybrid segregants were more highly methylated than the nonexpressing hybrid clones or than some GH_3 parent cells. We conclude that methylation of the cytosine residues in the DNA coding for growth hormone is not responsible for the failure of expression in hybrid cells.

What could account for the lack of RNA production and unresponsiveness to inducers in the hybrids? The preceding experiments seem to have reduced the number of possibilities. Still possible are controls that could be at several levels. We have begun to investigate these, and even at this early stage our other results have provided some further insights. One obvious and critical

than the GH_3 cells. The hybrid cells more closely resembled the L cell with respect to receptor number. An independent marker of their functionality is lacking; however, we do know that two nonexpressor clones and one expressor clone have similar numbers of nuclear triiodothyronine receptors. Thus a quantitative defect does not appear to explain the noninducibility of growth hormone in these hybrids.

Unlike the thyrolibrin and triiodothyronine receptors, steroid receptors, both for estrogen and glucocorticoid, are abundant in the hybrids. In fact, in many cases they are present in a larger number than in either parent. These receptors have the same affinity for steroid as do the parents. Thus the K_d of receptors for dexamethasone is approximately 2×10^{-8} M and that for estrogen is about 4×10^{-10} M. For estrogen receptors no functional marker was available; however, for glucocorticoids we could check the induction of glutamine synthetase, since that was inducible in both parental lines and in the hybrids. In the hybrids glutamine synthetase was inducible as well as it was in the parental cells, thus demonstrating that the glucocorticoid receptors are functioning in the hybrids. In summary, the steroid and triiodothyronine nonresponsiveness of the hybrids is probably not due to a lack of receptors. Furthermore, the biological activity of the glucocorticoid receptors has been proved.

We suggest that the following alternatives remain likely. First, there may be negative regulatory molecules produced by the dominant nonexpressing cell, or positive regulatory molecules produced by the GH_3 chromosomes may be diluted by the presence of the L cell genome. In either case, cell fusion results in the shutdown of the growth hormone and prolactin genes in the hybrid. Alternatively, there are subtle changes in the DNA of the structural or regulatory regions of the rat gene in the hybrids. In favor of the former argument are the experiments cited previously in which heterokaryons or cybrids appeared to show dominant negative regulation. It seems unlikely that at the very early times after heterokaryons are formed, before DNA replication has occurred, that mutations, DNA loss, methylations, etc., would have taken place. These limited experiments therefore suggest that at that stage there is already dominant negative control, and they also suggest that molecules from the cell must be causing failure of gene expression. What these putative regulatory chemicals are and how they act remains to be seen. What is required now in order to define them further are systems that will allow their identification and isolation. The kinds of techniques for analysis that are under development in our laboratory to test this theory include cell-free transcription systems, single-cell microinjection systems, direct analysis of protein fractions from each of the cell types, and studies employing various nucleases.

References

Barta A, Richards RI, Baxter JD, Shine J (1981) Proc Natl Acad Sci USA 78: 4867–4871

Bourgeois S, Newby RF, Huet M (1978) Cancer Res 38: 4279–4284

Chen M, Nienhuis AW (1981) J Biol Chem 256: 9680–9681

Chien Y, Thompson EB (1980) Proc Natl Acad Sci USA 77: 4583–4587

Coleclough C, Perry RP, Karjalainen K, Weigert M (1981) Nature 290: 1372–1378

Cooke NE, Coit D, Weiner RI, Baxter JD, Martial JA (1980) J Biol Chem 255: 6502–6510

Gehring U, Thompson EB (1979) In: Baxter JD, Rousseau GG (eds) Glucocorticoid Hormone Action. Springer-Verlag, Berlin, pp 399–421

Gopalakrishnan TV, Thompson EB, Anderson WF (1977) Proc Natl Acad Sci USA 74: 1642–1646

Grove JR, Dieckmann BS, Schroer TA, Ringold GM (1980) Cell 21: 47–56

Gubbins EJ, Maurer RA, Hartley JL, Donelson JE (1979) Nucleic Acids Res 6: 915–930

Gubbins EJ, Maurer RA, Lagrimini M, Erwin CR, Donelson JE (1980) J Biol Chem 255: 8655–8662

Harmon JM, Thompson EB (1981) Mol Cell Biol 1: 512–521

Hieter PA, Korsmeyer SJ, Waldmann TA, Leder P (1981) Nature 290: 368–372

Lippman ME, Thompson EB (1974) J Biol Chem 249: 2483–2488

Lipsich LA, Kates JR, Lucas JJ (1979) Nature 281: 74–76

Lyons LB, Thompson EB (1977) J Cell Physiol 90: 179–191

Orly J, Schramm M (1976) Proc Natl Acad Sci USA 73: 4410–4414

Page GS, Smith S, Goodman HM (1981) Nucleic Acids Res 9: 2087–2104

Wade P, Parks ES, Hubbell RJ, Goldberg RJ, O'Neill FJ, Scolnick EM (1976) Cell 8: 87–93

Pfahl M, Bourgeois S (1980) Som Cell Genet 6: 63–74

Razin A, Riggs AD (1980) Science 210: 604–610

Ringertz NR, Savage RE (1976a) In: Cell Hybrids, Academic Press, New York, 1976, pp 87–118

Ringertz NR, Savage RE (1976b) In: Cell Hybrids, Academic Press, New York, 1976, pp 180–212

Seeburg PH, Shine J, Martial JA, Baxter JD, Goodman HM (1977) Nature 270: 486–494

Sibley CH, Tomkins GM (1974) Cell 2: 221–227

Spofford JB In: Ashburner M, Novitski E (eds) The Genetics and Biology of Drosophila, Vol. lC. Academic Press, New York, 1976, pp 955–1018

Strobl JS, Dannies PS, Thompson EB (1982) J Biol Chem, 257: 6588–6594

Tashjian AH, Jr., Yasamura Y, Levine L, Sato GH, Parker ML (1968) Endocrinology 82: 342–352

Thompson EB, Dannies PS, Buckler CE, Tashjian AH, Jr. (1980) J Steroid Biochem 12: 193–210

Thompson EB, Gelehrter TD (1971) Proc Natl Acad Sci USA 68: 2589–2593

van der Ploeg LHT, Flavell RA (1980) Cell 19: 947–958

Wolf SF, Migeon BR (1982) Nature 295: 667–671

Yamamoto KR, Chandler VL, Ross SR, Ucker DS, Ring JC, Feinstein SC (1980) Cold Spring Harbor Symp Quant Biol 45: 687–697

Discussion of the Paper Presented by E. B. Thompson

O'MALLEY: It's striking that while you're looking at shut-off of some of these genes that the dexamethasone receptor has gone way up, seven- or eight-fold, and it might be interesting to keep your eye on that. Receptors are generally inducible gene products in most cells, or at least their levels can be regulated by hormones and other chemicals. What's known about the normal regulation of dexamethasone receptor in GH₃ cells?

THOMPSON: That's an interesting comment. The estrogen receptor also went up; I didn't show the data for that observation. The reason why up to this point we have not studied this phenomenon is because one could only measure the ligand binding sites. With the recent advent of antisera for receptors it's now possible to ask whether there is really more receptor protein present, and maybe in a few years we'll be able to study whether messenger RNA for it is being altered. Examples of positive control exist in hybrid cells. Albumin, for example, is a classic case. The regulation of glucocorticoid receptor sites is not as well studied as that of progesterone- and estrogen-binding sites. There is some evidence, however, that there is downregulation or repression of binding sites by glucocorticoids themselves. Studies in humans correlated levels of steroid and subsequent decrease in receptor sites. In addition, glucocorticoid receptor sites increase during the S phase of the cell cycle.

Dr. Baxter, I would like to ask a question about an experiment that you mentioned. In your data there were two forms of growth hormone messenger RNA. It appeared that stationary phase cells made the smaller, presumably under-adenylated form. Now dexamethasone slows the growth of these cells, and it has been suggested in other systems that in fact it tends to produce stationary-phase differentiated-type cells or even phenotypic recovery from oncologic transformation. Therefore, it seems strange to me that when you looked at growth hormone induction by dex, you saw only the larger mRNA form. Could that be because the induction by dex in fact is a relatively low one and the sensitivity of the method would not permit resolution of the two mRNA's?

BAXTER: Under the particular conditions of the experiment, the growth hormone mRNA is predominatly one form. However, there are experiments under different conditions where dex will increase both mRNA forms. We have not yet worked out all the conditions for understanding the differential expression of these two mRNA forms.

Discussants: B.W. O'MALLEY, J.D. BAXTER AND E.B. THOMPSON

Chapter 8

Studies of the Mechanism of Glucocorticoid Hormone Action

NANCY C. LAN, THAI NGUYEN, GUY CATHALA, STEVEN K. NORDEEN, MANFRED E. WOLFF, PETER A. KOLLMAN, SYNTHIA MELLON, TUAN NGUYEN, MICHAEL KARIN, NORMAN EBERHARDT, AND JOHN D. BAXTER

I. Introduction

Within the past two decades, much information has been gained about the action of glucocorticoid hormones. These steroids penetrate the cell membrane (by mechanisms as yet unknown) and bind to soluble cellular receptors. There then occurs a conformational change (or changes) of the glucocorticoid–receptor complex (termed activation) such that the complex binds to the nuclear chromatin. After this association with chromatin, there are, in some cases, changes in the chromatin structure and, in most cases, effects on the levels of specific mRNA's. Regarding the details of these steps, however, a number of questions remain to be answered. For instance, it has not been at all clear whether the receptor is a single polypeptide chain or whether it contains several subunits. The nature of the glucocorticoid–receptor interaction is not understood in detail, and little is known about the mechanism(s) by which the mRNA increases occur. In this report we present our investigation of a number of these steps in glucocorticoid hormone action through studies with cultured rat hepatoma and pituitary tumor cells.

II. Glucocorticoid–Receptor Interaction as Revealed by Thermodynamic Analysis of the Binding Reactions

To understand the nature of the interaction of glucocorticoids with receptors, we previously (Wolff et al 1978) examined the thermodynamics of steroid binding by glucocorticoid receptors in cytosol prepared from rat hepatoma tissue culture (HTC) cells. We studied the temperature dependency of the binding of corticosterone to the glucocorticoid receptor. A plot of the $\ln K_a$ (the equilibrium association constant) as a function of $1/T$ (temperature in K) is shown in Fig. 1. Unlike other simple association reactions (Moore 1962), this relationship is not linear. The decreasing slope with higher $1/T$ values indicates that the enthalpy (ΔH) decreases as the temperature increases. The entropy (ΔS) for association is positive at 0°C, obtained

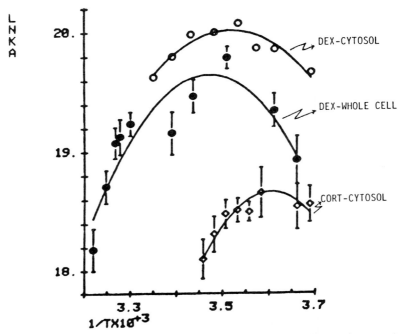

Fig. 1. Plot of the ln K_a versus $1/T$ for dexamethasone (dex) binding to glucocorticoid receptors in intact rat hepatoma cells (●) or cytosol (○) and for corticosterone (cort; △) binding to cytosol. [Data from Wolff et al. (1978).]

from the free energy of the binding and the enthalpy (ΔH). The entropy (ΔS) also decreases as temperature increases. These results led to the conclusion that the glucocorticoid–receptor interaction is driven predominantly by hydrophobic interactions.

In the intact cell, the steroid penetrates the cell membrane and then induces conformational changes in the receptor prior to, and in association with, activation. It is also conceivable that such changes are associated with nuclear binding. Any of these steps could affect the overall thermodynamics of the reaction. For instance, active transport or internalization processes or a barrier to steroid entry into the cell might affect the apparent affinity of the reaction measured in the intact cell, especially at lower temperatures. Further, because ligand-induced conformational changes can have major influences on the free energy of hormone–receptor interactions, these might also affect the overall driving force of the binding in the cell.

To answer these questions we compared the effect of temperature on the binding of dexamethasone to the glucocorticoid receptor in intact cells with binding in cytosol preparations.[1] In the latter studies, activation was blocked by sodium molybdate; in the intact cell studies, cell penetration of the steroid, activation and nuclear binding occurred, even at low temperature

[1] Nguyen T, Lan NC, Wolff ME, Kollman PA, Baxter JD, unpublished data.

(4°C). The K_as of dexamethasone for binding to glucocorticoid receptors in intact cells or cytosol preparations were measured at various temperatures under equilibrium conditions. The $\ln K_a$s were plotted as a function of $1/T$, as previously described (Wolff et al. 1978), and are shown in Fig. 1. The relationships between temperature and $\ln K_a$ for dexamethasone were found to be similar in both systems and were also similar to that previously found for corticosterone. This suggests that the initial glucocorticoid–receptor interaction dominates the overall binding reaction in intact cells. Thus, the cell membrane does not appear to affect binding in either a positive or a negative manner, arguing against any role for transport processes in the reaction. The data therefore give further support to the hypothesis that steroid entry into the cell is a passive phenomenon. The data also imply that any conformational changes in the receptor–glucocorticoid complex associated with activation or nuclear binding do not have a major overall effect on the thermodynamics of the binding in the cell, suggesting that the initial hormone–receptor interaction is the major force in driving the reaction in the intact cell.

III. Mechanisms of Glucocorticoid Antagonist Action

Certain steroids act as glucocorticoid antagonists. These compounds bind to the receptors but, in contrast to estrogen and androgen antagonists, do not promote the activation of the receptor–steroid complex necessary for nuclear binding (for review, see Baxter and MacLeod 1980). We previously found that the agonist dexamethasone associated with, and dissociated from, the receptors much more slowly than the antagonist progesterone (Rousseau et al. 1972). At 0°C and 10^{-8} M steroid concentration, the time for half-maximum binding of progesterone and dexamethasone to the receptors was 4 and 22 min, respectively; conversely, the dissociation rate constants for progesterone and dexamethasone were 3.4×10^{-2} and 3.0×10^{-3} min^{-1}, respectively. Because the differences in kinetics between dexamethasone and progesterone might be attributable to the higher affinity of dexamethasone for the receptors, we recently[1] compared the kinetics of receptor binding of the agonist aldosterone with those of progesterone, as these two steroids have nearly equal affinities for the receptors. We found similar results; the rates of association and dissociation of progesterone for the receptors are five times those of aldosterone. These results may imply that antagonists are able to recognize more readily the prevalent conformation of the receptors as they exist in the unactivated state. This could reflect the fact that the agonist undergoes high-affinity binding only after conformational changes occur in the receptor. An alternative hypothesis—that the ability of agonists to act as such requires that they remain bound for a certain period of time to elicit the subsequent conformational changes in the receptors—is probably not valid because receptors that are photoaffinity-labeled with the antagonist dexamethasone mesylate still are not active in eliciting glucocorticoid responses (Simons and Thompson 1981).

IV. Photoaffinity Labeling of Glucocorticoid Receptors

In order to understand the nature of glucocorticoid receptors, we have labeled them with photoactive ligands. Such specific and covalent "tagging" allows examination of the receptors under denaturing conditions. To accomplish this we have taken advantage of the fact that the synthetic progestin, 17α,21-dimethyl-19-norpregna-4,9-diene-3,20-dione (R5020), can bind to glucocorticoid receptors. R5020 has previously been used in photoaffinity labeling of progesterone receptors (Dure et al. 1980).

We first characterized the binding of R5020 to cytosol prepared from rat hepatoma tissue culture (HTC) and mouse S49 lymphoma cells (Nordeen et al. 1981). Scatchard analysis of R5020 binding and studies of the competition by a variety of steroids for R5020 binding indicate that, in both cell lines, R5020 binds to a single class of sites that are the glucocorticoid receptors. The affinity of R5020 for these receptors is about one-fifteenth that of dexamethasone (Nordeen et al. 1981). When cytosol from HTC cells is incubated with [^{3}H]R5020 for 2 h, irradiated at 350 nm, desalted, and subjected to electrophoresis on 7.5% SDS–polyacrylamide gels, three peaks of labeling are routinely observed (Fig. 2). However, labeling of only one peak can be blocked by the inclusion of excess unlabeled dexamethasone or R5020 in the initial binding reaction. Thus, labeling of only this peak, which corresponds to a molecular weight of 87,000 (average of four determinations), can be attributed to the glucocorticoid receptor. When cytosol incubated with unphotolyzed R5020 or R5020 prephotolyzed for 5 min is taken through an identical procedure, virtually all activity migrates with the dye (Fig. 2).

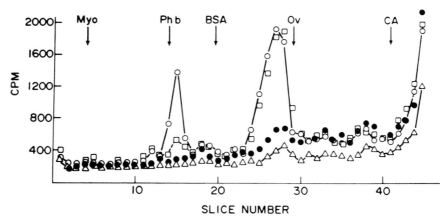

SLICE NUMBER

Fig. 2. Photolabeling of HTC cytosol with radiolabeled R5020. Cytosol was incubated with 35 nM [^{3}H]R5020 alone (○—○) or with 100 nM dexamethasone (□——□), irradiated, and then electrophoresed on discontinuous SDS–polyacrylamide gels. A third sample (△——△) was not irradiated, and a fourth (●——●) was incubated with 35 nM [^{3}H]R5020 that had been irradiated for 5 min before incubation with the cytosol. Arrows indicate the mobility of standard proteins run on a parallel gel: Myo = myosin 200,000; Phb = phosphorylase b, 94,000; BSA = bovine serum albumin, 68,000; Ov = ovalbumin, 43,000; CA = carbonic anhydrase, 30,000. [Reprinted from Nordeen et al. (1981).]

These results suggest that, in HTC cells, the glucocorticoid receptor consists of a single polypeptide chain. We find no evidence for dissimilar subunits, as has been reported for progesterone receptors (Dure et al. 1980).

For comparative purposes we also studied another cell line (S49 mouse lymphoma cells) with well-characterized glucocorticoid receptors (Sibley and Yamamoto 1979). Cytosol from S49 cells photolabeled with [^{3}H]R5020 gave results identical to those seen with HTC cells (Nordeen et al. 1981). A single peak of specific labeling (which can be blocked by dexamethasone), corresponding to a molecular weight of 87,000, is observed (Fig. 3). Thus, glucocorticoid receptors from cells derived from two different species and tissues appear quite similar and photolabel with the same efficiencies (2.5–5.0%). Even the major peak of nonspecific labeling has the same apparent molecular weight.

S49 lymphoma cells are particularly useful because of the variety of receptor mutants available. When cytosol prepared from S49R$^-$ cells (a variant line that has lost glucocorticoid-binding activity and responsiveness to hormone) is photolabeled with [^{3}H]R5020, no specific labeling can be seen (data not shown), whereas the nonspecific labeling is unchanged (Nordeen et al. 1981). These data further support the notion that the peak of specific labeling in the wild-type S49 cells is due to the glucocorticoid receptor.

We (Nordeen et al. 1981) have also photolabeled another S49 variant, S49nt^i, which is glucocorticoid-resistant and exhibits normal levels of total binding activity and increased binding of hormone-receptor complex to the

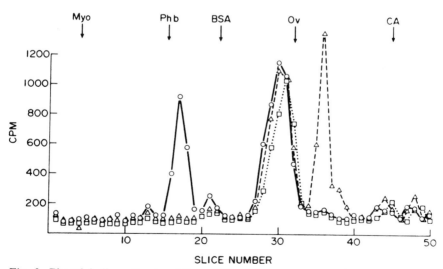

Fig. 3. Photolabeling of cytosol from S49 wild-type and S49nt^i cells. Cytosol from S49nt^i cells was incubated with 40 nM [^{3}H]R5020 alone (△——△) or with 1 μM dexamethasone as well (□——□). Parallel incubations utilized cytosol from wild-type S49 cells (○—○) containing [^{3}H]R5020. Irradiation and electrophoresis were performed as described. See legend to Fig. 2 for the key to the standards run on a parallel gel (arrows). [Reprinted from Nordeen et al. (1981).]

nucleus (Yamamoto et al. 1976). The photolabeled variant "receptor" from these cells is indeed labeled by [^{3}H]R5020 and has an apparent molecular weight of only 39,000, compared with the wild type's 87,000 (Fig. 3). Because receptors from S49nt^i retain the ability to bind hormone normally and to bind to DNA (although altered slightly), it appears that there is a domain on the receptor peptide that contains its "effector" function that is separate from the steroid- and DNA-binding domains and is lost by the mutation. A straightforward defect, which could result in the synthesis of a truncated protein, is a mutation creating a stop codon in the receptor gene. If this is the case (as is likely), then the hormone-binding domain is located in the amino-terminal portion of the receptor and the carboxy-terminal portion contains the receptor's effector function or at least would be required for its activity. In the future, photoaffinity labeling should continue to be of use in exploring the relation between receptor structure and function.

V. Complexity of Glucocorticoid Hormone Responses

The complexity of the glucocorticoid hormone response has been examined in cultured pituitary cells (GH$_3$ subline) in which growth hormone production is regulated by these steroids and by thyroid hormones (Ivarie et al. 1981). This study was conducted by pulse-labeling proteins with [^{35}S]methionine and examining the products on two-dimensional gels (Ivarie et al. 1981). It was found that glucocorticoids regulate gene expression by either increasing or decreasing the rate of synthesis of a small subset ($<1\%$) of the detected proteins in the cells. The domain of glucocorticoid hormone control overlapped to some extent with that of thyroid hormone, although most glucocorticoid-regulated gene products were unaffected by thyroid hormone and vice versa. Further, when both hormones affected the expression of the same gene, all of the possible combinations (synergisms, antagonisms, additive influences, etc.) were observed. These results imply that glucocorticoids regulate only a few specific genes and that their interactions with thyroid hormone do not occur through any central element common to all the actions of thyroid hormone (e.g., the receptors) but instead occur through gene-specific responses.

VI. Actions of Glucocorticoids on Growth Hormone mRNA

To pinpoint further the site of action of glucocorticoids in cultured pituitary cells, we have measured the levels of GH mRNA in response to the steroids. With the use of a cell-free translation assay, we earlier found that changes in GH production are paralleled by similar changes in GH mRNA (Martial et al. 1977a and 1977b). In these studies, GH mRNA copy numbers were assessed by hybridizing mRNA to a cDNA probe prepared from partially purified GH mRNA (Martial et al. 1977a).

Recently, with the use of a pure probe prepared from cloned cDNA to rat growth hormone (rGH) mRNA, we studied the kinetics of glucocorticoid effects on the relative changes of GH mRNA by Northern blotting techniques. To do this, GC cells were grown in a hypothyroid or serum-substitute medium and induced with hormones for 4, 8, and 24 h. The total RNAs were prepared as described elsewhere (Thomas 1980). The RNAs were electrophoresed on 1% agarose gels and transferred to nitrocellulose paper. The paper was then hybridized with ^{32}P-labeled cloned cDNA to rGH mRNA. The cDNA was labeled only on the strand complementary to GH mRNA with a specific activity of 1.5×10^8cpm/μg, using T_4 polymerase as described

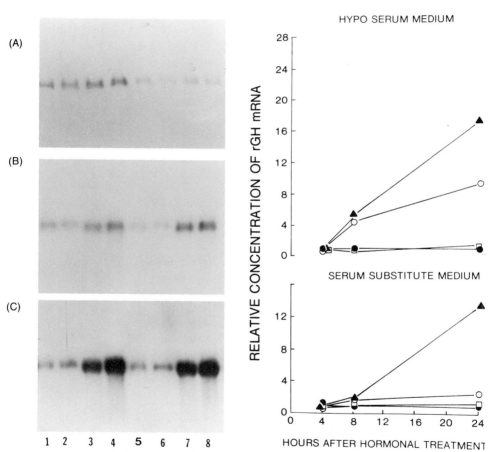

Fig. 4. Effect of glucocorticoids on GH mRNA levels in GC cells. RNAs were prepared from control cells (lanes 1 and 5) and cells were treated with dexamethasone (10^{-7} M; lanes 2 and 6), T_3 (lanes 3 and 7), or dexamethasone plus T_3 (lanes 4 and 8) grown in serum-substitute (lanes 1–4) or hypothyroid serum medium (lanes 5–8) for 4 h (a), 8 h (b), or 24 h (c). The RNA was run on 1% agarose gels, transferred to nitrocellulose paper, and hybridized with ^{32}P-labeled cDNA to rGH mRNA according to the method described by Thomas (1980). (A) Radioautogram of the RNA blots; (B) The relative concentrations of GH mRNA sequences, quantified by scanning the radioautogram shown in (A).

previously (O'Farrell 1981; O'Farrell et al. 1980). The relative concentrations of GH mRNA sequences in control and hormone-treated cells were quantified by scanning the radioautogram shown in Fig. 4A and plotted on Fig. 4B. In all cases examined, when a glucocorticoid (dexamethasone) was given to the cells alone, no significant effect on rGH mRNA levels was observed within 24 h of incubation. On the contrary, the glucocorticoid potentiated markedly the effects of thyroid hormone on rGH mRNA after 24 h of incubation. The data suggest that the glucocorticoid affects the expression of the rGH gene and that this effect is dependent on thyroid hormone.

VII. Effects of Glucocorticoids on Growth Hormone Pre-mRNA

Glucocorticoids could affect GH mRNA by increasing its pre-mRNA or by stabilizing mRNA—in which case there might be no effect on the pre-mRNA. In a preliminary study using a cDNA probe for hybridization, we found only a minute quantity of GH pre-mRNA (about 0.3% of GH mRNA) in these cells. To visualize pre-mRNA on a radioautogram without a great background influence of mRNA, we used a GH gene intron as a probe (GH gene intron D, prepared by stepwise restriction enzyme digestion, as indicated in Fig. 5). When such a probe was used to hybridize an RNA gel blot in

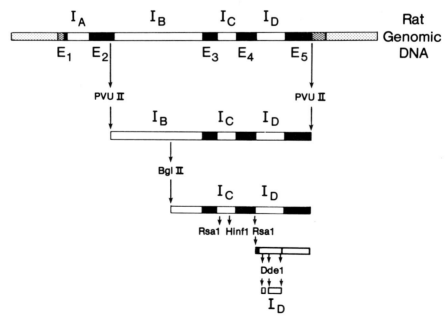

Fig. 5. Preparation of rGH chromosomal gene unlabeled intron D. The structure of the rGH chromosomal gene from prGHeh 5.8 is shown (■ , coding structure; ▭, introns; ▨ flanking structure). Intron D was obtained by double digestion with Rsa I and Hinf I of the PVU-II–PVU-II restriction fragment of the prGHeh 5.8 clone, followed by digestion with DdeI.

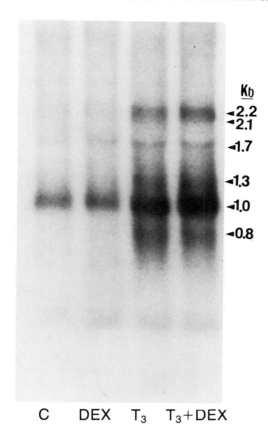

C   DEX   T₃   T₃+DEX

Fig. 6. Effect of glucocorticoids on rGH pre-mRNA. Total cellular RNA was prepared from control or 4-h hormone-treated GC cells grown in hypothyroid serum medium. Poly(A)-containing RNAs were prepared by binding total RNA to oligo (dT)-cellulose in $0.4\ M$ KC1 and eluting with water. Poly(A)–RNA (40 μg) was run on the gel and transferred to nitrocellulose paper, as described in the legend to Fig. 4. The blot was hybridized with ^{32}P-labeled intron D probe prepared by nick translation (Taylor et al. 1976).

which RNAs were prepared from control and 4-h hormone treatment, an mRNA band corresponding to GH pre-mRNA (2.2 kb) and several intermediates were observed, as shown in Fig. 6. Consistent with the observed effect of the glucocorticoid on GH mRNA levels, the glucocorticoid does not increase GH pre-mRNA when it is given alone to the cells, but it exerts an effect when thyroid hormone is present. These results suggest that (at least when thyroid hormone is present) glucocorticoids do increase GH pre-mRNA levels, and the influence is rather rapid.

VIII. Effects of Glucocorticoids on Growth Hormone Gene Transcription

We have attempted to determine if the effect of glucocorticoids on increasing GH pre-mRNA is due to influences on transcription by quantifying the relative number of RNA polymerase molecules actively engaged in transcribing the GH gene. To do this, nuclei prepared from control and hormone-treated (4 h) GC cells were incubated with ^{32}P-labeled ribonucleoside triphosphates under conditions in which the endogenously bound RNA polymerase molecules can proceed with transcription and chain termination but do not reinitiate RNA synthesis (Stallcup et al. 1978). The radiolabeled RNAs were isolated and the quantity of newly synthesized rGH mRNA was assessed by hybridizing it to nitrocellulose filters containing unlabeled cDNA to rGH mRNA. The filters were washed extensively and the specifically bound RNA was quantified by measuring the radioactivity on the filters (McKnight and Palmiter 1979). The results from an experiment are shown in Fig. 7. Consistent with the observed glucocorticoid effects on GH mRNA and pre-mRNA, the glucocorticoid increased GH gene transcription significantly only when it was given together with T_3 but did not have a clear influence when it was given alone. The data suggest that glucocorticoids can increase the rate of transcription of the GH gene and that this increase is at the level of initiation of transcription.

IX. Glucocorticoid Effects on Transferred rGH Genes

The regulation of certain genes can also be studied by transferring these genes to cells that ordinarily do not harbor them. The rGH gene has been transformed into mouse L cells with the use of herpes virus TK gene (Mantei et al. 1979; Pellicer et al. 1980; Wigler et al. 1977; Wold et al. 1979). Plasmid pTK (Wagner et al. 1981), containing the herpes simplex virus (HSV) TK gene, and prGHeh 5.8, harboring a 5.8-kb fragment containing the rGH gene,

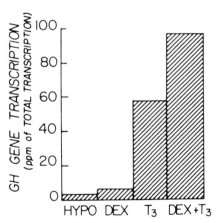

Fig. 7. Transcription of the rGH gene in isolated nuclei. GC cells were grown in medium containing 10% calf serum and then maintained for 7 days in medium containing 10% thyroidectomized calf serum. Cells were induced with dexamethasone (1 μM) or T_3 (10 nM), alone or in combination, for 4 h. Nuclei were then isolated and incubated in the presence of ^{32}P-ribonucleoside triphosphate. The ^{32}P-labeled RNA was isolated and hybridized to filters containing the cloned rGH cDNA strand. Values are represented as cpm of ^{32}P-labeled RNA hybridized to the rGH cDNA per million cpm in the hybridization reaction (ppm). [Data from Spindler et al. (1982).]

were used for the transformations; $CaCl_2$ was used to precipitate a mixture of pTK, prGHeh 5.8, and salmon sperm DNA at room temperature. The calcium precipitate was then added to mouse L cells at a density of about 5×10^5/plate grown in the complete DME medium for 16 h. The medium was changed and a mixture of hypoxanthine, amethopterin, and thymidine was then added. The colonies were transferred to a multiwell plate and propagated until they reached mass culture.

We obtained a number of L cells and 15 of these were chosen at random for analysis. These cells were treated with dexamethasone (10^{-6} M) or T_3 (10^{-7} M) alone or in combination for 2 days. The hormones did not affect the overall cell density during this time. The cell cytosol and nuclei were separated by lysis in hypotonic buffer containing NP40. The nuclei were used for DNA extraction and cytosol was used to prepare total cellular RNA (Anderson et al. 1974). The RNAs were analyzed by Northern blotting techniques, using ^{32}P-rGH cDNA as a probe, as described earlier.

Although there was substantial variation in the overall level of expression, eight of the 15 clones were found to have detectable rGH mRNA; in all eight cases, the mRNA levels were increased about four-fold by dexamethasone.

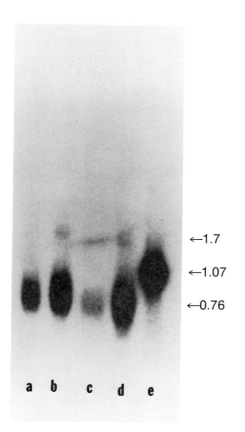

←1.7

←1.07

←0.76

a b c d e

Fig. 8. The GH mRNA levels in L cells transformed with the rGH gene with and without hormonal treatment. Results from one clone are shown. Lane a, no added hormone; lane b, dexamethasone; lane c, T_3; lane d, T_3 plus dexamethasone. For comparison, lane e shows RNA from cultured GH_3D_6 cells. (Data from Karin M, Eberhardt NL, Richards RI, Barta A, Malich N, Martial JA, Baxter JD, Cathala G, manuscript in preparation.)

With the use of densitometry and RNA from GH_3 cells, in which GH mRNA levels had been previously determined using R_0t analysis (Wegnez et al. 1982) as a calibration standard, we quantified the amount of GH transcripts in one clone (shown in Fig. 8). This clone contained 9–10 GH mRNA molecules/ cell before induction and 40–50 molecules/cell after treatment with dexamethasone. This compares favorably with about 5 and 50 molecules/cell of GH mRNA in GH_3D_6 and GC cells, respectively, grown in thyroid hormone-free medium (Wegnez et al. 1982).

Thyroid hormone does not increase the GH mRNA in these transformed cells. This may be explained by the lack of appreciable levels of T_3 receptors in L cells. (The specific T_3 binding capacity and the affinity of T_3 for these sites were found to be one-tenth of the values found in GH_3 cells.)

The major growth hormone-specific transcripts produced in the transformed L cells are shorter (750–780 bp) than the mature product (1070 bp) in GH_3 cells. However, there is a weaker band originating around 1.1 kb that could reflect a small amount of mature rGH mRNA. In addition, there are rGH gene transcripts migrating at 1.7, 1.55, and 1.4 kb. It is not currently known whether the shorter mRNA is produced by aberrant processing of a pre-mRNA whose transcription was initiated at the normal start site or by processing of a unique primary transcript initiated at some other site.

In any case, these data demonstrate that the information necessary for responsiveness to glucocorticoids is contained within the nucleotide sequence of the gene introduced into the L cells. It is hoped that by manipulating these sequences with the use of deletions, for example, it should be possible to identify the nucleotide sequence(s) responsible for the hormonal regulation and to determine whether those sequences are directly or indirectly the sites of action of glucocorticoid receptors.

Summary

We have examined several aspects of the mechanism of glucocorticoid hormone action. Thermodynamic studies of glucocorticoid binding by the receptor in the intact cell provide further support for the hypothesis that, under physiologic conditions, the glucocorticoid–receptor interaction is driven primarily by hydrophobic interactions. Kinetic studies of glucocorticoid agonist and antagonist interactions with the receptors indicate that agonists such as dexamethasone and aldosterone require longer to bind to receptors and remain bound longer than the antagonist progesterone—a finding that may imply that antagonists are able to recognize more readily the prevalent conformation of the receptors as they exist in the unactivated state and that the agonist undergoes high-affinity binding only after conformational changes occur. Photoaffinity labeling reveals that the glucocorticoid receptor is a single polypeptide chain of 87,000 MW, and that the domain required for receptor action is probably on the carboxy-terminal portion of the molecule and is distinct from the steroid- and DNA-binding domains.

We investigated the effect of glucocorticoids on specific gene activity by examining their ability to influence the expression of the rat growth hormone (rGH) gene in cultered rat pituitary cell lincs. We found that glucocorticoids increase GH pre-mRNA and the rate of GH gene transcription. These effects are rapid and thyroid hormone-dependent and can be attributed, at least in part, to the ability of glucocorticoids to increase rGH mRNA. The data suggest that glucocorticoids have direct influences on rGH gene activity. The effect of glucocorticoids on the rGH gene after it has been transferred to cells that ordinarily do not harbor them has also been studied. We found that glucocorticoids increase rGH mRNA levels in these transformed cells, but it is not clear whether the effect is at the level of transcription.

Acknowledgments. This work was supported in part by NSF grant PCM 8004-735 and NIH grant AM 19997-05. Dr. Baxter is an Investigator of the Howard Hughes Medical Institute. Dr. Cathala is a recipient of an NIH-French CNRS Program Award for Biomedical Scientific Collaboration.

References

Anderson CW, Lewis JB, Atkins JF, Gesteland RF (1974) Proc Natl Acad Sci USA 71: 2756–2760

Baxter JD, MacLeod KM (1980) In: Bondy PK, Rosenberg LE (eds) Metabolic Control and Disease, 8th ed, Saunders, Philadelphia, pp 104–160

Dure LS IV, Schrader WT, O'Malley BW (1980) Nature 283: 784–786

Ivarie RD, Baxter J, Morris JA (1981) J Biol Chem 256: 4520–4528

McKnight GS, Palmiter RD (1979) J Biol Chem 254: 9050–9058

Mantei N, Boll W, Weissmann C (1979) Nature 281:40–46

Martial JA, Baxter JD, Goodman HM, Seeburg PH (1977a) Proc Natl Acad Sci USA 74: 1816–1820

Martial JA, Seeburg PH, Guenzi D, Goodman HM, Baxter JD (1977b) Proc Natl Acad Sci USA 74: 4293–4295

Moore WJ (1962) Physical Chemistry, 3rd ed, Prentice-Hall, Englewood Cliffs, NJ

Nordeen SK, Lan NC, Showers MO, Baxter JD (1981) J Biol Chem 256: 10503–10508

O'Farrell PH (1981) Focus 3: 1–3

O'Farrell PH, Kutter E, Nakanishi M (1980) Molec Gen Genet 179: 421–435

Pellicer A, Robins D, Wold B, Sweet R, Jackson J, Lowy I, Roberts JM, Sim GK, Silverstein S, Axel R (1980) Science 209: 1414–1422

Rousseau GG, Baxter JD, Tomkins GM (1972) J Mol Biol 67: 99–115

Sibley CH, Yamamoto KR (1979) In: Baxter JD, Rousseau GG (eds) Glucocorticoid Hormone Action, Springer-Verlag, Berlin pp 357–376

Simons SS Jr, Thompson EB (1981) Proc Natl Acad Sci USA 78: 3541–3545

Spindler SR, Mellon SH, Baxter JD (1982) J Biol Chem 257: 11627–11632

Stallcup MR, Ring J, Yamamoto KR (1978) Biochemistry 17: 1515–1521

Taylor JM, Illmensee R, Summers J (1976) Biochim Biophys Acta 442: 324–330

Thomas PS (1980) Proc Natl Acad Sci USA 77: 5201–5205

Wagner MJ, Sharp JA, Summers WC (1981) Proc Natl Acad Sci USA 78: 1441–1445

Wegnez M, Schachter BS, Baxter JD, Martial JA (1982) DNA 1: 145–153

Wigler M, Silverstein S, Lee L-S, Pellicer A, Cheng Y, Axel R (1977) Cell 11: 223–232

Wold B, Wigler M, Lacy E, Maniatis T, Silverstein S, Axel R (1979) Proc Natl Acad Sci USA 76: 5684–5688

Wolff ME, Baxter JD, Killman PA, Lee DL, Kuntz ID, Bloom E, Matulich DT, Morris J (1978) Biochemistry 17: 3201–3208

Yamamoto KR, Gehring U, Stampfer MR, Sibley CH (1976) Rec Prog Horm Res 32: 3–32

Discussion of the Paper Presented by J. D. Baxter

THOMPSON: In these very interesting studies, where you looked at the size of the mRNAs from the transformed cells, do you have any other evidence besides their size that they are starting at that intron?

BAXTER: Yes. Preliminary nuclease S_1 mapping studies suggest that the 5'-terminus of the predominant 1.7 kb RNA species arises from intron B. We also know that there are no intron A and exon 2 transcripts in the 1.7 kb fragment. Thus, the 1.7-kb fragment could be generated either by initiation of transcription within the intron or by some type of aberrant RNA processing.

THOMPSON: Just a quick one; part of the same question is: What is the percentage in GH cells of regular transcript to the 1.7 kb?

BAXTER: It varies. In deinduced cells there is more 1.7-kb than 2.2-kb RNA, whereas in induced cells there is more 2.2-kb RNA.

THOMPSON: Do you include the mature mRNA?

BAXTER: No, if you include the mature mRNA, the ratio is 100 : 1 or greater.

O'MALLEY: What is the genesis of the large effect you obtain in the presence of serum? Would it occur if you use a mixture of hormones instead of serum in your cell culture?

BAXTER: We do not know what the serum is doing. We assume there is some as yet unidentified factor or factors that stimulate(s) rat growth hormone gene expression. Since we find differences between complete media and media from a hypothyroid calf that are not made up by adding thyroid hormone, we also speculate that in the animal, thyroid hormone is inducing some factor or factors that may affect rat growth hormone gene expression.

O'MALLEY: Is the internal start capped?

BAXTER: We do not know that. It is difficult to sort that out because the DNA is repetitive.

SHAPIRO: One control experiment that I think is important to do, although it can be quite difficult, is to demonstrate that your observation that the transcript remains under glucocorticoid control following deletion of a large portion of the 5' end of the gene actually reflect regulation of the initiation of transcription, and not control at the level of processing or stability in the nucleus.

BAXTER: That's an easy experiment. In fact, that's one of the ones we plan to do.

SCHRADER: You showed that you could pick up some fragments that appear to be terminated at the start of various introns and you thought that they were defective processing. Is that correct?

BAXTER: We do not know this. These could be aberrantly processed products.

SCHRADER: Is anything known about the preferred pathway of processing in vivo of this gene, say by R looping analysis or something to indicate whether or not it's the final introns that are removed last?

BAXTER: Yes, we have a lot of data on that but simply haven't digested it at this point.

MOUDGIL: In your synergistic effects of the hormones, i.e., T_3 and dexamethasone, am I correct to say that dexamethasone itself did not increase the message for growth hormone, but only when you give it with T_3?

BAXTER: That's what we thought earlier. However, it now appears that under all conditions dexamethasone alone will increase growth hormone mRNA.

MOUDGIL: To what proportion compared with T_3?

BAXTER: T_3 alone almost invariably has a greater effect than dexamethasone.

MOUDGIL: Is it a possibility that dexamethasone may be increasing T_3 receptor population?

BAXTER: Herb Samuels tooked at that and it does not.

MOUDGIL: And if so would it in parallel then increase the growth hormone message?

BAXTER: No, that doesn't happen. In fact, what happens is (mostly data of Samuels) that T_3 downregulates its own receptors. In fact, the rate of transcription goes up very quickly in response to T_3 and then it declines somewhat with the decrease in T_3 receptors.

Discussants: E.B. THOMPSON, J.D. BAXTER, B.W. O'MALLEY, W. SCHRADER AND V.K. MOUDGIL

Chapter 9

Corticosteroid Binder IB, A Potential Second Glucocorticoid Receptor

GERALD LITWACK, MICHAEL MAYER, VIRGINIA OHL, AND BERNARD SEKULA

I. Introduction

In 1973 (Litwack et al. 1973) we discovered a second liver glucocorticoid binding protein separable from transcortin and from the glucocorticoid receptor that we named binder II. We named the new protein IB. The nomenclature developed from the sequence of elution of glucocorticoid or metabolite binding proteins from DEAE–Sephadex columns at pH 7.5. Since ligandin, a steroid metabolite binding protein eluted first (pI ~8.9), it was named IA and the new binding protein was eluted just after it, hence IB. The receptor eluted next (II) and subsequently other proteins including Transcortin (IV) were eluted. Initially we thought, since IB was a binder of unmetabolized potent glucocorticoids, it might function as a "storage" protein in the cytosol, perhaps analogously to the cytosolic thyroid hormone binding protein or that it might be a second receptor (Litwack and Rosenfield 1975). Recently, there has been much emphasis on the proteolytic degradation of the glucocorticoid receptor into forms that retain the steroid-binding function and either retain or do not retain the DNA-binding site (Wrange and Gustafsson 1978; Sherman et al. 1979). In this paper we review and extend the information on corticosteroid binder IB. We try to draw some conclusions about its possible function in light of the recent emphasis on artifactual proteolytic digestion products of the glucocorticoid receptor. This protein can be viewed as a test of the hypothesis that there exist multiple hormone receptors or "isoreceptors" that, if they truly exist, could help to determine the specificity of the hormonal response when a particular form is present in a specific cell. At our current state of knowledge, when monoclonal antibodies and other specific probes are becoming available, the definitive answer to the question of glucocorticoid receptor polymorphism cannot be completely decided. However, the information presented here will favor that hypothesis over artifactually produced forms.

II. Purification and Properties of Liver IB

We reported the purification of this protein in 1975 (Litwack and Rosenfield 1975) about 2500-fold over liver cytosol. This was accomplished by a series of column chromatographic steps on DEAE–Sephadex, Sephadex G–100, and CM–Sephadex. Some characteristics are distinctly different from the traditional glucocorticoid receptor (binder II). The Stokes radius is about 20–30 Å, whereas that for II separated rapidly on DEAE–Sephadex columns is 50–60 Å. The in vitro saturation curves with [^{3}H]dexamethasone, cortisol, or corticosterone are sigmoidal, whereas with binder II this curve is always hyperbolic (Fig. 1). The ligand specificity is somewhat different between IB and II. For binder II the order of ligand potency is dexamethasone > corticosterone > cortisol > progesterone $\geqslant$ cortisone $\geqslant$ deoxycorticosterone determined in vivo (Litwack et al. 1973), whereas the order for IB determined in vitro is dexamethasone > cortisol = corticosterone = estradiol-17β $\geq$ deoxycorticosterone = dihydrotestosterone > aldosterone = cortexolone > testosterone (Litwack and Rosenfield 1975). Thus, some significant differences exist in ligand binding specificity. Moreover, we have repeatedly

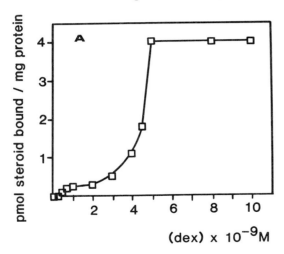

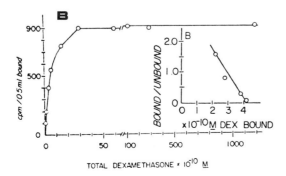

Fig. 1. Titration with [^{3}H]dexamethasone of (A) binder IB and (B) binder II. Since both components were isolated by DEAE–Sephadex chromatography in the absence of added steroid, saturation is accomplished by exchange of endogenous ligand. [Figure 1B reproduced from Litwack et al. (1973).]

confirmed that multiple fractionation procedures applied to liver IB are possible only when the in vivo ligand is corticosterone; dexamethasone and triamcinolone acetonide produce complexes that are not nearly as stable. This is in marked contrast to the traditional receptor (binder II). When we learn more about ligand exchangeability of IB the steroid binding specificity will be reinvestigated. The molecular weights of the two forms are different. A majority of laboratories now consider that the traditional receptor is about 90,000–120,000 MW. This value is equivalent to the 50–60-Å form of receptor on SDS–gel electrophoresis (Eisen et al. 1981). The consensus is that the 50–60-Å form represents a single polypeptide chain with one steroid binding site. IB is in excess of 40,000 MW as calculated from an S value of 5–6 and a Stokes radius of 20–28 Å. It has yet to be visualized on gels under denaturing conditions. Thus, most of the phenomenology of the IB and II proteins and the steroid binding site appears to be different.

III. Comparison of IB with Proteolytic Digestion Products of II

Wrange and Gustafsson (1978) have studied the generation of proteolytic cleavage products of the liver glucocorticoid receptor (50–60 Å-form = binder II) and functionalities that are either retained or lost. Their data can be pictured as in Fig. 2. At least two cleavage planes produced by trypsin occur, and the resulting fragments have been visualized in tissue cytosols (Carlstedt-Duke et al. 1979). Thus, the native activated receptor, in the current view, aligns with the 50–60-Å form, as shown. A single cleavage by

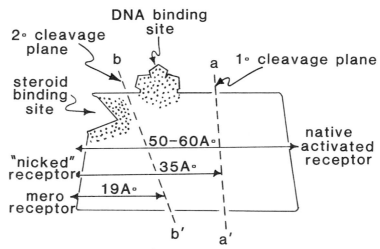

Fig. 2. Scheme depicting activated binder II showing two cleavages, the first by trypsin (or chymotrypsin) and the second by extended incubation with trypsin. a—a' refers to the first cleavage and b—b' refers to the second cleavage. [From data of Wrange and Gustafsson (1978).]

trypsin or chymotrypsin or lysosomal proteases (Carlstedt-Duke et al. 1979) at a–a′ yields a putative fragment that lacks an identifiable active site and the 35-Å product that retains both steroid- and DNA-binding sites. Further proteolysis (along b–b′) produces "meroreceptor" (Wrange and Gustafsson 1978) of about 19 Å which retains the steroid binding function but not the DNA-binding site. A major consideration has been whether the 35-Å cleavage product of II is identical to IB (20–28 Å). Originally we fractionated binder II on long columns of DEAE–Sephadex A-50, which required about 14 h. A typical chromatogram obtained from livers of adrenalectomized male rats that had been injected intraperitoneally with [³H]corticosterone for 10 min is reproduced in Fig. 3. The load was subjected to prior chromatography on Sephadex G-25 to remove free steroid. The relevant peaks are those labeled IB and II. The slight leading shoulder on the binder II peak is steroid dissociated during the long run. Binder II eluted from this chromatogram as a 35 Å Stokes radius, a pI of approximately 6.7, and a molecular weight ~67,000 (Litwack et al. 1973). These characteristics align with the 35-Å proteolytic cleavage product depicted in Fig. 2 (Wrange and Gustafsson 1978). Clearly, proteolysis takes place during the long experiments done in the early 1970s. Since then we have developed methods for rapid ion exchange chromatography in the presence of stabilizing agents that produce

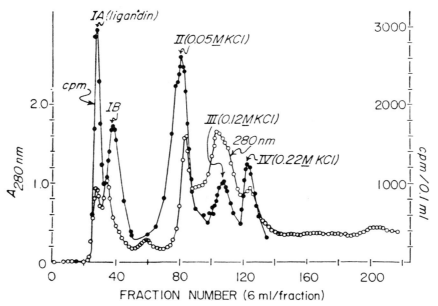

Fig. 3. Chromatography of cytosol of male, adrenalectomized rat 10 min after in vivo injection of 300 μCi [³H]corticosterone. Free steroid was removed by gel filtration prior to ion exchange chromatography. Solid circles, dpm; open circles, A_{280nm}. A long column (3 × 76 cm) was used requiring about 14 h to complete; consequently, all the binder II is the proteolyzed 35-Å form. Here it is shown to be separated distinctly from IB. [Reproduced from Litwack et al. (1973).]

binder II with a 50–60 Å Stokes radius (Parchman and Litwack 1977). Obviously, IB, the second binding peak in the chromatogram (Fig. 3), is eluted before the "nicked" 35-Å binder II and is thus separated from it on the basis of ion exchange chromatography, leading to the conclusion that IB is not identical to the 35-Å proteolytic cleavage product of binder II. It is possible that there is further processing of the 35-Å proteolytic product to IB but we have no evidence in support of this and there is some evidence, to be presented later, against this possibility based on proteolytic inhibitors.

We reasoned that if activation were physiologically important, it might manifest itself in a time-dependent fashion, at a rate not so rapid as to exclude experimental observation. By injection of [^{3}H]triamcinolone acetonide intraperitoneally and demonstrating unactivated and activated forms by ion exchange chromatography from resulting cytosols, the results should show either identical distributions at all time points or a sequence in which

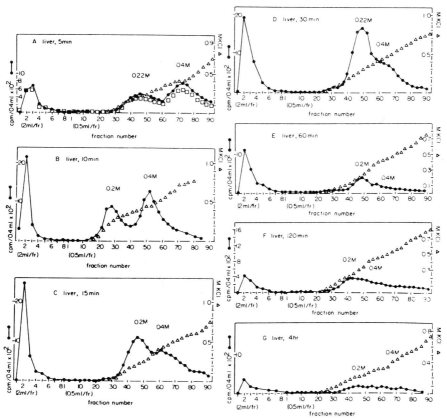

Fig. 4. Progressive in vivo appearance of unactivated and then activated forms of the liver cytosol glucocorticoid receptor (binder II) and IB following intraperitoneal injection of [^{3}H]triamcinolone acetonide. [Reproduced from Marković and Litwack (1980).]

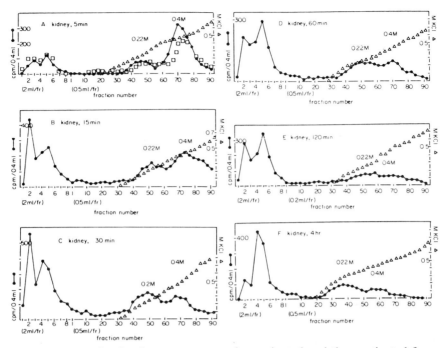

Fig. 5. Progressive in vivo appearance of unactivated and then activated forms of kidney cytosol glucocorticoid receptor, mainly IB and II, following intraperitoneal injection of [^{3}H]triamcinolone acetonide. [Reproduced from Marković and Litwack (1977).]

unactivated forms predominate and decline in concert with appearance of activated forms. The former would lead to a negative conclusion, whereas the latter would justify the conclusion that activation was physiologically relevant. Because the time of preparation of cytosols and ion exchange chromatography would be identical, the variable would be the time of exposure to steroid within the adrenalectomized rat. Thus, if activation were not a time-dependent process in the animal, fluctuation in unactivated and activated forms would not be expected. The alternative proved to be the case (Figs. 4 and 5) for both tissues examined, liver and kidney. In either case, unactivated forms are prominent, decline with time in alignment with the appearance of activated binders II and IB. A single form of IB appears in liver, but a doublet appears in kidney. Activated forms decline at longer times presumably because of translocation to the nucleus. These results are taken to confirm the idea that activation of receptor precursors is a physiological process, and our results (Marković and Litwack 1980) are similar to those obtained with thymus cells under tissue culture conditions (Munck and Foley 1979).

IV. Polyclonal Antibody to Glucocorticoid Receptor

Howard Eisen developed a rabbit polyclonal antibody to the activated binder II (Eisen 1980). This antibody has been used to study the activation process in vitro in liver and kidney (Marković et al. 1980). The antibody was connected to Sepharose, enabling column chromatography. In liver cytosol substantial binding of activated steroid–receptor complex occurred while a portion did not bind. Analysis of the nonbinding wash-through by rapid ion exchange chromatography showed that it behaved exclusively like IB. A summary of the data appears in Table 1.

These results are presented graphically in Fig. 6. The following generalizations can be made from these experiments. The polyclonal antibody crossreacts with activated binder II, probably with antigenic groups in the nonactive site region (see Fig. 2), since the antibody does not interfere with steroid binding or with DNA binding by the activated steroid–receptor complex. The antibody cross-reacts with the unactivated precursor of binder II. The antibody does not cross-react with activated IB or with its unactivated precursor in liver or in kidney cortex. The loss of cross-reaction between II and IB could be explained by proteolysis of the antigenic site (Fig. 2), although we have reason to believe this is not the case. However, the failure of crossreaction with the unactivated precursor of IB is a more formidable challenge. Here it seems less likely that selective proteolysis of the antigenic site has occurred. Stevens and his collaborators have described the receptors of glucocorticoid-sensitive or resistant P1798 murine lymphosarcoma (Stevens et al. 1981a; 1981b). The sensitive cell contains a receptor resembling binder II, while the resistant cell receptor resembles IB. Interestingly, the same antigenic behavior toward the Eisen antibody obtained in this system, namely, that the antibody cross-reacts with sensitive cell receptor but not with resistant cell receptor. Stevens observes that the unactivated precursor of sensitive receptor is 9 S and 80-Å Stokes radius producing a molecular weight of about 300,000 while the data for resistant cell receptor are 10 S, 70-Å Stokes radius producing a similar molecular weight. Thus, it would appear that the unactivated precursors of either form do not differ much in molecular weight. It will be important to know if the unactivated precursors of

Table 1. Summary of Results with Polyclonal Antibody to Binder II[a]

	Binding to Ab column	Not binding to Ab column	Subsequent analysis of wash-through
Activated liver cytosol	$+++$	$+$	IB
Unactivated liver cytosol	$+++$	$+$	IB unactivated precursor
Activated kidney cortex	$-$	$+++$	IB
Unactivated kidney cortex	$-$	$+++$	IB unactivated precursor

[a] Derived from Marković et al. (1980).

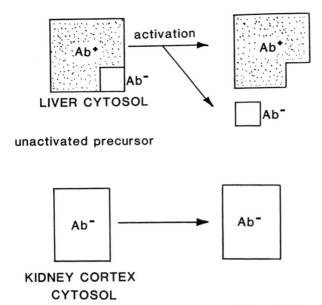

Fig. 6. Graphic summary of cross-reactivity of polyclonal anti-rat liver cytosol-activated binder II with activated glucocorticoid receptors and their unactivated precursors in liver and kidney cortex cytosols.

sensitive and resistant receptors share antigenic determinants or not. We need monoclonal antibodies against several sites on IB and II for a thorough examination of the unactivated precursors in order to conclude whether they are genetically separate isoreceptors or whether a putative separate gene for IB (or resistant P1798 receptor) is a truncated form of the gene-encoding message for binder II.

V. Protease Inhibitors and the Appearance of Binder IB

A wide variety of protease inhibitors has been tested for their effects on the appearance of IB. When IB is formed by in vivo injection of [^{3}H]triamcinolone acetonide, inclusion of micromolar amounts of molybdate, leupeptin, antipain, or phenylmethylsulfonylfluoride in the homogenizing medium failed to alter the appearance of IB and II by rapid ion exchange chromatography of cytosols from which unbound excess steroid was removed during the chromatography (Parchman and Litwack 1977). When generation of IB was studied in vitro in kidney cytosols, profound inhibition of protease activity by leupeptin, antipain, N,α-p-tosyl-L-lysine chloromethyl ketone, L-1-tosylamide-2-phenylethyl chloromethyl ketone, or iodoacetamide occurred, but there was no discernible effect on appearance of IB during activation as examined by ion exchange chromatography. Protease inhibitors that have smaller effects on general proteolysis, such as sodium vanadate, phenylmethanesulfonylfluoride, soybean trypsin inhibitor, and lima bean trypsin inhibitor also were without effect on appearance of IB during activation (Ohl VS, Mayer M, Sekula BC, Litwack G, unpublished experiments). Protease inhibitors were generally without effect whether they were added prior to, during, or after the specific binding of radioactive steroid. The obvious conclusion from these experiments is that proteases are not responsible for the appearance of IB from its unactivated precursor.

VI. Receptor Properties of IB

It is possible to examine the properties of IB separately from II by first separating the two components batchwise on DEAE–Sephadex. The utility of this separation is visualized by inspection of Fig. 4. Additionally, the activity of binding of IB can be measured by difference between whole cytosol and one treated with DNA–cellulose or nuclei compared to cellulose controls. What specifically bound steroid remains after binding to an acceptor can be detected easily by ion exchange chromatography. When acceptors are used, it is frequently better to employ the indirect method, since we are not able to visualize stable steroid–IB complexes after extraction from acceptors and we do not obtain, so far, a form of steroid–IB complex after ion exchange chromatography that binds efficiently to DNA or to nuclei. It is possible that some component of the system, present in cytosol but fractionated away by ion exchange chromatography, is required for binding to DNA or nuclei. Using those techniques that give reliable data, we have studied binding of IB complexes to nuclei, DNA–cellulose, and homodeoxypolymers. Using sucrose nuclei (Blobel and Potter 1966) from liver we find that binding of binder II is rather salt sensitive, whereas binding of IB is more stable to salt. Thus, binding of binder II to nuclei is reduced from 30% of specific steroid binding to 10% in the presence of 0.3 M KCl, a drop of 68%. Under the same conditions IB binding to nuclei falls from 27% to 17%, a drop of 38%, or only half the sensitivity to salt as exhibited by binder II complexes. At a KCl level of 0.15 M, IB is virtually unaffected in its binding to nuclei, whereas the binding of II is reduced by 30% compared to the control. Both IB and II bind well to DNA–cellulose, but II is more easily extracted after it is bound than is IB. With 1mM spermidine, 25% of bound II complexes are extracted, whereas only 4% of IB complexes are extracted. Five mM MgCl$_2$ extracts 26% of bound II but only 2% of bound IB. Ten mM pyridoxal-P (Dolan et al. 1980) extracts 95% bound II and 70% IB, and 0.45 M NaCl extracts 79% bound II and 64% bound IB. Similar experiments have been carried out with binding the complexes to nuclei followed by extraction. One mM spermidine is ineffective for either complex but 5 mM pyridoxal-P (Dolan et al. 1980) extracts 32% bound II and 36% bound IB; 0.2 M NaCl extracts 78% bound II and 55% bound IB. Either complex can be extracted from nuclei by the action of pancreatic DNase I but to a negligible extent by micrococcal nuclease. DNAse at 284 Kunitz units extracts 56% II and 43% IB; doubling the amount of the enzyme extracts 74% II and 57% IB under our conditions (Ohl VS, Mayer M, Sekula BC, and Litwack G, unpublished data). Thus, steroid–IB complexes seem to bind more firmly to DNA cellulose and to nuclei than do steroid –II complexes, as judged by their relative efficiency of extraction by various agents.

Binding experiments conducted with homodeoxypolymers gave similar results. Olig(dA), oligo(dC), oligo(dG), and oligo(dT) bound to cellulose were used. The average chain length was 12 nucleotides. Binder IB complexes bound more extensively than binder II complexes in terms of percentage of specifically bound steroid. Again binder II bound to oligodeoxynucleotides more extensively in low salt, and this binding was much more salt-sensitive than binding of IB complexes. In 0.15 M NaC1 the binding preference of binder II complexes for oligodeoxynucleotides was oligo(dG) $\geq$ oligo(dT) > oligo(dC) $\gg$ olig(dA), whereas the preference of binder IB complexes was oligo(dC) $\geq$ oligo(dG) > oligo(dA) $\gg$ oligo(dT). This result suggested marked differences between the two binding proteins. These differences must accrue from the different structures of IB and II as well as differences in their DNA binding sites.

VII. Conclusions

IB is a lower-molecular-weight protein than the traditional receptor, binder II, and it has different properties in terms of binding to nuclei, DNA, and oligodeoxypolymers. The appearance of IB, clearly, is unrelated to the activity of proteases as judged by exhaustive experiments with proteolytic inhibitors. This leads to the conclusion that IB probably does not arise directly from II during activation (transformation). The activation process, availing IB in its active form, seems to have physiological significance judging from time-course experiments after in vivo labeling with [³H]triamcinolone acetonide. Polyclonal antibody prepared against activated binder II does not cross-react with IB. Results with this antibody also suggest that there are two antigenically distinct unactivated precursors, one giving rise to binder II and one giving rise to binder IB, upon activation. Thus at our current level of understanding, the remaining possibility for connecting these two "receptors" is that the precursor for IB may be derived from the precursor of II. That such an interconversion could involve protease activity has not been ruled out. Future experiments will be directed to this possibility especially using monoclonal antibodies against II or IB. If it proves that the unactivated precursors are interrelated by a processing event, whether or not it involves proteolytic activity, it is possible that such an interconversion is physiologically important and leads to active receptors that have distinctive functions.

Acknowledgments. Research on this subject is currently supported by Research Grant BC-361 from the American Cancer Society and by grant CA-12227 from the National Cancer Institute to the Fels Research Institute.

References

Blobel G, Potter VR (1966) Science 154: 1662–1665

Carlstedt-Duke J, Wrange Ö, Dahlberg E, Gustafsson J-A, Högberg B (1979) J Biol Chem 254: 1537–1539

Dolan KP, Diaz-Gil JJ, Litwack G (1980) Arch Biochem Biophys 201: 476–485

Eisen HJ (1980) Proc Nat Acad Sci 77: 3893–3897

Eisen HJ, Schleenbaker RE, Simons Jr SS (1981) J Biol Chem 256: 12920–12925

Litwack G, Filler R, Rosenfield SA, Lichtash N, Singer S (1973) J Biol Chem 55: 977–984

Litwack G, Rosenfield SA (1975) J Biol Chem 215: 6799–6805

Marković RD, Eisen HJ, Parchman LG, Barnett CA, Litwack G (1980) Biochemistry 19: 4556–4564

Marković RD, Litwack G (1980) Arch Biochem Biophys 202: 374–379

Munck A, Foley R (1979) Nature 278: 752–753

Parchman LG, Litwack G (1977) Arch Biochem Biophys 183: 374–382

Sherman MR, Barzilai D, Pine PR, Tuazon, FB (1979) In: Leavitt WW, Clark JH (eds) Steroid Hormone Receptor Systems, Plenum Press, New York and London, pp 357–375

Stevens J, Stevens YW (1981a) Cancer Res 41: 125–133

Stevens J, Eisen HJ, Stevens YW, Haubenstock H, Rosenthal RL, Artishevsky A (1981b) Cancer Res 41: 134–137

Wrange Ö, Gustafsson J-A (1978) J Biol Chem 253: 856–865

Discussion of the Paper Presented by G. Litwack

SCHRADER: As I pointed out in my talk last night, these protease inhibitors that people have been using may or may not be able to protect receptors from proteolysis. I could think of at least a couple of experiments that might bear on the question of binder IB and its relationship. You mentioned you had iodinated it, I believe. I wonder if you have tried to do the sort of tryptic mapping that we've done for the progesterone receptor?

LITWACK: No. Our preparations aren't as homogeneous as yours.

SCHRADER: A second possibility would be to look in cell variants that lack the binder II, the classical glucocorticoid receptor, and ask whether or not binder IB is detectable.

LITWACK: Apparently certain mutants are reputed to lack it. In fact, what you measure is a lack of binding of steroid. Ringold mentioned to me certain experiments he has done with some of his glucocorticoid-resistant mutants in which he's looked at cross-reactivity with antibody. He finds just as much there as in the glucocorticoid-sensitive strains. There may be a mutation in the steroid-binding site to produce a floppy site. John Stevens also has done some very interesting studies with glucocorticoid-resistant P1798 mouse lymphosarcoma. Dr. Gustafsson has collaborated with him on it. The sole receptor there does have properties that are similar to the IB we are looking at in terms of its tighter binding to nuclei and to DNA. Some of our extraction data from nuclei are different from Stevens' but the general properties are similar, and I would tend to believe that his receptor and our IB may be similar. We need to have the antibodies to answer that question.

SCHRADER: I would like to follow up on one thing you said. You pointed out that you thought that the binding of the ligand to IB has sigmoidal binding kinetics. Is that correct?

LITWACK: Yes. Let me qualify that again. What has happened is either a rebinding or an exchange. What we've done is to isolate it in the activated position in the chromatogram and hit it with a competition experiment. But we haven't studied the exchange properties of this protein at all.

SCHRADER: That implies the possibility of multiple binding sites for the ligand on the protein.

LITWACK: It could imply that. Our calculations have been made on the basis of one-to-one binding of ligand to protein, and those calculations give us an extremely low concentration. In vitro experiments indicate the amount of IB, based on binding, is 0.1 of binder II, which would put it at 0.0005% of the cytosol proteins (requiring 200,000-fold purification to homogeneity). Now we know there is probably a lot more there for two reasons: (1) At 2500-fold purity we know there's antigenecity there. So there's a lot more protein than we're able to measure in vitro for some reason; (2) the other point is that when we inject radioactive steroid, we see quite a bit more IB. I don't understand the phenomenology going on there yet.

GUSTAFSSON: I think that you have shown quite nicely that the Binder II and 1B are separate proteins. But perhaps I'm not quite convinced yet that the 1B may not in fact represent a proteolytic fragment of the receptor or the form II. You mentioned that you have seen a proteolytic fragment of binder II. Have you tried to study that further? Have you tried to isolate it? Because I think that it will be very nice if you can compare that proteolytic form with binder IB. By such a comparison I think that you can make a stronger case.

LITWACK: Yes, we will do that. Also I think we need the monoclonal antibody. The other point I should make clear is that in our rapid ion exchange experiments we see little or no degradation of II. That chromatogram I showed you where II is all in the "nicked form" (35 Å) is after 20 h or more of manipulation. In our hands, endogenous proteolysis doesn't go that fast. If you do a 15- or 16-h incubation in the cold we see it, but it's all in the 54 Å form when we do rapid ion exchange or minicolumns (~45 min).

PECK: Perhaps I missed this, but have you studied the level of IB as a function of previous exposure to glucocorticoids of your target tissues?

LITWACK: You mean do glucocorticoids induce it?

PECK: Right. By analogy with the type II receptor in the estrogen system.

LITWACK: We have done a time course study looking at this and what we see is a conversion of unactivated to activated forms. So we don't know anything about the half-life of this protein. I doubt if we could do satisfactory experiments until we have a very good probe to do it. We need the monoclonal antibody. I am hopeful we can get them, since we now have monoclonal antibodies for binder II; I think we can do the same with IB.

PECK: The properties are so strikingly similar except for the nomenclature. Type II in the estrogen system seems similar to IB in the glucocorticoid system. You almost are suspicious that they are the same phenomenon. I am not sure inducible is the right word either. "Exposable" by previous treatment or whatever might be a better word to use.

LITWACK: Yes. The only difference we can cite is the K_d values where the potent glucocorticoids are the same with the traditional receptor and IB. I think you have a 30-fold difference or so between type I and type II. I don't know what to say on that. Have you physically separated those?

PECK: No. Well, except through extraction.

LITWACK: If we get some antibody we'd be glad to try that experiment.

MOUDGIL: My inquiry relates to some of your and Dr. Pratt's earlier work on effects of phosphatase inhibitors and molybdate. Have you looked into the in vitro steroid binding, transformation, or DNA binding characteristics of binder IB? Is it stabilized in the presence of molybdate at high temperature or does it block its activation?

LITWACK: Actually, not. The only thing we've done is to test pyridoxal phosphate, which does two things to IB. By some mechanism it obliterates half of the steroid receptor complex. Of what's left, it completely inhibits its DNA binding. But we haven't done a molybdate study.

MOUDGIL: But that's when you break the cells.

LITWACK: Break the cells, yes, but we haven't studied stabilization. With corticosterone it is quite stable. With triamcinolone acetonide it is surprisingly unstable to fractionation procedures.

MOUDGIL: Have you looked at the effect of O-phenathroline that you were recently studying?

LITWACK: No, we haven't.

SCHRADER: Could I get you to extend to something that you did not directly address in your talk? This is something that many people in the steroid receptor area banter about a great deal. What do you think the term *receptor activation* means? People use that term constantly and nobody ever puts a physical hypothesis to it. It could be proteolysis, it could be subunit assembly, it could be subunit disassembly, it could be none of the above. It could be (as I prefer to think) an in vitro artifact. There are all kinds of possibilities. What is your operating definition of receptor activation and do you think it occurs in vivo?

LITWACK: First of all, activation, in my mind, is the conversion of a steroid receptor complex from a form that is not DNA-binding protein or to one that is or is a nuclear-binding protein, by some mechanism. We provided some indirect evidence on this point, and that's all it is, which suggests that dephosphorylation (and Bill Pratt has also provided some) may be a part of this process. We still believe that. We haven't any reason not to believe it. And we are trying to do direct experiments of the type you eluded to last night: direct phosphorylation in vitro and in vivo. We think it's not proteolysis. But I'm not sure now. I'll tell you why I'm not sure. We think it's not proteolysis in vitro because a wide variety of proteolytic inhibitors do not affect the conversion of unactivated to activated forms as we visualize them on chromatography. Brad's (Thompson) lab has done a lot of work on this (with Sakaue) that has never been published and maybe he'd be willing to make a comment. Want to make a comment, Brad?

O'CONNOR: Before he does, is that because that work was not accepted for publication or not submitted for publication?

TOMPSON: The work has not been submitted. Yoshi Sakaue tried a number of proteolytic inhibitors and none prevented activation. We also found that you can identify the form which binds nuclei and DNA after in vitro activation and it is the same size as the unactivated form. Finally, there are the Foley and Munck experiments, in which they did whole cell studies using pulse labeling methods and showed that when the steroid first enters the cell it binds to what one calls the unactivated form. Then in the cell, the complex shifts progressively to the activated form; so I think activation is a physiologic step (Munck and Foley, J Ster Biochem (1980) 12:225).

LITWACK: May I say one more thing in answer to part of Bill Schrader's question. In the last few weeks Dr. Grandics in my lab has been examining the receptor isolated by affinity chromatography, gel filtration, and subsequent activation and binding to DNA and elution with pyridoxal phosphate. We have found a strong caseinolytic activity that goes through all these fractionation steps. On the surface it appears to bind to an affinity column which has a steroid on it, so it may be a steroid-binding protein. And it is not inhibited by any of the inhibitors we've tried. Nothing inhibits it. And we've tried everything from iodoacetate to soybean trypsin inhibitor and nothing touches it. Now I believe that this could have some function in hormone action somewhere, but I'm not sure that it has any function in activation.

O'CONNOR: I'd like to make the point that in cultured *Drosophila* cells and in imaginal disc of *Drosophila* most of the receptor is resident in the nuclei of naive cells. In addition, cytosol receptor has the same S values and elution from phosphocellulose as the material that one isolates from nuclei. It may be in this system that there is no need of "activation" to get binding to DNA.

MOUDGIL: I would like to make a comment on the possibility of phosphorylation and dephosphorylation, involvement in activation. I think Dr. Schrader's data that was presented yesterday has an activated receptor that they claim to have phosphorylated. However, Dr. Litwack's work indicates that dephosphorylation of the receptor (since most receptors become activated on purification) would be phosphorylated rather than dephosphorylated. Second, molybdate and tungstate can extract nuclear and DNA-bound receptor. They can work as low as 0.1 mM concentration to extract 80–90% DNA bound receptor. This is the case with rat liver glucocorticoid as well as chick oviduct progesterone receptor. Some of our data that we have submitted for publication suggests that molybdate or tungstate or molecules similar to them may be working via a more direct action rather than through phosphatases alone. At this point it is not clear that dephosphorylation or phosphorylation may be necessary for activation, although these possibilities exist.

LITWACK: I think that the alkaline phosphatase experiments done by Schmidt and Barnett in my laboratory, where we showed that it stimulated the rate of activation, indicate that dephosphorylation is the most direct guess at what is happening. I suppose we ought to keep in mind that phosphatases can catalyze other kinds of reactions, such as transphosphorylation. We are just continuing on the basis of what we think is a sensible idea, either to prove it or disprove it if we can. It may be that it's not direct phosphorylation of the receptor protein but that some other regulatory protein is the one that is phosphorylated or dephosphorylated.

SCHRADER: As far as the chick progesterone receptor is concerned, the most pronounced and most basic coomassie blue staining spot is what we are tentatively assigning to be the dephosphoreceptor. We are tentatively assuming that the most basic is the one that has no phosphate on it. That material also binds to DNA, just as the more acidic forms do. I wasn't trying to raise a hornet's nest here about which things weren't directly covered; it's just that I think it's time that we get some hypotheses out on the table and start doing some experimenting to determine what these mechanisms really are.

ROY: Don't you people think that the degree of phosphorylation will decide whether it will bind to DNA, that too much phosphate will artificially repel the DNA and if there is a little bit or just enough it might change the conformation in such a way that it can more easily bind to DNA?

LITWACK: We provided some evidence using chemical probes that what changes on the surface of the receptor in a conversion from unactivated to activated form is the emergence of a segregated group of positive charges, consisting minimally of lysine, arginine, and histidine residues. We think it's possible that a phosphate group resides adjacent to that group of positive charges neutralizing them and that if you remove that phosphate group you then expose those charges. It's something nice to think about but we don't have any direct evidence.

LIAO: I would like to note the confusion in the use of the word *activation*. Munck was the first to suggest that a cortisol receptor may exist as a form (R") that is unable to bind cortisol. His hypothesis that R" is converted to the steroid binding form through an energy-dependent process was supported by the recent work of Pratt and his co-workers, who used the term *activation* for the process. On the other hand, Jensen, Munck, and others initially used the term *transformation* to describe the change in the steroid–receptor complex from the form that does not bind to nuclei (DNA or chromatin) to the nuclear binding form. The term *transformation* was subsequently replaced by the term *activation* by many investigators. As we know more about the chemical nature of these two processes we probably will use better terms, but meanwhile we should minimize the confusion in the terminology. (Additional note: Since many people prefer to use *activation* for the DNA-binding activity, it may not be a bad idea to use the term *conversion* for the activation of the steroid-binding activity of a receptor protein.

LITWACK: Sounds like you are writing a letter to *Science*.

MOUDGIL: To my knowledge there are only a few laboratories working in the general-area of steroid receptors (e.g., Dr. Pratt's and some others) who have used the term *activation* for a form of receptor that binds steroid. Most other laboratories have used the term *activation* as representing a cytoplasmic receptor form transforming to a nuclear binding form, although I agree with Dr. Liao's comments that this has been used variably and it should be more consistent.

Discussants: W. SCHRADER, S. LIAO, G. LITWACK, J.A. GUSTAFSSON, E.J. PECK, V.K. MOUDGIL, J.D. O'CONNOR, A.K. ROY, and E.B. THOMPSON.

Chapter 10

Functional Analysis of the Glucocorticoid Receptor by Limited Proteolysis

JAN CARLSTEDT-DUKE, ÖRJAN WRANGE, SAM OKRET, JOHN STEVENS, YEE-WAN STEVENS, AND JAN-ÅKE GUSTAFSSON

I. Introduction

Steroid hormones, like many other low-molecular-weight hormones and vitamins, exert their biological effect(s) via an intracellular receptor protein. Receptor proteins for each of the different groups of steroid hormones have been demonstrated. Thus, there are separate receptor proteins for androgens, estrogens, glucocorticoids, mineralocorticoids, progestins, and vitamin D, respectively. The current model for the mechanism of action of steroid hormones is shown in Fig. 1. This model has been used for the past 15 years and the reader is referred to the abundant review articles on this topic for the details originally supporting the model.

The steroid, in this case a glucocorticoid, passes through the cell membrane of the target cell, presumably by passive diffusion, into the cytoplasm. Here the steroid binds to specific receptor proteins and the complex is then translocated into the nucleus, where it interacts with DNA. Before the nuclear translocation or interaction with DNA can take place in vitro, the glucocorticoid receptor complex (GR) must be activated. Activation can be achieved by incubating the complex at 20°C or by incubation at 0°C in the presence of 0.12 M NaCl or KCl. Activation probably incurs a conformational change of the steroid–receptor complex and presumably occurs immediately after the binding of steroid to the receptor in vivo. Activation of the complex greatly stabilizes the interaction of the steroid with the receptor protein.

After the translocation of the complex into the nucleus, it binds to the genome, presumably binding to the DNA of the specific genes regulated by the hormone. This interaction results in the specific transcription of these genes with a resulting synthesis of the specific enzymes induced. For instance, the addition of glucocorticoids to liver cells results in the induction of tyrosine aminotransferase and tryptophan oxygenase activities. As stated earlier, this model for the mechanism of action of steroid hormones has held true over the past 15 years and the evidence for this model has been greatly strengthened over the years.

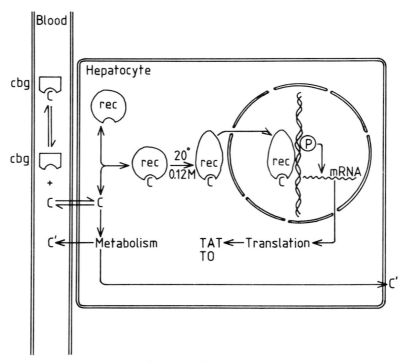

Fig. 1. Model for the mechanism of action of glucocorticoids. Abbreviations: C, corticosteroid; cbg, corticosteroid-binding globulin; rec, glucocorticoid receptor protein; P, DNA-dependent RNA-polymerase; TO, tryptophan oxygenase; TAT, tyrosine aminotransferase, C', corticosteroid metabolites (inactive). [From Carlstedt-Duke (1979).]

II. Physicochemical Characterization of the Glucocorticoid Receptor

A large variety of physicochemical characteristics has been described for the glucocorticoid receptor with a very great range in the molecular weight described. This great variation is partly due to the variety of methods used for the preparation and characterization of GR.

A. Cytosolic Receptor

Following the injection of [^{3}H]dexamethasone in rats and the isolation of the liver cytosol, a peak of radioactivity on Sephadex G-200 was found with a Stokes radius of 6.1 nm (Carlstedt-Duke et al. 1977). This peak of radioactivity with an identical Stokes radius was also found if the cytosolic fraction was labeled with the radioactive synthetic glucocorticoid in vitro. The peak of radioactivity could be saturated by incubation of the cytosol with the radioactive steroid in the presence of a 100-fold concentration of unlabeled dexamethasone or triamcinolone acetonide.

Complexes of similar size have been described in other target tissues for glucocorticoids. Giannopoulos (1973) described a GR in rabbit fetal lung that had a sedimentation rate of 4 S and that chromatographed on Sephadex G-200 between the void volume and bovine serum albumin. This corresponds approximately to the elution volume found by us for the rat liver GR. Similarly, GR from mouse fibroblast cytosol was found to have a Stokes radius of 5.9 nm (Aronow 1979; Middlebrook and Aronow 1977). A similar size has been reported for the cytosolic GR in corticosteroid-sensitive mouse lymphoma cells S49 (Yamamoto et al. 1974; Yamamoto et al. 1976) and P1798 (Stevens and Stevens 1979; Stevens and Stevens 1981) as well as in human leukemic lymphocytes (Stevens et al. 1979). In all of these cases, the cytosolic GR was found to have a Stokes radius of 5.7–6.2 nm, a sedimentation rate of 4.0 S, and a calculated molecular weight of about 90,000. The rat liver cytosolic GR (Carlstedt-Duke et al. 1977) had a sedimentation rate of 4.0 S, and this, together with the Stokes radius, gave a calculated molecular weight of 102,000.

Thus, although initial studies described a large variation in the size of the cytosolic GR from various different target organs, there is now a great similarity in the physicochemical characteristics described for the GR from a variety of species, from man to mouse, and in a variety of target tissues.

B. Nuclear Receptor

Following the injection of [^{3}H]dexamethasone in rats, as described earlier, and the extraction of the purified nuclear fraction with 0.4 M KCl, chromatography on Sephadex G-200 revealed a peak of radioactivity corresponding to a Stokes radius of 3–3.6 nm (Carlstedt-Duke et al. 1977). Thus, the nuclear form of GR appeared to have a smaller size than the cytosolic GR in rat liver cytosol. Similar findings were reported for mouse pituitary adenocarcinoma AtT-20 cells, where the nuclear GR had a sedimentation rate of 3.2–3.6 S, whereas the cytosolic GR had a sedimentation rate of 4.0 S (Garroway et al. 1976), and for mouse fibroblasts, where the nuclear GR was described as having a Stokes radius of 3.2 nm and a calculated molecular weight of 54,000 (Aronow 1979; Middlebrook and Aronow 1977). In contrast, the cytosolic GR in mouse fibroblasts was reported to have a Stokes radius of 5.9 nm and a calculated molecular weight of 109,000. However, Giannopoulos et al. (1973) found no difference in size for the nuclear GR in rabbit fetal lung when compared to the cytosolic form, both sedimenting at 4–5 S and eluting on Sephadex G-200 closely after the void volume.

If cytosol incubated in vitro with [^{3}H]dexamethasone was further incubated with purified nuclei after heat activation, the salt extract of these nuclear fractions was sometimes found to contain small amounts of a complex with Stokes radius 5–6 nm as well as the usual nuclear GR with Stokes radius 3–3.6 nm (Carlstedt-Duke et al. 1977). Stevens et al. reported the stabilization of nuclear GR in mouse lymphoma P1798 cells (1978a) and in human chronic lymphatic leukemia lymphocytes (1978b) by carbobenzoxy-

L-phenylalanine (CBZ-Phe), an inhibitor of chymotrypsin. When [^{3}H]triamcinolone acetonide-labeled cytosol was incubated with rat liver cytosol in the presence or absence of CBZ-Phe together with purified nuclei, the nuclear extract was found to contain considerably larger amounts of the larger 5–6 nm form of GR in the presence of the protease inhibitor CBZ-Phe (Fig. 2). Carter and Chae (1976) have previously reported protease activity in extracts of highly purified nuclei. This proteolytic activity is activated by high ionic strength such as that used for the extraction of GR from nuclear preparations. Thus, it would appear that the apparently smaller size of the nuclear GR reported in various tissues as described earlier is the result of intranuclear proteolysis. Under suitable conditions with minimal proteolytic activity, it is possible to demonstrate the occurrence of the larger 5–6 nm form of GR in the nuclear fraction as well as in cytosol.

C. The Effect of Molybdate

The addition of the chaotropic ion molybdate to GR preparations prevents the heat activation or salt activation of GR (Dahmer et al. 1981; Murakami and Moudgil 1981a; Murakami and Moudgil 1981b). No effect of molybdate is seen on the DNA-binding when it is added to the activated GR, however. Paradoxically, molybdate can be used for the reversible extraction of GR from DNA–cellulose or from nuclei. However, if the weaker chaotropic ion tungstate is used instead, the extraction of GR from DNA–cellulose is irreversible (Murakami and Moudgil 1981b).

The addition of molybdate to cytosol containing GR has a profound effect on its size. Sherman et al. (1982) found that the presence of molybdate during fractionation of cytosols from various tissues stabilizes forms of the GR with stokes radii of 7.1–8.2 nm. They inferred that smaller forms, e.g., the 2.0–3.4 nm complexes in rat kidney cytosol, are the products of subunit dissociation and proteolysis. Stevens et al. (1981a) have studied the effect of molybdate in more detail. Cytosolic GR in corticosensitive (CS) P1798 cells or human leukemia CEM-CH6 cells has a Stokes radius of 6.0 nm. The addition of molybdate increases the size of the GR in these cells to 8.0 nm. In corticosteroid-resistant (CR) P1798 cells or cytosol from CS P1798 cells that has been treated with α-chymotrypsin (cf. later), the GR has a stokes radius of 2.8 nm. The addition of molybdate to these cytosol preparations results in a GR with Stokes radius 7.0 nm. If cytosol from CS or CR P1798 cells is treated with trypsin (discussed later) the GR is reduced to a form with Stokes radius 1.9–2.1 nm. The addition of molybdate to these preparations has no effect on the size of this form of GR. Heat activation prior to the addition of molybdate reduces the effect of molybdate in CS or CR cytosol.

We have found that the addition of molybdate to rat liver cytosol results in a transposition of the cytosolic 6.1 nm GR to the void volume when chromatographed on Sephadex G-150 (Stokes radius >7.0 nm). Addition of molybdate to liver cytosol treated with α-chymotrypsin (discussed later) results in a transposition of GR from a volume corresponding to Stokes radius 3.6 nm to elution in or close to the void volume when chromatographed on Sephadex G-150 (Carlstedt-Duke et al. 1982). The increase in size of the nonactivated GR in the presence of molybdate is presumably due to a specific aggregation of the receptor. Alternatively, the size of GR in the presence of molybdate may represent the size of the native non-activated GR.

III. Limited Proteolysis of the Glucocorticoid Receptor

The varying size of the glucocorticoid receptor, particularly the difference in size between the cytosolic and nuclear forms of GR (see earlier) led us to investigate the effect of proteolysis on the glucocorticoid receptor. The receptor is affected both by endogenous proteases, such as the nuclear proteolytic activity described earlier and by exogenous proteases.

A. Endogenous Proteases

As described earlier, the nuclear form of GR in rat liver was initially found to have a smaller Stokes radius then the cytosolic form. This difference could be partially abolished by the addition of the chymotrypsin inhibitor CBZ-Phe (Fig. 2). When the liver cytosol was prepared in hypotonic buffers rather than the hypertonic buffers usually used, the GR in these preparations was found to have a Stokes radius very similar to that of the nuclear form, 3.6 nm. A similar effect was seen if the cytosol was prepared as usual using hypertonic buffer to which a hypotonic extract of the fraction of liver sedimenting between 1000 and 10,000 g was added. Increasing amounts of this extract resulted in an increasing conversion of the 6.1-nm form of GR into the 3.6-nm form (Carlstedt-Duke et al. 1977; Wrange et al. 1979a).

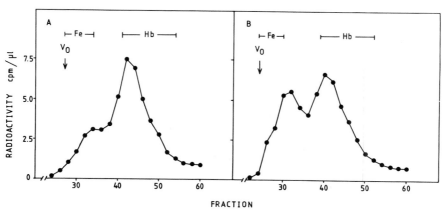

Fig. 2. [³H]Triamcinolone acetonide-labeled rat liver cytosol was incubated in the (A) absence or (B) presence of 10 mM carbobenzoxy-L-phenylalanine (CBZ-Phe) together with purified rat liver nuclei for 30 min at 25°C. After incubation the nuclei were washed and the intranuclear receptor complex was extracted by incubation with 0.4 M KCl and analyzed by chromatography on Sephadex G-150. [From Wrange et al. (1979a).]

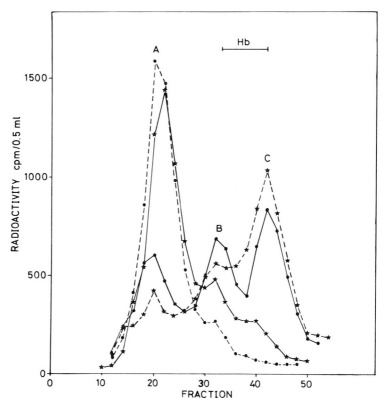

Fig. 3. The effect of concentration of lysosomal extract on the 6.1 → 3.6 nm complex conversion. Two milliliters of [^{3}H]dexamethasone-labeled rat liver cytosol containing the 6.1-nm complex was incubated with 0 ml (●---●), 0.1 ml (*——*), 0.3 ml (●——●), or 0.5 ml (*---*) of lysosomal extract at 0°C for 30 min. The incubations were analyzed by gel filtration on Sephadex G-150. Hemoglobin (Hb) was added to the samples as the internal marker. Peak A is the 6.1-nm complex, B is the 3.6-nm fragment, and C is the 1.9 nm fragment of GR. [From Carlstedt-Duke et al. (1979).]

The fraction of liver homogenate that was extracted and added to the cytosol as described earlier contains, among other cell organelles, the lysosomes. As these are rich in proteolytic activity, the effect of extracts of purified lysosomes on GR was tested. Lysosomes were highly purified by loading with iron–sorbitol–succinate complex for several weeks and then separation by ultracentrifugation. The lysosomes were extracted with hypotonic buffer and the extract added to labeled rat liver cytosol containing only the 6.1-nm form of GR (Fig. 3). Increasing amounts of the lysosomal extracts resulted in a conversion of the GR from the 6.1-nm form (peak A) first to the 3.6-nm form (peak B) and subsequently to an even smaller form with Stokes radius 1.9 nm (peak C) (Carlstedt-Duke et al. 1979). Thus, endogenous proteases normally occurring in the rat liver can reduce the native cytosolic form of GR with Stokes radius 6.1 nm to smaller fragments. These lysosomal proteases responsible for this conversion are released when the cytosol is prepared in hypotonic buffer. Furthermore, the lysosomes are damaged in varying degrees, depending on the conditions used during homogenization. We have found that the best conditions for the preparation of rat liver cytosol with the least amount of proteolytic degradation of the receptor are obtained when using a glass–teflon Potter–Elvehjem homogenizer with a speed of rotation of the pestle of 1300 r.p.m. The homogenizer is cooled on ice prior to homogenization and an ice jacket around the homogenizer is used during homogenization. Homogenization using a Dounce homogenizer with either the tight- or loose-fitting pestle or using a Waring blender greatly increases the amount of lysosomal rupture with subsequent proteolysis of the receptor. Also, inadequate cooling of the homogenizer during homogenization results in an increased proteolysis of the receptor.

Limited proteolysis of the glucocorticoid receptor by endogenous proteases has also been described in other target tissues. Sherman et al. (1978) showed that the conversion of the 70,000 form of GR in rat kidney into a form with molecular weight 32,000 could be blocked by the cathepsin inhibitor leupeptin. They also found a similar effect of leupeptin on the conversion of the 130,000 form of GR in mouse mammary tumor into a form with molecular weight 41,000. Further work on the mouse mammary tumor GR by Costello and Sherman (1979) showed that the larger form had a Stokes radius of 4.8–5.0 nm and a calculated molecular weight of 87–93,000 and the smaller form had a Stokes radius of 2.9 nm and a calculated molecular weight of 46,000. The conversion of the larger form into the smaller was blocked by leupeptin and induced by calcium. In a subsequent work (Sherman et al. 1982) the rat kidney GR was reported to have a Stokes radius of 3.4 nm and was reduced into a smaller form with Stokes radius 2.6 nm. This reduction was also blocked by leupeptin as previously described. As the addition of molybdate to the rat kidney cytosol resulted in a complex with Stokes radius 7.1 nm, the 3.4-nm form of GR described in this study probably represents a proteolytic fragment (cf. earlier). The 2.8–3-nm proteolytic fragment of CS P1798 GR is also converted into a form with Stokes radius 7.0 nm after the addition of molybdate (Stevens et al. 1981a).

Endogenous proteolysis of GR in lymphocytes from both rat and human has been described. Stevens et al. (1979) reported the reduction of the 5.6–6.0-nm form with a calculated molecular weight of 90–96,000 in human leukemic lymphocytes into a form with Stokes radius 3.5 nm. Cidlowski (1980) found that the 5.6-nm form in rat thymocytes was reduced to a form with Stokes radius 1.8 nm by endogenous proteases.

B. Exogenous Proteases

As the proteolytic degradation of GR by endogenous proteases appeared to be so specific, further studies were carried out using purified proteases. Rat liver cytosol was prepared under optimal conditions to minimize the endogenous proteolysis, as described earlier, and incubated with either [³H]dexamethasone or [³H]triamcinolone acetonide. These preparations contained only the 6.1-nm form of GR. The addition of increasing amounts of trypsin (Fig. 4) or papain to these cytosol preparations resulted in the conversion of the 6.1-nm form of GR into the 3.6-nm form and then further to the smaller 1.9-nm form. The addition of α-chymotrypsin, however, resulted only in the conversion of the 6.1-nm form into the 3.6-nm form of GR. There was no conversion of GR into the 1.9-nm form by α-chymotrypsin. No further decrease in size was seen when incubating with large amounts of the proteases (Carlstedt-Duke et al. 1977; Wrange and Gustafsson, 1978).

The effect of various protease inhibitors with different specificities was also tested on the endogenous proteolytic activity. As described earlier, CBZ–Phe had a partial effect on the nuclear proteolytic activity (Fig. 2). However, no effect was seen on the proteolytic activity in either the cytosol or lysosomal extract with any of the following protease inhibitors: soybean trypsin inhibitor, lima bean trypsin inhibitor, diisopropylfluorophosphate, antipain, leupeptin, pepstatin, Trasylol, N-ethylmaleimide, or p-chloromercuribenzoate (Carlstedt-Duke et al. 1977; Carlstedt-Duke et al. 1979; Wrange and Gustafsson 1978; Wrange et al. 1979a).

The proteolytic degradation pattern of GR when studied by isoelectric focusing in polyacrylamide gel slabs following limited proteolysis by trypsin

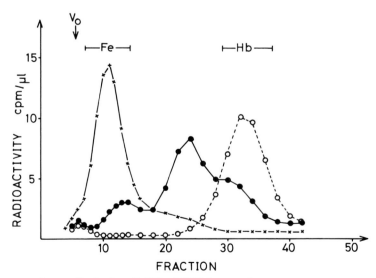

Fig. 4. Chromatography on Sephadex G-150 of [³H]dexamethasone–receptor complex after incubation of rat liver cytosol with [³H]dexamethasone. The labeled cytosol was divided into three fractions. To one fraction was added 0.1 μg (●——●), and to another fraction 0.5 μg (○– – –○) of trypsin per $A_{280-310}$ nm. The third fraction (×——×) served as control. Each sample was incubated at 10°C for 30 min and then analyzed by Sephadex G-150 chromatography using ferritin (Fe) and hemoglobin (Hb) as internal standards. [From Carlstedt-Duke et al. (1977).]

was identical for rat liver cytosol with corticosterone, dexamethasone, or triamcinolone acetonide as ligand (Wrange et al. 1979a). Furthermore, limited proteolysis of GR in rat kidney and hippocampus by trypsin gives rise to identical patterns (Wrange 1979; Wrange et al. 1979a). There is also a similarity between the proteolytic degradation of GR between species. Normal human lymphocyte GR gave rise to the same fragments as rat liver GR on isoelectric focusing following incubation with trypsin (Hansson et al. 1981).

Similar findings have been described for steroid receptors in other systems. Stevens and Stevens (1981) have described the limited proteolysis of the glucocorticoid receptor in mouse lymphoma P1798 cells. In the corticosteroid-sensitive (CS) cells the GR has a Stokes radius of 6.1 nm. Chymotrypsin-treatment of this cytosol results in the conversion of GR into a form with Stokes radius 2.8 nm. This is identical to the size of GR in cytosol from corticosteroid-resistant (CR) cells. Trypsin-treatment of cytosol from either CS or CR cells results in the formation of a fragment with Stokes radius 1.9 nm. Both the 6.1-nm and the 2.8-nm forms of GR in these cells will aggregate in the presence of molybdate to complexes with size 8.0 and 7.0 nm, respectively (Stevens et al. 1981a). The 1.9-nm fragment, however, is unaffected by the addition of molybdate.

Wilson and French (1979) have described the proteolytic degradation of the androgen receptor by trypsin, as well as by endogenous proteases. In both cases the larger 5.8-nm form is converted into the two smaller fragments with Stokes radius 3.7 nm and 2.3 nm.

Thus, steroid receptors appear to be particularly susceptible to limited proteolysis by either endogenous or exogenous proteases. The result of this proteolysis gives rise to specific fragments of the receptor, although there appears not to be any specificity with regard to the proteolytic activity. Thus, the steroid receptors contain several sites that are highly susceptible to proteolytic cleavage by a large variety of proteases.

IV. Purification of the Glucocorticoid Receptor

Recently, several groups have been successful in purifying the glucocorticoid receptor using a variety of different techniques. We have purified the rat liver GR by sequential chromatography on DNA–cellulose with an intermediate step involving heat activation (Wrange et al. 1979b). A flow diagram for the principle of the purification is shown in Fig. 5. After preparation of the rat liver cytosol and incubation with [^{3}H]triamcinolone acetonide, the cytosol is passed through a phosphocellulose and a DNA–cellulose column to remove all DNA-binding proteins. During these steps it is important to work as fast as possible and to keep the receptor preparation well cooled in order to minimize proteolysis and activation of the GR. After passage through the first DNA-cellulose column, the receptor preparation is heat-activated and rechromatographed on DNA–cellulose to which it will now bind. After thorough washing of the column, the GR can be eluted from the

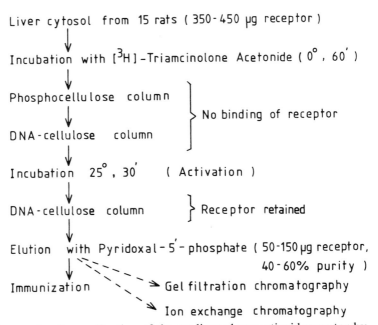

Liver cytosol from 15 rats (350-450 µg receptor)

↓

Incubation with [^{3}H]-Triamcinolone Acetonide (0°, 60')

↓

Phosphocellulose column

↓

DNA-cellulose column

} No binding of receptor

↓

Incubation 25°, 30' (Activation)

↓

DNA-cellulose column } Receptor retained

↓

Elution with Pyridoxal-5'-phosphate (50-150 µg receptor, 40-60% purity)

Immunization ➤ Gel filtration chromatography

➤ Ion exchange chromatography

Fig. 5 Flow diagram for the purification of the rat liver glucocorticoid receptor by sequential chromatography on DNA–cellulose with intervening activation. After elution of the purified GR from the second DNA–cellulose column using either pyridoxal 5′-phosphate or 0.18 M NaCl, the purity was approximately 50%. Further purification could be achieved by gel filtration on Sephadex G-200, by ion exchange chromatography on DEAE–Sepharose or by SDS–polyacrylamide gel electrophoresis. [For further details see Wrange et al. (1979b).]

column using either 10 mM pyridoxal 5′-phosphate or at least 0.2 M NaCl. The purity of the receptor preparation after elution with pyridoxal phosphate is about 50% based on the radioactivity eluted and the protein quantitation. The stability of the GR eluted is greater when elution is performed using pyridoxal 5′-phosphate. However, this blocks the DNA-binding site of the GR, although the pyridoxal phosphate can be removed by incubation with 100 mM dithiothreitol (Payvar et al. 1981).

After elution from DNA–cellulose, the GR can be further purified to about 80% purity by chromatography on DEAE–Sepharose with elution by a linear NaCl gradient.

Chromatography of the purified GR on Sephadex G-200 gives a Stokes radius of 6.0 nm. The sedimentation rate of the purified GR was 3.6 S in low ionic strength and 3.4 S in the presence of 0.15 M KCl. Thus, in contrast to GR in crude cytosol, the purified GR does not show any tendency to aggregate in low ionic strength.

These data gave a calculated molecular weight of 85,000 and a frictional ratio of 1.9, which was in good agreement with the molecular weight determined by SDS–polyacrylamide gel electrophoresis, namely, 89,000. From the radioactivity in the purified preparation of GR and assuming a molecular weight of 89,000, it was calculated that there were 1.2 ligand binding sites per receptor molecule. This supports the previous hypothesis that there is only one steroid-binding site on each receptor molecule.

Govindan and collaborators (Govindan 1979; Govindan, 1980a; Govindan and Manz 1980; Govindan and Sekeris 1976; Govindan and Sekeris 1978; Tsawdaroglou et al. 1981) have purified GR from various tissues using affinity chromatography or a combination of affinity chromatography and sequential chromatography on DNA–cellulose. The initial affinity matrices used had low affinity for the receptor and the procedure was very slow with a resulting proteolysis of the receptor. In these studies the purified forms of GR from rat liver had molecular weights of both 45,000 and 90,000. Antibodies raised against the two forms cross-reacted with both forms, thus further strengthening the hypothesis that the 45,000 form was a proteolytic fragment. If the purification was performed using rat liver nuclear GR (Govindan, 1980a) or using rat liver cytosol GR but purifying using a combination of sequential chromatography on DNA–cellulose and affinity chromatography (Govindan and Manz, 1980), only the larger form of GR with molecular weight 90,000 was obtained. Purification of GR from rat thymocytes resulted in three different forms with molecular weights 90,000, 72,000, and 45,000, respectively (Tsawdaroglou et al. 1981). The 72,000 form was also found in the nuclear fraction. Antibodies raised against either the 90,000 or the 45,000 form cross-reacted with both these forms but did not cross-react with the 72,000 form. The function of this third form and its derivation is unclear. Westphal and Beato (1980) have described the purification of GR from rat liver cytosol with a molecular weight of 40,000 and a Stokes radius of 2.7 nm. Their purification was also based on sequential chromatography with intermediate heat activation, but chromatography was performed on phosphocellulose instead of on DNA–cellulose. Presumably this preparation also represents the proteolytic fragment of the native form of GR.

Recently, two reports have been presented concerning the affinity labeling of GR in hepatoma tissue culture cells. Following analysis by SDS–polyacrylamide gel electrophoresis a band of radioactivity with molecular weight 85,000 (Simons and Thompson 1981) or 87,000 (Nordeen et al. 1981) was obtained. Thus, there is very strong evidence that the native GR has a molecular weight of 85,000–90,000 but that it is very susceptible to proteolytic degradation to a form with molecular weight 40,000–45,000.

V. Antibodies Against the Glucocorticoid Receptor

Antibodies against the purified rat liver GR were raised in two rabbits by immunization of the rabbits with trichloroacetic acid-precipitated GR eluted from the second DNA–cellulose column (Okret et al. 1981). Antibodies were also raised in five other rabbits by immunization with GR further purified by SDS–polyacrylamide gel electrophoresis. The GR band was localized by the staining of parallel tracks, and the corresponding band from the unstained tracks was cut out and homogenized in Freund's complete or incomplete adjuvant. The rabbits were injected subcutaneously at 10–15 sites with a total of 10 μg GR in Freund's complete adjuvant. After 4 and 6 weeks, the rabbits were boostered with 10 μg GR in Freund's incomplete adjuvant. The specificity of the resulting antisera was the same as previously described (Okret et al. 1981) and the titers were of the same order of magnitude.

A. Specificity

The specificity of the antisera against rat liver GR was shown by incubation of labeled cytosol with the antiserum, and the binding of specific antibodies to GR was shown by (1) adsorption of the GR–antibody complex to protein A linked to Sepharose, (2) an increased sedimentation rate of the GR–antibody complex when compared to that of the GR, and (3) an increased molecular size of the GR–antibody complex when compared to GR by gel filtration (Okret et al. 1981). Specificity of the antisera was tested by analysis by glycerol gradient centrifugation (Fig. 6) or protein-A–Sepharose chromatography. The gradients consisted of 12–30% (w/v) glycerol layered on top of 0.4 ml 87% glycerol. When experiments were performed with the rat liver GR, the GR–antibody complex centrifuged down through the 87% glycerol and pelleted at the bottom of the tube. Using this method it was possible to show that antisera did not cross-react at all either with the androgen, estrogen, or progestin receptors or with transcortin. The antisera do not bind to [^{3}H]triamcinolone acetonide. The binding of the antibodies to the receptor does not interfere with the binding of the steroid by the receptor or with the ability of the activated GR to bind to DNA–cellulose. No difference in affinity was observed between the antibodies and activated or nonactivated GR.

Four of the antisera were further tested with regard to cross-reactivity to GR from other species and tissues. All four of these antisera cross-reacted with mouse liver GR as well as with rabbit lung GR. The antibodies cross-reacted with GR from various rat tissues (liver, thymus, and hippocampus). Two of the four antisera tested cross-reacted with GR from human normal lymphocytes, chronic lymphatic leukemia cells (Fig. 6), and human hippocampus. The same two antisera cross-reacted with chick embryo liver GR.

The seven antisera were also tested with regard to specificity for the three different forms of rat liver GR described earlier (Fig. 7). The 3.6-nm form of GR was obtained by treating labeled cytosol with α-chymotrypsin and the 1.9-nm form was obtained by treatment with trypsin. When analyzed by glycerol gradient centrifugation or protein A–Sepharose chromatography, all seven of the antisera were found to interact only with the intact 6.1-nm form of GR. None of the seven antisera interacted with either the 3.6-nm or 1.9-nm forms of GR (Fig. 7). However, when these three cytosol preparations were tested using ELISA (see later) following proteolytic treatment, the immunological activity was found to be identical in all three cytosol

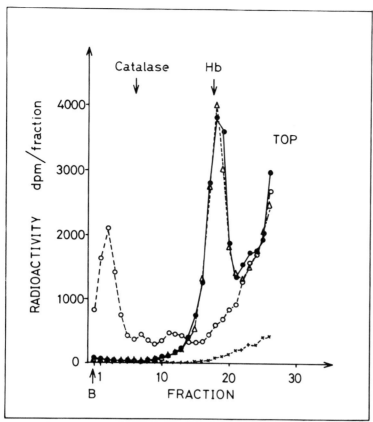

Fig. 6. Density gradient centrifugation of antiglucocorticoid–receptor-antiserum-treated GR from human chronic lymphatic leukemia lymphocyte cytosol. Aliquots of [^{3}H]triamcinolone acetonide-labeled cytosol, 0.15 ml containing 1.36 pmol GR, were incubated with 50 μl (5 mg protein/ml) anti-GR IgG (O----O), nonimmunized rabbit IgG (●——●), or sodium phosphate buffer (△----△) for 2.5 h at 4°C in the presence of 0.15 M NaCl. Cytosol incubated in the presence of a 100-fold concentration of unlabeled triamcinolone acetonide was also analyzed (×----×). After treatment with dextran-coated charcoal, the incubations were analyzed by density gradient centrifugation. B indicates that the radioactivity adhered to the bottom of the tube after emptying of the gradient and gentle washing. [From Okret et al. (1981).]

preparations. Thus, the antigenic determinant(s) were not destroyed by treatment with trypsin or α-chymotrypsin.

In contrast, Govindan (1979) raised antibodies against both the 90,000 and 45,000 forms of rat liver GR that he had purified and showed that both types of antibodies cross-reacted to an equal degree with both forms of GR. The antiserum against the 45,000 form must therefore be directed toward completely different determinants from our seven antisera. Eisen (1980) has also raised antibodies against the purified rat liver glucocorticoid receptor.

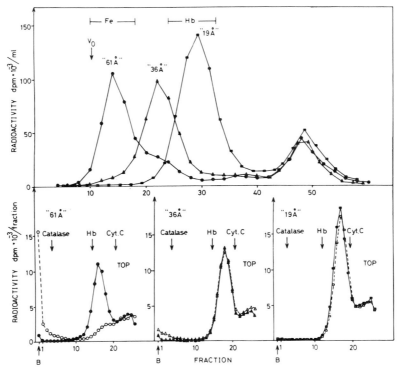

Fig. 7. Density gradient centrifugation and gel filtration chromatography of antiglu-cocorticoid–receptor-antiserum-treated native 6.1-nm GR and 3.6-nm and 1.9-nm fragments from rat liver cytosol. The proteolytic fragments were generated by treating labeled cytosol with α-chymotrypsin (3.6 nm) or trypsin (1.9 nm). After dextran-coated charcoal treatment, 3.5 ml of each preparation was analyzed by chromatography on Sephadex G-150. One hundred fifty μl of each preparation was incubated with 50 μl of anti-GR IgG (open symbols) or 50 μl of nonimmunized rabbit IgG (closed symbols for 2.5 h at 4°C in the presence of 0.15 M NaCl and analyzed by glycerol density gradient centrifugation. B indicates the radioactivity adhered to the bottom of the tube after emptying the gradient and gentle washing. [From Okret et al (1981)].

B. ELISA

An indirect competitive ELISA (enzyme-linked immunosorbent assay) was used for the detection of immunological activity in various preparations (Okret et al. 1981; Carlstedt-Duke et al. 1982). Samples of 0.2 ml were incubated with 0.05 ml of antiserum, purified on protein A–Sepharose, diluted 1 : 40–1 : 70 at 4°C overnight. After incubation, the amount of unconjugated anti-GR antibodies was measured on micro-ELISA plates coated with 40 ng purified GR in each well. Thus, color development in the well was inversely proportional to the amount of GR in the test sample.

Using this technique, the ELISA color-inhibition profile of chromatograms following gel filtration, density gradient centrifugation, ion exchange chromatography, isoelectric focusing, and DNA–cellulose chromatography was shown to follow identically the specifically bound radioactivity representing GR (Okret et al. 1981). Thus, this method appears to be highly specific for the detection of GR immunoactivity.

C. Immunofluorescence

Using specific antibodies against the glucocorticoid receptor, it has been possible to prove another stage of the hypothetical model for the mechanism of action of steroid hormones, namely, the nuclear translocation of the steroid–receptor complex following the binding of the steroid to the receptor in the cytosol. Two studies have been carried out using specific antibodies against GR. Papamichail et al. (1980) described a shift of immunofluorescence from the cytosol to the nucleus of rat thymocytes, HeLa cells, and human mononuclear cells from peripheral blood following incubation of the cells with 10^{-7} M dexamethasone at 37°C for 15 min. Govindan (1980b) reported similar findings for hepatoma tissue culture cells following incubation with dexamethasone using his antibody against the 45,000 form of GR.

Using several of our specific antisera against rat liver GR we have been able to show specific immunoflourescence in cell nuclei of the mediobasal hypothalamus, the periventricular posterior hypothalamus, and the posterior arcuate nucleus in the rat brain (Gustafsson et al. 1981). No fluorescence was seen in these regions when serum from nonimmunized rabbits was tested. It should be possible to use this method for the further investigation for the mechanism of action of steroid hormones in the brain using the previously described antisera and antibodies raised against other steroid receptors.

VI. Functional Domains of the Glucocorticoid Receptor

Limited proteolysis of rat liver GR has been shown to give rise to two specific fragments of GR as described earlier. These fragments have been analyzed with regard to several biological functions common to steroid receptors, namely, steroid- and DNA-binding (Wrange and Gustafsson 1978) and also immunoactivity (Carlstedt-Duke et al. 1982). Using limited proteolysis of GR with trypsin or α-chymotrypsin, it was possible to define three separate functional domains containing the three previously described functions. Similar findings have been described for the androgen receptor (Wilson and French 1979).

A. Steroid-Binding Site

The classical method for the detection and analysis of steroid receptor proteins during the past 20 years has involved the employment of ^{3}H-labeled ligands with high specific activity. Thus, it has not been possible previously to detect the receptor protein itself directly, but one has only been able to follow it indirectly by studying specifically bound radio-labeled steroid. All of the three forms of GR first described in rat liver cytosol by us (Carlstedt-Duke et al. 1977), namely, the 6.1-, 3.6-, and 1.9-nm forms of GR, were defined by the elution of radioactivity from the gel filtration columns; thus, all three forms contain the steroid binding site. The equilibrium constant for dissociation measured by analysis according to Scatchard was found to be identical for the 6.1- and the 3.6-nm forms (Wrange and Gustafsson 1978). However, the 1.9-nm form of GR was found to be much more unstable. In fact, it was not possible to relabel this fragment after removal of the ligand. Neither was it possible to prepare the 1.9-nm fragment by trypsinization of liver cytosol prior to incubation with steroid. This part of the receptor appears to be completely unstable in the absence of ligand. It was, however, possible to treat cytosol with α-chymotrypsin and subsequently label the receptor fragment with radioactive steroid without any loss of activity.

B. DNA-Binding Site

One of the typical characteristics of steroid receptor proteins is that, following activation, the steroid–receptor complex is able to bind to purified nuclei or to DNA. The DNA-binding capabilities of the three different forms of rat liver GR were tested by binding to DNA–cellulose (Wrange and Gustafsson 1978). Both the 6.1- and the 3.6-nm forms of GR bound to DNA–cellulose. However, the 3.6-nm form was eluted at a higher concentration of NaCl than the 6.1-nm form was, 0.15–0.2 M and 0.11–0.13 M NaCl, respectively. The 1.9-nm form of GR did not bind to DNA–cellulose at all.

The binding of the two larger forms of GR to DNA–cellulose represents nonspecific binding of the receptor complex to DNA. There is no species or tissue specificity with regard to DNA used and one can use either single- or double-stranded DNA.

Treatment with trypsin enables the steroid-binding site to be separated from the DNA-binding site. The 1.9-nm fragment of GR retains the ability to bind the steroid but has lost the ability to bind DNA. Stevens and Stevens (1981) have described similar findings for GR from corticosensitive P1798 mouse lymphoma cells. The intact GR has a Stokes radius of 6.1 nm. Following treatment of the cytosol with α-chymotrypsin, the GR is reduced to a fragment with a Stokes radius of 2.8 nm. This fragment binds more tightly to DNA than the original 6.1-nm form of GR. Treatment of the cytosol with trypsin results in a GR fragment with Stokes radius 1.9 nm that does not bind to DNA. In contrast, the DNA-binding properties of the proteolytic fragments of the androgen receptor were somewhat different (Wilson and French, 1979). All three of the forms described retained the capability of binding both androgens and DNA.

C. Immunological Determinant(s)

As mentioned earlier, the classical method for detection of steroid receptor proteins has been to follow the radioactively labeled ligand. The advent of specific antibodies directed against the receptor protein enables the detection of the receptor itself for the first time rather than the detection of the bound ligand. Using the specific antibodies raised against the purified rat liver GR, we can detect unlabeled glucocorticoid receptor using an indirect competitive ELISA (see earlier discussion). Using this assay, the immunoactivity detected follows exactly the specifically bound radioactivity corresponding to GR when analyzed by various different chromatographic methods (Okret et al. 1981). As described previously, the anti-GR antibodies only interact with the 6.1-nm form of GR and not with either the 3.6-nm or 1.9-nm forms. Thus, treatment with α-chymotrypsin separates the immunological determinant(s) from the steroid- and DNA-binding sites.

When labeled cytosol is analyzed by gel filtration on Agarose A-0.5m and the fractions are assayed for radioactivity and immunoactivity using ELISA, both the radioactivity and immunoactivity are eluted together close after the void volume (Fig. 8). The elution volume is above the linear part of the standard curve and corresponds to a Stokes radius of 5–6 nm. If the labeled cytosol is incubated with α-chymotrypsin prior to chromatography on

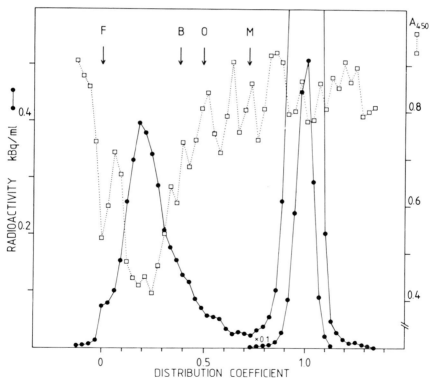

Fig. 8. Gel filtration of cytosol labeled with [³H]triamcinolone acetonide on Agarose A-0.5m. After incubation with triamcinolone acetonide, the concentration of NaCl was adjusted to 0.15 M and the sample applied on the column that was eluted with EPG buffer containing 0.15 M NaCl and 0.02% NaN₃. After chromatography, the fractions were analyzed for radioactivity and for immunoactivity by ELISA (A_{450}) (Okret et al. 1981; Carlstedt-Duke et al. 1982). F = ferritin; B = bovine serum albumin; O = ovalbumin; M = myoglobin.

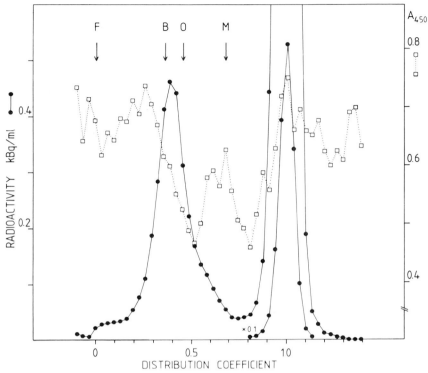

Fig. 9. Gel filtration of [³H]triamcinolone acetonide-labeled cytosol treated with α-chymotrypsin. After incubation of the cytosol with triamcinolone acetonide, the labeled cytosol was incubated with α-chymotrypsin and the incubation was terminated by the addition of lima bean trypsin inhibitor. Chromatography on Agarose A-0.5m was performed as described in the legend to Fig. 8.

Agarose A-0.5m, the immunoactivity elutes at a later volume than the radioactivity (Fig. 9). The radioactivity elutes at a volume corresponding to a Stokes radius of 3.3 nm, whereas the immunoactive fragment elutes at a volume corresponding to a Stokes radius of 2.6 nm. Thus, treatment of GR with α-chymotrypsin appears to cleave the complex into two specific fragments. The larger, with a Stokes radius of 3.3 nm (3.6 nm on Sephadex G-150), contains the steroid- and DNA-binding sites. The smaller, with a Stokes radius of 2.6 nm, contains the immunological determinant(s) (Carlstedt-Duke et al. 1982). That the DNA-binding site is separated from the immunological determinant(s) can be seen in Fig. 10. Labeled cytosol that is heat-activated and analyzed by DNA–cellulose chromatography results in the simultaneous elution of radioactivity with the immunoactivity, eluting at 0.17 M NaCl (Fig. 10A). However, treatment of the labeled heat-activated cytosol with α-chymotrypsin results in the separation of the peak of immunoactivity, eluting at 0.06 M NaCl, from the peak of radioactivity, eluting at 0.25 M NaCl (Fig. 10B).

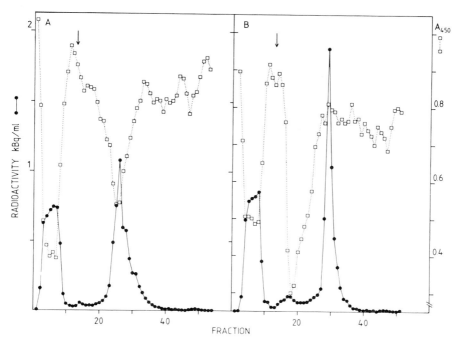

Fig. 10. DNA-cellulose chromatography of labeled cytosol treated (B) or not treated (A) with α-chymotrypsin. After application of the dextran-coated charcoal-treated samples, the columns were washed with EPG buffer and then eluted with a linear 0–0.5 M NaCl gradient. The arrow marks the start of the gradient. In A both the peaks of immunoactivity and of radioactivity were eluted at 0.17 M NaCl. In B the peak of immunoactivity was eluted at 0.06 M NaCl and the peak of radioactivity at 0.25 M NaCl.

Further proteolysis of the cytosol with α-chymotrypsin appears to further reduce the immunoactive fragment to a smaller form with Stokes radius 1.4 nm (Fig. 9). Prolonged proteolysis results in a reduction of the first of the immunoactive peaks on Agarose A-0.5m and an increase in the second immunoactive peak. Chromatography of the immunoactive peak eluted from DNA–cellulose at 0.06 M NaCl on Agarose A-0.5m results in a peak of immunoactivity eluting at a volume corresponding to a Stokes radius of 2.8 nm (Carlstedt-Duke et al. 1982). The immunoactive peak eluting at 0.06 M NaCl on DNA–cellulose following α-chymotrypsin treatment of the cytosol is recovered irrespective of prior heat treatment.

VII. Specific Binding of the Glucocorticoid Receptor to DNA

After removal of the pyridoxal phosphate from the purified GR, it was possible to show the specific interaction of the receptor complex with a cloned fragment of the murine mammary tumor virus (MTV) gene (Payvar et al. 1981). Expression of the MTV gene is regulated by glucocorticoids (Ringold et al. 1977; Varmus et al. 1979; Grove et al. 1980). Addition of glucocorticoids to hepatoma tissue culture cells containing the MTV gene leads to a 50–1000-fold induction of synthesis of MTV RNA. However, this induction is dependent on the site of insertion of the MTV gene in the hepatoma cell genome. If the MTV gene is inserted at a site where there is a tight packing of the DNA, the MTV gene is not induced by glucocorticoids (Yamamoto et al. 1981). However, this appears to be a function of the chromatin proteins at the site of insertion and does not have anything to do with the regulatory sites on the DNA at the site of insertion.

Using a nitrocellulose filter assay, it was possible to show that the purified rat liver GR specifically retains a cloned fragment of the MTV gene, labeled pMTV2 (Payvar et al. 1981). Recently, the long terminal repeat sequence from the MTV gene has been fused with dihydrofolate reductase cDNA and in this way the expression of dihydrofolate reductase in cells infected with this plasmid is regulated by glucocorticoids (Lee et al. 1981). The further characterization of the interaction of the purified GR with the MTV gene should provide great insight in the regulation of mammalian genes.

VIII. Corticosteroid Resistance

The presence of glucocorticoid receptor in the cell is the key requirement for expression of glucocorticoid effects in a variety of target tissues. However, this does not always guarantee corticosteroid sensitivity (CS). Sibley and Tomkins (1974) isolated variant clones of mouse lymphoma S49 that displayed levels of glucocorticoid binding but were unaffected by concentrations of dexamethasone that normally caused extensive lysis of wild-type S49 cells. Several groups have later described corticosteroid-resistant (CR) cells, mainly lymphocytes, both murine and human lymphoproliferative cell lines, that contain nuclear GR but do not respond to glucocorticoids in a normal manner (Bourgeois et al. 1978; Crabtree et al. 1978; Gailani et al. 1973; Gehring 1976; Lippman et al. 1974; McPartland et al. 1977; Schmidt et al. 1980; Stevens et al. 1978a; Thompson et al. 1979).

One or more steps in the mechanism of action of glucocorticoids may be altered giving rise to corticosteroid resistivity. The possible defects that may occur are (cf. Fig. 1) alterations in

1. Steroid binding
2. Receptor structure
3. Activation of GR
4. The translocation and nuclear binding of the activated GR
5. Interaction of the activated GR with the chromatin
6. The subsequent reactions leading to the effects on the cell
7. Substance present intracellularly modulating the effects

Yamamoto and collaborators (Sibley and Yamamoto 1979; Yamamoto et al. 1974; Yamamoto et al. 1976) have isolated and characterized several mutant CR strains of S49 mouse lymphoma cells. Apart from the normal, wild type (wt), there are four types of CR variants: receptor deficient (r^-), nuclear transfer deficient (nt^-), increased nuclear transfer (nt^i), and deathless (d^-). Each of these variants represents a defect at a different level in the mechanism of action of glucocorticoids. However, in this review we shall concentrate our interest on the CR strain that is characterized by an increased nuclear transfer of the activated GR (nt^i). Cells of this strain contain a normal number of receptor binding sites but the receptor is in some way altered. Physicochemical characterization of the receptor in this strain (nt^i) reveals a calculated molecular weight of 50,000, a sedimentation rate of 3.5 S, and an increased affinity for DNA with elution from DNA–cellulose at 0.21 M NaCl. In contrast, the receptor in normal cells (wt) has a calculated molecular weight of 90,000, has a sedimentation rate of 4.0 S, and elutes from DNA–cellulose at 0.17 M NaCl.

Similar results have been described for mutant receptor-positive CR P1798 mouse lymphoma cells (Stevens and Stevens 1979; Stevens and Stevens 1981). Characterization of GR from CS P1798 cells reveals a Stokes radius of 6 nm and a calculated molecular weight of 90,000. However, GR from CR P1798 cells has a Stokes radius of about 3 nm and a calculated molecular weight of 40,000. Furthermore, GR from CR P1798 cells has an increased affinity for DNA when compared to GR from CS P1798 cells. A similar "resistantlike" form of GR could be obtained from CS P1798 cytosol by limited proteolysis with α-chymotrypsin (Stevens and Stevens 1981; Stevens et al. 1981a). This gave rise to a receptor fragment with Stokes radius 3 nm and tighter DNA binding, identical to the GR from CR P1798 cells. No effect of α-chymotrypsin was seen on GR from CR cells.

Nordeen et al. (1981) have affinity-labeled the glucocorticoid receptor from various cells using the synthetic gestagen, R5020. They found that GR from wt S49 cells had a molecular weight of 87,000, whereas GR from nt^i cells only had a molecular weight of 39,000, when analyzed by SDS–polyacrylamide gel electrophoresis.

The results of the limited proteolysis of GR from CS P1798 cells might suggest that the receptor in CR cells may be synthesized as a larger precursor that then undergoes rapid intracellular processing (i.e., proteolysis) to the 3-nm form. This, however, is not very likely, as experiments consisting of the mixing of CR and CS P1798 cytosol have shown no conversion of the 6-nm CS GR to a 3-nm form (Stevens and Stevens, 1979; Stevens and Stevens, 1981). Furthermore, hybrids between CS (wt) S49 cells and CR EL4 cells showed dominance of the CS properties, as well as assigning the murine GR gene to chromosome 18 (Francke and Gehring, 1980), strongly indicating that intracellular modulating components (i.e. inhibiting substances or proteolysis) do not exist (Yamamoto et al. 1976).

Alternatively, the CR P1798 receptor may be synthesized directly as a polypeptide corresponding to the 3-nm form because of a defect at the genome level permitting transcription of only a smaller receptor mRNA. A third alternative is that different domains of the glucocorticoid receptor are synthesized as separate polypeptides but fail to link up with each other.

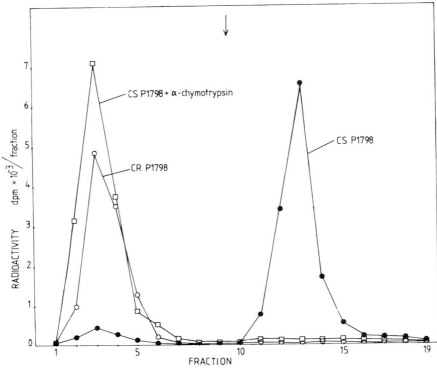

Fig. 11. Protein-A–Sepharose chromatography of GR–antibody complexes from corticosteroid-sensitive (CS; ●), α-chymotrypsin-treated CS (□) and corticosteroid-resistant (CR; ○) P1798 lymphoma cytosol. Preparation of cytosol, labeling with [³H]triamcinolone acetonide and treatment with α-chymotrypsin was performed as described by Stevens et al. (1981b) using EPG buffer. One hundred μl of labeled cytosol was incubated with 50 μl purified antiserum for 1.5 h at 4°C. The incubation mixtures were then applied on 2-ml columns of protein A–Sepharose and fractions of 0.5 ml were collected (Okret et al. 1981). The arrow indicates the start of elution with 0.1 M acetic acid. Incubation of CS cytosol with normal nonimmunized rabbit serum gave no retention of radioactivity on the protein A–Sepharose column.

Since our antibodies bind to a domain of the rat liver glucocorticoid receptor that is removed by limited proteolysis with α-chymotrypsin (see earlier Carlstedt-Duke et al. 1982), an immunocharacterization of the receptor from CS and CR cells was deemed of great value. It has previously been reported that GR from CS P1798 cells interacts with specific antibodies against rat liver GR, whereas GR from CR P1798 cells does not bind to these antibodies (Stevens et al. 1981b). The same is observed with our antibodies (Fig. 11).

In contrast to CS P1798 cytosol, both before and after limited proteolysis by α-chymotrypsin, very little immunoactivity could be found in the cytosol from CR P1798 cells (Figs. 12 and 13). It is unclear what the low amount of immunoactivity found in P1798 cells at low dilutions represents (Fig. 12). This was assayed by the indirect ELISA based on the specific anti-GR antibodies

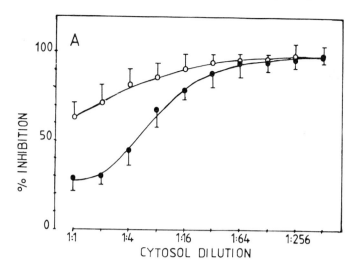

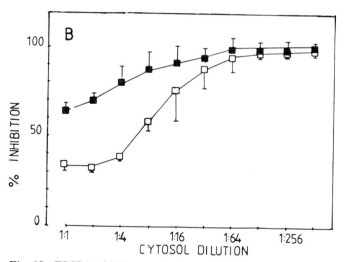

Fig. 12. ELISA of (A) cytosol dilutions from corticosteroid-sensitive (CS; (●), corticosteroid-resistant (CR; ○) P1798 cells and of (B) α-chymotrypsin-treated CS (□) and CR (■) cytosol. Cytosol preparation from P1798 tumors was performed in EPG buffer, pH 7.4, followed by centrifugation at 170,000 g. Treatment with α-chymotrypsin was carried out as previously described (Wrange and Gustafsson 1978). Cytosol (5 mg protein/ml) was diluted in EPG buffer and 200-μl aliquots of each dilution was incubated with 50 μl of purified antiserum solutions and assayed by indirect competitive ELISA (see above; Okret et al. 1981; Carlstedt-Duke et al. 1982). The bars represent 1 SD; CS (●) $n = 3$, CR (○) $n = 5$, α-chymotrypsin-treated CS (□) $n = 2$, and α-chymotrypsin-treated CR (■) $n = 4$.

and purified GR as described earlier, both in whole cytosol and after chromatography by gel filtration on Agarose A-0.5m. These results suggest that the domain normally removed by limited proteolysis by α-chymotrypsin appears to be missing in CR P1798 cells. It would seem, therefore, that this domain plays an important role in the mechanism of action of glucocorticoids. However, another possibility that cannot be ruled out is a complete change of the immunoactive domain resulting in loss of immunoactivity. The evidence presented here for the apparent lack of the immunoactive domain in CR P1798 cells and the molecular weight of GR from S49 nt^i cells of 39,000 strongly suggested that this domain is completely missing in these cells and that a mutation has occurred affecting the genome resulting in a defect

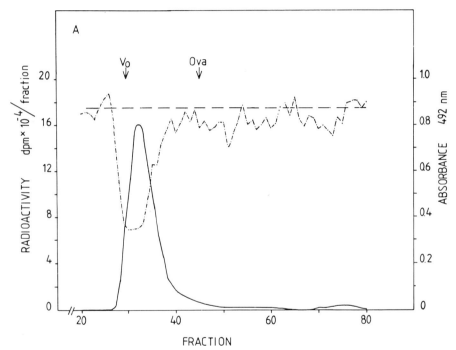

Fig. 13. Gel filtration of cytosol from (A) corticosteroid-sensitive (CS) cytosol, (B) α-chymotrypsin-treated CS cytosol, and (C) corticosteroid-resistant (CR) cytosol followed by analysis by indirect competitive ELISA. Preparation of cytosol from P1798 tumors, incubation with [^{3}H]triamcinolone acetonide and treatment with α-chymotrypsin was performed as described by Stevens et al (1981b) using EPG buffer containing 0.15 M KCl, pH 7.4. Three ml of labeled cytosol (5 mg protein/ml) was applied to columns of Agarose A-0.5m (90 × 1.6 cm) in EPG buffer containing 0.15 M KCl and 0.02% NaN$_3$ and fractions of 3 ml were collected. Two hundred μl from each fraction was assayed for radioactivity (——); 150 μl from each fraction was incubated with purified antiserum and assayed by indirect competitive ELISA (–·–·–·–). Background values (––––) were obtained by assaying fractions after chromatography of only elution buffer, using indirect competitive ELISA. V$_0$ signifies the exclusion volume and Ova, ovalbumin.

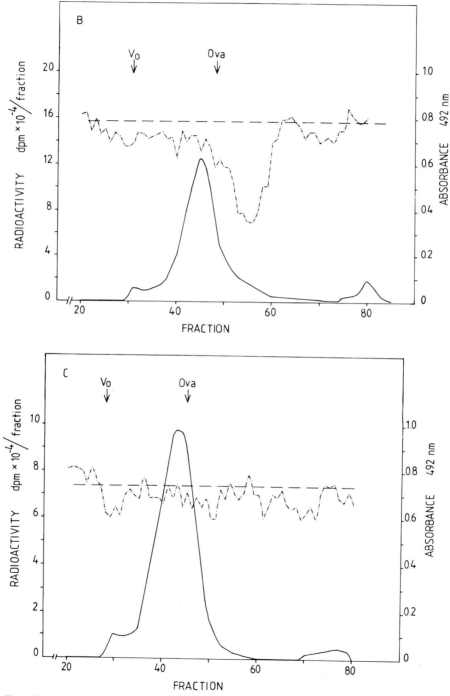

Fig. 13. (continued)

transcription of the receptor gene. Our data further substantiate the hypothesis raised by Nordeen et al. (1981) that the hormone-binding and DNA-binding domains are located in the NH_2-terminal of the protein.

IX. Summary and Future Perspectives

Limited proteolysis of the glucocorticoid receptor has proved to be a valuable tool for a functional analysis of the receptor protein. With the help of these analyses, it has been possible to describe three functional domains of the receptor protein (Fig. 14). The native glucocorticoid–receptor complex contains a steroid-binding domain (A), a DNA-binding domain (B), and an immunoactive domain (C). This form of GR has a Stokes radius of 6.1 nm and a molecular weight of 89,000 when purified. Two steroid-binding proteolytic GR fragments can be found. The larger has a Stokes radius of 3.3–3.6 nm and a calculated molecular weight of 46,000 and contains both the steroid- and DNA-binding sites (A + B). The smaller steroid-binding GR fragment, with a Stokes radius of 1.9 nm and a calculated molecular weight of 19,000, contains only the steroid-binding domain (A). Analysis of the proteolytic fragments of GR using the specific anti-GR antibodies revealed the occurrence of a proteolytic fragment with Stokes radius 2.6 nm following limited proteolysis of GR by α-chymotrypsin. This fragment contains neither the steroid-binding nor the DNA-binding domains but consists only of the immunoactive domain (C). Further proteolysis of this fragment results in an even smaller form with Stokes radius 1.4 nm.

The apparent identity of the larger of the two proteolytic forms of GR (the 3.3–3.6-nm form) with GR isolated from certain corticosteroid-resistant cells, together with the lack of the immunoactive domain in these cells appears to indicate an important function of this domain with regard to biolog-

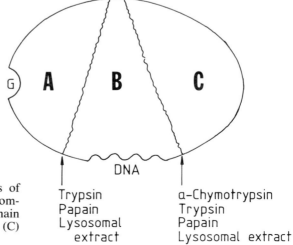

Fig. 14. Functional domains of the glucocorticoid-receptor complex. (A), steroid-binding domain (B) DNA-binding domain, (C) immunoactive domain.

ical activity of the receptor. Although this domain does not bind to DNA with any affinity, it seems most likely that it plays an important role in the specific interaction of the receptor complex with the genome. One of the most studied systems with regard to interaction of regulatory proteins with the genome is the *lac* repressor protein and its binding to the *lac* operon. This protein also contains several functional domains that can be separated by limited proteolysis (Dunaway et al. 1980). Ligand-binding occurs in one domain but DNA binding occurs in the other domain. However, the protein must be intact in order to elicit its effect. Binding of the ligand appears to abolish the relatively weak interdomain contacts, which leads to increased motional freedom of the DNA-binding domain (Schnarr and Maurizot, 1981). It will be of great interest to see whether similar functions occur within the glucocorticoid receptor with regard to its specific interaction with those genes that are regulated by glucocorticoids such as the MTV gene. Further work on this is at present under progress.

Acknowledgments. We are very greatful for the expert technical assistance from Anita Andersson, Ulla-Britt Harnemo, Marianne Nilsson, Ing-Marie Nilsson, and Ingalill Ramberg over the years these studies have been performed. This work was supported by grants from the Swedish Medical Research Council (13X-2819 and 13X-06245), LEO Research Foundation, Harald Jeanssons Stiftelse, and the Swedish Society of Medical Sciences. Jan Carlstedt-Duke and Örjan Wrange are recipients of research fellowships from the Swedish Cancer Society and from the Swedish Medical Research Council, respectively.

References

Aronow L (1979) In: Baxter JD, Rousseau GG (eds) Glucocorticoid Hormone Action. Springer-Verlag, Heidelberg, pp 327–340

Bourgeois S, Newby RF, Huet M (1978) Cancer Res 38: 4279–4284

Carlstedt-Duke J (1979) Thesis, Karolinska Institute, Stockholm

Carlstedt-Duke J, Gustafsson J-Å, Wrange Ö (1977) Biochim Biophys Acta 497: 507–524

Carlstedt-Duke J, Wrange Ö, Dahlberg E, Gustafsson J-Å, Högberg B (1979) J Biol Chem 254: 1537–1539

Carlstedt-Duke J, Okret S, Wrange Ö, Gustafsson J-Å (1982) Proc Natl Acad Sci USA 79: 4260–4264

Carter DB, Chae C-B (1976) Biochemistry 15: 180–185

Cidlowski, JA (1980) Biochemistry 19: 6162–6170

Costello MA, Sherman MR (1979) The Endocrine Society 61st Annual Meeting abstract 273, p 141

Crabtree GR, Smith KA, Munck A (1978) Cancer Res 38: 4268–4272

Dahmer, MK, Quansey, MW, Bissen, ST, Pratt, WB (1981) J Biol Chem 256: 9401–9405

Dunaway M, Manly SP, Matthews KS (1980) Proc Natl Acad Sci USA 77: 7181–7185

Eisen HJ (1980) Proc Natl Acad Sci USA 77: 3893–3897

Francke U, Gehring U (1980) Cell 22: 657–664

Garroway NW, Orth DN, Harrison RW (1976) Endocrinology 98: 1092–1100

Gehring U (1976) In: James VHT (ed) Proceedings of the Fifth International Congress of Endocrinology, Excerpta Medica, Amsterdam, pp 536–541

Gialani S, Minowada J, Silvernail P, Nussbaum A, Kaiser N, Rosen F, Shimaoka K (1973) Cancer Res 33: 2653–2657
Giannopoulos, G (1973) J Biol Chem 248: 3876–3883
Giannopoulos G, Mulay S, Solomon S (1973) J Biol Chem 248: 5016–5023
Govindan MV (1979) J Steroid Biochem 11: 323–332
Govindan MV (1980a) Biochim Biophys Acta 631: 327–333
Govindan MV (1980b) Exptl Cell Res 127: 293–297
Govindan MV, Manz B (1980) Eur J Biochem 108: 47–53
Govindan MV, Sekeris CE (1976) Steroids 28: 499–507
Govindan MV, Sekeris CE (1978) Eur J Biochem 89: 95–104
Grove JR, Dieckmann BS, Schroer TA, Ringold GM (1980) Cell 21: 47–56
Gustafsson J-Å, Carlstedt-Duke J, Fuxe K, Carlström K, Okret S, Wrange Ö (1981) In: Fuxe F, Wetterberg L, Gustafsson J-Å (eds) Steroid hormone regulation of the brain. Pergamon Press, Oxford, pp 31–39
Hansson L-A, Gustafsson SA, Carlstedt-Duke J, Gahrton G, Högberg B, Gustafsson J-Å (1981) J Steroid Biochem 14: 757–764
Lee F, Mulligan R, Berg P, Ringold G (1981) Nature 294: 228–232
Lippman ME, Perry S, Thompson EB (1974) Cancer Res 34: 1572–1576
McPartland RP, Milholland RJ, Rosen F (1977) Cancer Res 37: 4256–4260
Middlebrook JL, Aronow L (1977) Endocrinology 100: 271–282
Murakami N, Moudgil VK (1981a) Biochem J 198: 447–455
Murakami N, Moudgil VK (1981b) Biochim Biophys Acta 676: 386–394
Nordeen SK, Lan NC, Showers MO, Baxter JD (1981) J Biol Chem 256: 10503–10508
Okret S, Carlstedt-Duke J, Wrange Ö, Carlström K, Gustafsson J-Å (1981) Biochim Biophys Acta 677: 205–219
Papamichail M, Tsokos G, Tsawdaroglou N, Sekeris CE (1980) Exp Cell Res 125: 490–493
Payvar F, Wrange Ö, Carlstedt-Duke J, Okret S, Gustafsson J-Å, Yamamoto KR (1981) Proc Natl Acad Sci USA 78: 6628–6632
Ringold GM, Yamamoto KR, Bishop JM, Varmus HE (1977) Proc Natl Acad Sci USA 74: 2879–2883
Schmidt TJ, Harmon JM, Thompson EB (1980) Nature 286: 507–510
Schnarr M, Maurizot J-C (1981) Biochemistry 20: 6164–6169
Sherman MR, Pickering LA, Rollwagen FM, Miller LK (1978) Federation Proc 37: 167–173
Sherman MR, Moran MC, Neal RM, Niu E-M, Tuazon FB (1982) In: Lee HJ, Fitzgerald TJ (eds) Progress in Research and Clinical Applications of Corticosteroids. Heyden, Philadelphia, pp 45–66
Sibley CH, Tomkins GM (1974) Cell 2: 221–227
Sibley CH, Yamamoto KR (1979) In: Baxter JD, Rousseau GG (eds) Glucocorticoid Hormone Action. Springer-Verlag, Heidelberg, pp 357–376
Simons SS, Thompson EB (1981) Proc Natl Acad Sci USA 78: 3541–3545
Stevens J, Stevens Y-W (1979) Cancer Res 39: 4011–4021
Stevens J, Stevens Y-W (1981) Cancer Res 41: 125–133
Stevens J, Stevens Y-W, Rhodes J, Steiner G (1978a) J Natl Cancer Inst 61: 1477–1485
Stevens J, Stevens Y-W, Sloan E, Rosenthal R, Rhodes J (1978b) Endocrine Res Comm 5: 91–108
Stevens J, Stevens Y-W, Rosent' ⸳ RL (1979) Cancer Res 39: 4939–4948
Stevens J, Stevens Y-W, Haubenstock H (1981a) The Endocrine Society 63rd Annual Meeting abstract 483, p 203
Stevens J, Eisen HJ, Stevens Y-W, Haubenstock H, Rosenthal RL, Artishevsky A (1981b) Cancer Res 41: 134–137
Thompson EB, Granner DK, Gelehrter TD, Hager GL (1979) Cold Spring Harbor Conferences on Cell Proliferation 6: 339–360

Tsawdaroglou NG, Govindan MV, Schmid W, Sekeris CE (1981) Eur J Biochem 114: 305–313

Varmus HE, Ringold G, Yamamoto KR (1979) In: Baxter JD, Rousseau GG, (eds) Glucocorticoid Hormone Action. Springer-Verlag, Heidelberg, pp 253–278

Westphal HM, Beato M (1980) Eur J Biochem 106: 395–403

Wilson EM, French FS (1979) J Biol Chem 254: 6310–6319

Wrange Ö (1979) Biochim Biophys Acta 582: 346–357

Wrange Ö, Gustafsson, J-Å (1978) J Biol Chem 253: 856–865

Wrange Ö, Carlstedt-Duke J, Gustafsson J-Å (1979a) In: Agarwal MK (ed) Proteases and Hormones. Elsevier/North-Holland Biomedical Press, Amsterdam pp 141–157

Wrange Ö, Carlstedt-Duke J, Gustafsson J-Å (1979b) J Biol Chem 254: 9284–9290

Yamamoto KR, Stampfer MR, Tomkins GM (1974) Proc Natl Acad Sci USA 71: 3901–3905

Yamamoto KR, Gehring U, Stampfer MR, Sibley CH (1976) Rec Prog Hormone Res 32: 3–32

Yamamoto KR, Chandler VL, Ross SR, Ucker DS, Ring JC, Feinstein SC (1981) Cold Spring Harbor Symp Quant Biol 45, in press

Discussion of the Paper Presented by J.-Å. Gustafsson

ROY: In your studies for specific binding, did you load the receptors with the steroid or can the unoccupied receptor also bind with the DNA fragment?

GUSTAFSSON: That is a very interesting question. We have only tested the glucocorticoid–receptor complex, not the receptor without ligand.

ROY: This brings us back to the old question of what is the role of the hormone in hormone action? Because in one of the earlier slides you showed that the fragment that does not bind the hormone can bind to the DNA.

GUSTAFSSON: Yes, very slightly.

SCHRADER: From your DNA-binding experiment, what is your estimate of the distances between the start of the gene and the location at which DNA binding of the receptor seems to occur?

GUSTAFSSON: Well, like I said, it seems as if the receptor interacts with several sites on the MTV genome; one of these sites seems to lie within the LTR region. Other sites are located at some distance from the LTR region.

SCHRADER: Upstream?

GUSTAFSSON: To the right, downstream in the gene.

SCHRADER: Did I understand correctly that the length of these DNA fragments was in the order of between 0.49 and 6 kilobases?

GUSTAFSSON: That is correct. The whole pMTV2 DNA is about 4.5 kilobases.

SCHRADER: Have you done any footprinting studies either to see the size of the protected DNA fragment, or to see if a unique sequence on the DNA is protected?

GUSTAFSSON: Such studies are under way at the present moment.

PECK: I am interested in your immunofluorescent studies in the hypothalamus. You suggested that, in fact, those were neurons. Aren't they just as likely to be glial cells in that particular location, since glia are known to be a glucocorticoid target? That is question one. Two, have you looked at a better target, such as the hippocampus for neuronal glucocorticoid systems.

GUSTAFSSON: With regard to your first question, yes, you are right. It is not possible to distinguish fluorescent glial cells from fluorescent neurons. With regard to your second question, we have postponed further experiments until more suitable antibody preparations have been obtained.

SHAPIRO: In order to function in the cell in the kind of binding assay that you described here with DNA, the glucocorticoid receptor is going to have to seek out a unique MMTV sequence, diluted roughly a million-fold from genomic DNA. I think that one of the problems with other kinds of selectivity studies has been that the level of selectivity that has been obtained has been relatively modest. Can you give us an idea of the kind of dilution that you employed when you used these multiple restriction fragments? Is it anything close to the relation between the amount of MMTV DNA in cells and the genomic DNA that it is going to have to search through or is it of several orders of magnitude lower?

THOMPSON: The question really is: What did the affinity turn out to be? Have you done those studies?

GUSTAFSSON: In very recent experiments the discrimination between specific and nonspecific DNA binding appears to be about four orders of magnitude.

THOMPSON: By analogy, the *lac* repressor and others were only visible when it was possible to get the purified forms.

GUSTAFSSON: It seems that the selectivity is slightly less than the selectivity by which the lambda repressor may pick out the lambda operator. But on the other hand, it is known that the lambda operator contains several binding sites for the lambda repressor protein. So it is difficult to compare.

THOMPSON: I should probably mention the Ringold experiment, since nobody seems to be here representing that laboratory. He and his colleagues have taken the LTR region of the MMTV, hooked it up to the dihydrofolate reductase gene, and transformed cells, using the calcium phosphate technique, I believe. At any rate, they transformed cells with this, and showed that they can then induce the DHFR enzyme with glucocorticoids, whereas normally it is not inducible (Lee et al., Nature, 294: 228, 1981). That might suggest that of your two binding regions that you seem to have, that perhaps the LTR is the one that really is the active one.

PONGS: I will directly comment to that. There are apparently two initiation sites for transcription in the MMTV-proviral DNA: one lies inside the coding region. According to data from several laboratories, the LTR-binding site for glucocorticoid receptor has been shown to be within about 400 base pairs upstream from the TATAA box. I have a question actually, not just a comment. Did you try to do immunoprecipitations with your sera? Can you take cytosol and, specifically, can you immunoprecipitate glucocorticoid receptor?

GUSTAFSSON: Yes, I showed you such a dilution curve; it is possible.

PONGS: You didn't show gel data. For instance, if you use hot protein, do you precipitate just one protein or more than one?

GUSTAFSSON: No such data are yet available.

Discussants: J.-Å. GUSTAFSSON, A.K. ROY, W. SCHRADER, E.J. PECK, D.J. SHAPIRO, E.B. THOMPSON, and O. PONGS.

Chapter 11

Identification of the DNA-Binding Domain of the Chicken Progesterone Receptor A Subunit

PHILLIP P. MINGHETTI, NANCY L. WEIGEL,
WILLIAM T. SCHRADER, AND BERT W. O'MALLEY

Regulation of expression of egg white genes in chicken oviduct by the steroid hormones progesterone and estrogen has been firmly established (O'Malley et al. 1969; LeMeur et al. 1981; Roop et al. 1978; Swaneck et al. 1979). The progesterone receptor has been purified and characterized; it is a complex containing two nonidentical subunits A and B having molecular weights of 79,000 and 108,000, respectively (Kuhn et al. 1975; Schrader et al. 1977; Coty et al. 1979). Analysis of the structural organization of both subunits by limited proteolytic digestion of the native proteins produced two hormone-binding fragments, termed Form IV (43,000 g/mole) and meroreceptor (23,000 g/mole) (Sherman et al. 1976; Sherman et al. 1978; Vedeckis et al. 1980a; Vedeckis et al. 1980b). We now report the isolation of proteolytic fragments of the receptor A protein containing both hormone- and DNA-binding activity. Furthermore, by using the protein blotting technique of Bowen et al. (1980) we have been able to identify not only two large peptides containing both hormone-binding and DNA-binding activities, but also a small peptide that retains only DNA-binding activity.

Progesterone receptor subunit A of chick oviduct binds tightly to DNA (K_d 10^{-10} M) (Schrader et al. 1972; Hughes et al. 1981). To determine whether this DNA-binding domain could be excised from the A protein, a 30-fold purified mixture of [^{3}H]progesterone-labeled A and B receptor proteins was digested with *Staphylococcus aureus* V8 protease (Miles Laboratories) under the conditions described in the legend of Fig. 1. These conditions convert about 80% of the intact A and B proteins to smaller fragments as shown by gel filtration studies of the digests on Sephadex G-100 (data not shown). Since neither subunit B nor its digests show detectable DNA-binding activity at KCl molarities above 0.05 M, (Vedeckis et al. 1980b), it was unnecessary to remove receptor B before analysis. The digest was analyzed by DNA–cellulose chromatography as shown in the top panel of Fig. 1. About 40% of the input ^{3}H was recovered in the eluate. Although the peak appeared at an elution molarity characteristic of authentic receptor A (Coty et al. 1979; Vedeckis et al. 1980b), the extent of digestion suggested that a proteolytic fragment retaining hormone-binding and DNA-binding activity might have been produced.

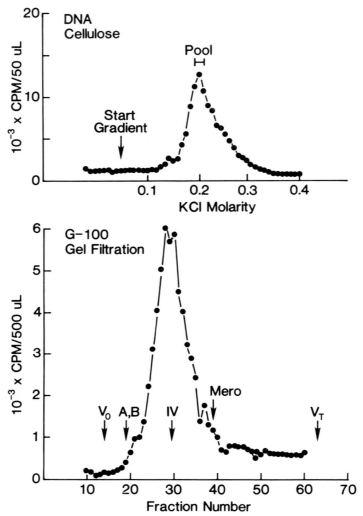

Fig. 1. Assignment of the DNA-binding domain of receptor A to its Form IV fragment. Oviducts from estrogenized immature chicks were homogenized in 10 mM Tris–HCl containing 0.001 M Na$_2$ EDTA and 0.012 M 1-thioglycerol pH 7.4 (Buffer A) and centrifuged to prepare 30 ml of cytosol (Coty et al. 1979). Cytosol was labeled with [^{3}H]progesterone, 2×10^{-8} M (50 Ci/mmole), for 2 h and then receptors were precipitated by bringing the cytosol to 35% ammonium sulfate, as described previously (Schrader et al. 1975). The precipitate was dissolved in 4 ml of Buffer A and dialyzed against Buffer A overnight. To this were added 75 μl of *Staphylococcus aureus* V8 protease (1 mg/ml). Incubation was at 22° for 10 min. An aliquot (3 ml) of the reaction mixture was loaded onto a 7-ml DNA–cellulose column prepared in Buffer A containing 0.05 M KCl. After extensive washing, adsorbed receptors were eluted using a 60-ml KCl gradient from 0.05 to 0.6 M in Buffer A. Fractions (1.0 ml) were collected and aliquots counted for ^{3}H to detect receptor-derived hormone-binding fragments as shown in the top panel. The two peak fractions were combined and analyzed by gel filtration on a Sephadex G-100 column (2.6 cm $\times$ 40 cm) run in Buffer A containing 0.3 M KCl. Fractions (3 ml) were collected and counted for ^{3}H as shown in the lower panel. The void volume (V_0), salt elution volume (V_T), and the elution positions for authentic receptor A and receptor B, Form IV, and meroreceptor are shown by the arrows.

To confirm this finding, the two peak fractions from the DNA–cellulose elution profile were analyzed by Sephadex G-100 gel filtration, as shown in the lower panel of Fig. 1. The column had previously been calibrated with a receptor digest obtained using an endogenous Ca^{2+}-activated neutral protease as described elsewhere (Vedeckis et al. 1980a; Vedeckis et al. 1980b). This enzyme yields predominantly two receptor hormone-binding proteolytic fragments, Form IV (M_r 43,000) and meroreceptor (M_r 23,000). When the DNA–cellulose eluate was analyzed, we found that the majority of the ^3H-receptor fragments eluted coincident with Form IV. The profile shows no contamination with intact A or B monomers and less than 5% of the material co-migrating with meroreceptors. Thus, we conclude that limited digestion with *Staphylococcus aureus* V8 protease yields a Form IV receptor fragment and that this fragment contains DNA-binding activity. We had previously reported (Coty et al. 1979) that receptor AB complexes have the DNA site on A occluded. In earlier digestion studies using crude cytosol, we found that Form IV did not bind to DNA (Vedeckis et al. 1980b). However, this was due to the fact that Form IVA and Form IVB can remain associated with each other after mild protease digestion if not treated with salt. In fact, trypsin, *S. aureus* protease and the endogenous Ca-activated neutral protease[11] all yield a DNA-binding Form IV activity from receptor A after salt treatment (data not shown).

We then examined the digestion products produced by *S. aureus* protease treatment of receptor A to determine whether a small DNA-binding peptide could be identified. These studies were carried out using highly purified receptor A protein from hen oviducts as the starting material. Our isolation of receptor A from hen oviducts utilized a modification of the protocol developed previously for purification of this protein from immature chicks (Coty et al. 1979) (see legend of Fig. 2). Yields are on the order of 1%; about 150 μg of receptor A is obtained from 2 kg of oviducts. The protein migrates as a single major Coomassie blue-stained spot on two-dimensional gel analysis by the procedure of O'Farrell (1975) as shown in Fig. 2. The protein band has an isoelectric point of pH 5.9 and an apparent molecular weight of 79,000. This spot has been positively identified as the receptor A protein by coincident migration of the protein, with chicken receptor A covalently coupled to the synthetic progestin R5020 (Dure et al. 1980). We concluded that the receptor A preparation can be purified to over 90% purity and is suitable for protease digestion studies.

To identify proteolytic fragments containing DNA-binding activity, we used a very sensitive protein blotting technique (Bowen et al. 1980). This technique involves a three-step procedure. The proteins are first separated by SDS–polyacrylamide gel electrophoresis. Second, the denaturant is removed from the gel to allow refolding of the proteins, and the proteins are transferred to sheets of nitrocellulose filter paper. Third, the proteins bound to nitrocellulose are analyzed for DNA binding by incubating the filter with [^{32}P]DNA. After washing the filter with appropriate salt solutions to remove unbound DNA, the filter is dried and autoradiographed to detect pro-

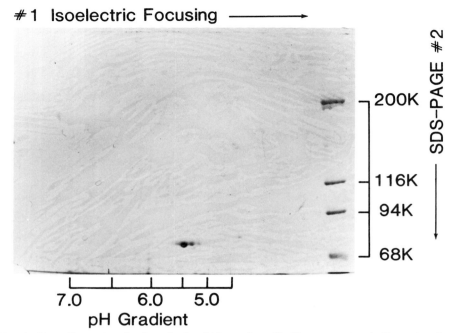

#1 Isoelectric Focusing ⟶

SDS-PAGE #2

200K
116K
94K
68K

7.0   6.0   5.0
pH Gradient

Fig. 2. Two-dimensional gel analysis of 10 μg of purified hen receptor A. Receptor A was purified by a modification of the procedure of Coty et al. (1979). Briefly, 2 kg of the magnum portion of the hen oviducts were homogenized in 2.5 vol of 10 mM Tris pH 7.4 containing 1 mM Na$_2$-EDTA and 12 mM 1-thioglycerol (Buffer A), centrifuged in a Beckman J-6 centrifuge at 4200 rpm, and the supernatant was passed through cheesecloth to remove floating fat. The receptor complexes were then precipitated at a final concentration of 10% polyethylene glycol 6000, centrifuged to collect the precipitate, redissolved in Buffer A with the aid of a tissue homogenizer, and recentrifuged to remove insoluble material. The receptor complexes were passed through a 700-ml phosphocellulose column equilibrated with Buffer A and then through a 500-ml DNA–cellulose column equilibrated with Buffer A containing 50 mM KCl. [The intact receptor A–B complexes do not bind to either of these columns (Coty et al. 1979).] The receptors were then labeled with 10^{-8} M [^{3}H]progesterone, precipitated, and dissociated into subunits A and B with ammonium sulfate at a final concentration of 40% and dialyzed to remove excess salt. The receptors were next applied to a 40-ml DNA–cellulose column equilibrated with Buffer A containing 0.1 M KCl and eluted with a 500-ml KCl gradient in Buffer A (0.1–0.5 M KCl). The receptor A peak eluting at 0.19 M KCl was applied to a 10-ml DEAE–cellulose column equilibrated with Buffer A containting 50 mM KCl. The receptor was eluted with a 150-ml KCl gradient in Buffer A (0.05–0.3 M KCl); receptor A eluted at 0.1–0.15 M. Finally, the receptor pool was applied to a 1-ml heparin Sepharose column equilibrated with 0.1 M NH$_4$HCO$_3$ (pH 8.5) and eluted with a 20-ml linear gradient (0.1–1 M NH$_4$HCO$_3$). Receptor A eluted at 0.34 M buffer. This purified receptor was pooled, lyophilized, and analyzed by two-dimensional gel electrophoresis (O'Farrell 1975). Horizontal dimension, first step isoelectric focusing (pH range 4–7); vertical dimension, electrophoresis in SDS (5% acrylamide). The proteins were stained with Coomassie blue.

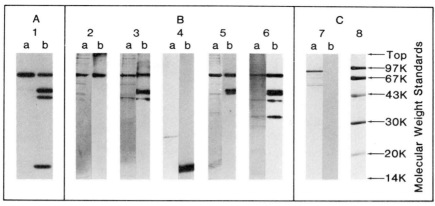

Fig. 3. Identification of the DNA-binding domain of the progesterone receptor A subunit by protein blotting. All enzymatic digestions were performed at room temperature. Proteins were separated electrophoretically on 1-mm-thick sodium dodecyl sulfate polyacrylamide slab gels containing 12.5% total acrylamide and 0.33% cross-linking (Laemmli 1970). Following electrophoresis, the slab was immersed in 200 ml of renaturation buffer containing 1% Triton X-100 (50 mM NaCl, 2 mM Na$_2$-EDTA, 4 M urea, 0.1 mM dithiothreitol, 10 mM Tris–HCl, pH 7.0, 1% Triton X-100) and rotated on a platform shaker for 1.5 h at room temperature. One fresh buffer change was made 45 min after the initial immersion. The slab was then washed an additional 1.5 h in renaturation buffer without Triton X-100. Fresh buffer was added at 30-min intervals during the 1.5-h wash period. After the entire 3-h wash, the slab was sandwiched between two nitrocellulose filters (BA 85, Schleicher and Schuell) while submerged in transfer buffer (50 mM NaCl, 2 mM Na$_2$-EDTA, 0.1 mM dithiothreitol, 10 mM Tris–HCl, pH 7.0) as described previously (Bowen et al. 1980). Fresh transfer buffer was added after 12 h and the protein diffusion process was allowed to continue for an additional 36 h with rotation at room temperature. Filters containing the transferred protein were then placed in 100 ml of binding buffer (50 mM NaCl, 1 mM Na-EDTA, 10 mM Tris HCl, pH 7.0, 0.04% bovine serum albumin, 0.04% ficoll, 0.04% polyvinyl pyrollidone) 20–30 min prior to incubation with the [^{32}P] DNA probe. All filters were incubated for 2 h at room temperature with 100,000 cpm/ml of a nick-translated [^{32}P]ov1.7 linear DNA fragment derived from a recombinant subclone of the natural chicken ovalbumin gene (Dugaiczyk et al. 1979; Tsai et al. 1981). Unbound [^{32}P]ov1.7 was washed from filters with six changes of binding buffer over a 3-h period. The filters were air-dried and autoradiographed with Kodak XRP-1 x-ray film. Overnight exposure of the x-ray film was sufficient to detect radioactive bands. Panel A: Lane 1a, DNA-binding activity of a typical receptor A preparation (5 μg) detected by autoradiography. Lane 1b, DNA binding of 5 μg of receptor following incubation with 0.15 μg of *S. aureus* V8 protease for 5 min (total reaction volume = 0.5 ml of 0.3 M NH$_4$HCO$_3$, pH 8.5). Panel B: Lanes 2–7 show the Coomassie blue protein staining pattern of gels (lanes a) and the resultant autoradiograph after [^{32}P]DNA filter binding (lanes b). Lane 2, 5 μg of a receptor preparation; lane 3, 5 μg of receptor incubated for 10 min with 0.05 μg of *S. aureus* protease (total volume = 0.5 ml of 0.3 M NH$_4$HCO$_3$, pH 8.5); lane 4, 10 μg receptor incubated 18 h with 1 μg *S. aureus* protease (total volume = 0.7 ml of 0.3 M NH$_4$HCO$_3$, pH 8.5); lane 5, 5 μg receptor incubated with 0.05 μg trypsin for 30 min (total volume = 0.5 ml of 0.3 M NH$_4$HCO$_3$, pH8.5); lane 6, 5 μg of receptor incubated with 0.15 μg of trypsin for 2 h (total volume = 0.35 ml of 0.3 M NH$_4$HCO$_3$, pH 8.5). Panel C: Lane 7, 5 μg of conalbumin incubated with 0.1 μg of *S. aureus* protease for 30 min (total volume = 0.1 ml of 0.1 M NH$_4$HCO$_3$). Lane 8, molecular weight markers obtained from Phar-

tein–[^{32}P]DNA interaction. Since only a fraction of the initial protein band is actually transferred to the nitrocellulose, the gel can be stained subsequent to transfer to reveal the protein migration pattern. The major technical advantage of this protein blot technique is that it combines the high resolving power of polyacrylamide gel electrophoresis with the use of a very small quantity (pico- to nanogram amounts) of a high specific activity DNA probe (10^7–10^8 cpm/μg DNA).

The key strategy of the technique takes advantage of the fact that proteins can be partially renatured after sodium dodecyl sulfate treatment. Enzymatic activity is recovered in some cases (Manrow and Dottin 1980; Lacks and Springhorn 1980), as is immunologic reactivity (Towbin et al. 1979). More important for the present study is the fact that DNA-binding activity can be retained, including recovery of nucleotide-sequence specificity (Bowen et al. 1980). The extreme sensitivity of the method allows detection of DNA binding even if only a few thousandths of a percent of the protein in question satisfactorily renatures.

Progesterone receptor A protein was prepared, and aliquots were digested with protease before SDS–polyacrylamide gel electrophoresis (Laemmli 1970), as shown in Fig. 3. In panel A, lane 1a shows the autoradiograph of a filter containing proteins transferred from an SDS gel loaded with purified receptor A protein and then incubated with [^{32}P]DNA. Under the conditions used, receptor A is the only protein that binds [^{32}P]DNA. Small amounts of contaminants apparently are not sufficiently avid DNA binders to be visible by this technique. As shown in lane 1b, digestion of receptor A with *S. aureus* protease generates three new DNA-binding peptides with molecular weights of 45,000, 40,000, and 15,000. The 45K and 40K proteins are the Form IV proteins. Apparently, the Form IV peak seen in Fig. 1 (lower panel) is a mixture of these fragments.

The three peptides observed in lane 1b (Fig. 3) are apparently generated sequentially by digestion of receptor A, as shown in panel B. Lane 2a shows the Coomassie blue staining pattern of a typical receptor preparation. The receptor is the predominant protein on the gel, migrating at 79,000 g/mole on these 12.5% acrylamide gels. Other faintly staining bands are evident, but none is a significant fraction of the A protein preparation. Lane 2b is the autoradiograph of the nitrocellulose filter after incubation with [^{32}P]DNA. The receptor A band is the only protein illuminated. Some DNA binding is seen at the top of the gel, which may be due to protein aggregation. In any case, we obtain identical DNA-binding fragments from preparations that contain this material and preparations that do not have it. The receptor A

---

macia Fine Chemicals (electrophoresis standards kit; low-molecular-weight range). Standards were 97K, phosphorylase b; 67K, bovine serum albumin; 43K, ovalbumin; 30K, carbonic anhydrase; 20.1K, soybean trypsin inhibitor; 14.4K, α-lactalbumin. The α-lactalbumin transfers quantitatively out of the gel during blotting and hence is not visible on this lane. Migration of this standard was determined in companion gels not subjected to blotting, and the R_f of this protein plotted as shown.

preparation shown in lane 2a was then digested for various times using *S. aureus* protease. Lanes 3a and 3b show the Coomassie blue staining and DNA-binding patterns, respectively, after digestion for 10 min. Lanes 4a and 4b are the results of exhaustive digestion for 18 h. The filter autoradiograph in lane 3b shows the 45K and 40K bands. More extensive digestion (lane 4b) results in elimination of these two intermediates, concomitant with the appearance of a pronounced 15K band. Since the 15K band is the only DNA-binding polypeptide detectable after 18 h, we believe that this fragment is the limit digest of the DNA-binding domain under these conditions. Furthermore, since the Form IV 45K and 40K peptides are produced and then consumed in the time course of digestion, we conclude that they are intermediates of digestion by this enzyme. The protein band at 23,000 g/mole, evident in lane 4a, which does not bind DNA, corresponds with the hormone-binding portion of the molecule, the meroreceptor, as shown by covalent labeling of this fragment with the synthetic progestin [^{3}H] R5020 (Birnbaumer, manuscript in press). It is notable that the sum of molecular weights of this 23,000 fragment and the 15,000 DNA-binding fragment are nearly equal to that of the 40K Form IV peptide.

The results of similar digestion studies using trypsin are shown in lanes 5 and 6. Mild digestion yielded two DNA-binding fragments at 46,000 and 44,000, as shown in lanes 5a,b. More extensive digestion produced additional DNA-binding bands at 38,000 and 30,500 (lanes 6a,b) and a non-DNA-binding fragment at 26,500 g/mole that corresponds to the hormone-binding fragment. Exhaustive digestion yielded no smaller DNA-binding fragments identifiable by this method (data not shown).

In panel C, lanes 7a and b show the results of *S. aureus* protease digestion of conalbumin, another chicken protein of similar molecular weight, which should not contain DNA-binding domains. Lane 7a shows the Coomassie blue staining pattern; lane 7b demonstrates that DNA-binding peptides are not present. Lane 8 contains stained protein standards. Not shown are lanes of the two proteases alone, which also contained no DNA-binding activity.

The protein blot method of Bowen et al. (1980) used here is a very useful one for further studies of receptors and other DNA-binding proteins. The method is extremely sensitive, since only a few micrograms of receptor A were needed to see ^{32}P-autoradiograms in 10–18 h exposure time, even though only a small fraction of receptor (or its fragments) transferred to the filters. Because of the nonquantitative nature of the transfer, the method allows only qualitative answers. We cannot assess from the present experiments whether all DNA-binding receptor fragments display equivalent equilibrium constants, for example.

Thus, we have found that the progesterone receptor A protein can be digested to liberate intermediate forms (40–45K) and using more exhaustive digestion conditions, a small fragment (15K), all of which contain the DNA-binding site. Under the conditions employed, this region of the protein is the only one displaying this activity as determined by the protein blot assay. Since the DNA-binding activity resists digestion under nondenaturing conditions (Fig. 1), and since the 15K fragment arises from a Form IV intermediate as the only detectable DNA-reactive fragment, we conclude that this region is at least a part of the native A protein's DNA-binding domain, and not merely an artifactual interaction of the fragment in the blotting method. The method does not allow us to exclude, however, the possibility that in the native protein other portions of the protein also play some role in DNA binding. This question awaits purification of the small fragment for direct DNA adsorption studies as we have done for the intact protein (Hughes et al. 1981).

References

Bowen B, Steinberg J, Laemmli UK, Weintraub H (1980) Nucleic Acids Res 8: 1–20
Coty WA, Schrader WT, O'Malley BW (1979) J Steroid Biochem 10: 1–12
Dugaiczyk A, Woo SLC, Colbert DA, Lai EC, Mace ML, O'Malley BW (1979) Proc Natl Acad Sci USA 76: 2253–2257
Dure LS IV, Schrader WT, O'Malley BW (1980) Nature 283: 784–786
Hughes MR, Compton JG, Schrader WT, O'Malley BW (1981) Biochemistry 20: 2481–2491
Kuhn RW, Schrader WT, Smith RR, O'Malley BW (1975) J Biol Chem 250: 4220–4228
Lacks SA, Springhorn SS (1980) J Biol Chem 255: 7467–7473
Laemmli UK (1970) Nature 227: 680–685
LeMeur M, Glanville N, Mandel JL, Gerlinger P, Palmiter R, Chambon P (1981) Cell 23: 561–571
Manrow RE, Dottin RP (1980) Proc Natl Acad Sci USA 77: 730–734
O'Farrell PH (1975) J Biol Chem 250: 4007–4021
O'Malley BW, McGuire WL, Kohler PO, Korenman SG (1969) Recent Prog Horm Res 25: 105–160
Roop DR, Nordstrom JL, Tsai SY, Tsai MJ, O'Malley BW (1978) Cell 15: 671–685
Schrader WT, O'Malley BW (1972) J Biol Chem 247: 51–59
Schrader WT, Toft DO, O'Malley BW (1972) J Biol Chem 247: 2401–2407
Schrader WT, Socher SH, Buller RE (1975) Meth Enzymol 36: 292–313
Schrader WT, Kuhn RW, O'Malley BW (1977) J Biol Chem 252: 299–307
Sherman MR, Pickering LA, Rollwagen FM, Miller LK (1978) Fed Proc Fed Am Soc Exp Biol 37: 167–173
Sherman MR, Tuazon FB, Diaz SC, Miller LK (1976) Biochemistry 15: 980–989
Swaneck GE, Nordstrom JL, Kreuzaler F, Tsai MJ, O'Malley BW (1979) Proc Natl Acad Sci USA 76: 1049–1053
Towbin H, Staehelin T, Gordon J (1979) Proc Natl Acad Sci USA 76; 4350–4354
Tsai SY, Tsai MJ, O'Malley BW (1981) Proc Natl Acad Sci USA 78: 879–883
Vedeckis WV, Freeman MR, Schrader WT, O'Malley BW (1980a) Biochemistry 19: 335–343
Vedeckis WV, Schrader WT, O'Malley BW (1980b) Biochemistry 19: 343–349

Discussion of the Paper Presented by B. Schrader and N. Weigel

GREENE: These very nice studies prompt me to ask a question, or a couple of questions, in keeping with the tradition of this meeting. Do you know anything about the specificity of DNA binding or of chromatin binding for your R-5020 photolabeled progesterone receptor? Have any of these studies been done or are they under way, and what is your feeling about this issue?

SCHRADER: We have now succeeded in identifying preferential receptor binding sites upstream of the ovalbumin gene in the flanking DNA. The prominent site is located between -135 and -249 bp upstream. This region shows about 10- to 20-fold better binding of purified receptor A subunit than does any other DNA fragment we have tested. However, we as yet have no proof that this region is the receptor binding site used in vivo. Nor have we any additional information on chromatin binding such as the R5020 method you propose.

GREENE: How about chromatin experiments?

SCHRADER: Chromatin experiments with R-5020 are just getting started and I have nothing to report on them at this time.

GREENE: Presumably this would be a good application of a covalently labeled receptor. You do have a tag, at least, for looking at some of these sites.

SCHRADER: That is one of the strategies we would like to pursue.

ROY: When you did the affinity labeling with the radiolabeled steroid, did the label always go to the same amino acid?

SCHRADER: We have not yet done that. We've tried to do that study by tryptic digestion of the labeled complex, followed by a simple tryptic map. The hormone-binding fragment that we get in that experiment is so hydrophobic that it runs at the solvent front in one of the dimensions of the tryptic map. That is why we had to go to HPLC; we are just starting those experiments. There is so far no evidence that the R-5020 is labeling anything except one unique site on the progesterone receptor. The specificity appears to be excellent for the progesterone receptor.

ROY: Don Coffey showed a couple of weeks ago that when he had the nuclei digested completely with DNase, most of the receptor remained bound to the nuclear matrix. How does your data reconcile with his results?

SCHRADER: I don't think I showed any DNA digestion experiments. Richard Buller in our group had done similar experiments on digestion a long time ago, with DNase I, and showed that he could digest up to 60 or 70% of the DNA with no loss of receptor binding activity in the nuclei. That work was published about 8 years ago. So I would say our results are comparable to his.

CLARK: But Arun, what you said is not quite right, because Coffey's residual receptor on the matrix was only a percentage of the total that could be driven there, right Don?

COFFEY: It's 30–60% of the total nuclear receptor, depending on the hormone treatment and which tissue and so forth.

BARDIN: What kind of receptor did you use in your studies reported in which you concluded that receptor might be a phosphoprotein in vivo. Was that an isolated dimer that was isolated and then phosphorylated?

WEIGEL: We used purified A and B subunits. We see charge heterogeneity in our purified material, suggesting that the receptors may be phosphorylated in vivo. The incorporated ^{32}P was a result of in vitro phosphorylation.

BARDIN: And that was the one where you showed the two binding sites, was that the isolated receptor?

WEIGEL: Receptor isolated either from chicks or from laying hens shows both binding sites. The experiment shown was done with partially purified chick B protein.

SCHRADER: Can I issue one disclaimer? As far as I'm concerned, there is only one definitive experiment to test whether steroid receptors are or are not phosphoproteins. That is to show in living cells the incorporation of ^{32}P into the protein. We have tried that experiment one time so far, and the experiment didn't work. There are no other data on that from our lab.

BARDIN: So how can you conclude that it is a phosphoprotein?

WEIGEL: I said that the data suggest that the receptors are phosphorylated. We are trying to test this possibility directly.

BARDIN: Mr. Schrader said it didn't work in the intact cell.

SCHRADER: Before you find that too damning, let me explain the experiment that I did. I took one chick oviduct, minced it, put it into tissue culture medium, and added some [^{32}P]orthophosphate to the medium. Then we took that one oviduct mince out of the culture and homogenized it. After an hour we threw the whole thing in with a 2-kg hen oviduct prep and purified the receptor B protein out of the preparation and asked if there was any ^{32}P in the receptor. There was none detectable. There was a mass ratio of about 100 mg to about 2 kg. That experiment is the only experiment that has been tried in that regard. I think that one needs to be cautious to conclude that these receptors are phosphoproteins, but the evidence we have from the purified protein is that one can explain the charge heterogeneities; we showed here that we can duplicate the charge heterogeneity by ^{32}P incorporation.

WEIGEL: We are going to look at the ^{31}P-NMR of the purified B to see if, in fact, there is phosphorous on the protein.

PONGS: In your blotting experiment, what kind of DNA did you use? Was it nick-translated or end-labeled?

WEIGEL: We used nick-translated DNA. The fragment was OV1.7 from the 5' end of the ovalbumin gene.

MOUDGIL: I was wondering whether you have looked at the effects of molybdate in relation to the conversion of B to A form of the receptor. We have been trying to purify the receptor in a nonactivated state and have seen that there is a considerable amount of B going into A if one homogenizes the tissue in presence of molybdate. What is the possibility that form A and form B are interconvertible?

SCHRADER: The possibility of conversion of B to A has been a possibility for a number of years. I think there are now several facts that now say, I think rather unequivocably, that that reaction cannot take place. One is the tryptic mapping data that show that B and A do not contain common tryptic peptides. There are tryptic peptides in A, which is the smaller protein, that are not present in B, and there are tryptic peptides in B that are not present in A. If A were a derivative of B, there can only be two possible tryptic peptides, which could be different, namely, the one at the N-terminus, where the clip occurred, and/or the one at the C-terminus. So I feel that the mapping experiment rules out that possibility. Second, I would caution that molybdate does not block proteolysis, when assayed under denaturing conditions. Thus, when people use nondenaturing conditions to assay for the integrity of receptors, the polypeptide chains may be held together by internal disulfides. But only by using a covalent labeling technique, as we have done, does one have sufficiently high resolution under denaturing conditions to address the question of whether or not the proteins are degraded. And as I showed in the experiments I presented there, those proteins can be considerably degraded and have many, many smaller-molecular-weight fragments. However, if you were to look at that stuff on sucrose gradients, it would be a nice garden variety 4 S progesterone receptor. I think it is necessary when addressing the question of whether or not B is being converted to A to be able to show by incubation with some enzymes that that reaction can occur. In our experiments we have been unable to perform the conversion. We've tried to do the enzymatic conversion of B to A and asked whether or not DNA-binding activity can be obtained by incubating radioactive B with anything, cytosol, other enzymes, or what have you. And in no experiments were we able to perform that, so I don't think that B can be converted to A.

Discussants: N. WEIGEL, W. SCHRADER, G.L. GREENE, A.K. ROY, J.H. CLARK, D.S. COFFEY, W.C. BARDIN, O. PONGS and V.K. MOUDGIL

Chapter 12

Immunochemical Studies of Estrogen Receptor

GEOFFREY L. GREENE

I. Introduction

The mechanism of action of steroid hormones in reproductive tissues has been studied extensively during the past two decades and it is generally accepted that specific intracellular receptor proteins mediate the biological action of steroids in these tissues. A model of steroid hormone action in target cells, first proposed for estrogens (Gorski et al. 1965; Jensen et al. 1968) and subsequently for all classes of steroid hormones (Muldoon 1980), has emerged from these studies. The most important features of this model are that the steroid, upon entering the cell, associates with an extranuclear protein and induces its conversion to a form that has increased affinity for chromatin and DNA. This temperature- and steroid-dependent process is accompanied by translocation of the hormone–receptor complex to the nucleus, where it interacts with the chromatin and in some way stimulates the production or accumulation of specific RNAs. The net result is the production of specific intracellular and secreted proteins that are involved in the growth and/or regulation of the target tissue and possibly other tissues.

Despite the accumulation of a considerable amount of data on the intracellular interactions and behavior of steroid hormone complexes, most of the processes involved in receptor activation, nuclear uptake, biosynthesis, degradation, and movement within cells are still not understood. In addition, knowledge of the composition, sequence, and physical and chemical properties of all classes of steroid hormone receptors is still very limited. Much of this lack of knowledge is due to the low availability of steroid receptors and to difficulties encountered in the handling and purification of the receptor proteins, particularly in mammalian systems. The presence of multiple forms of these receptors in cell-free extracts of target tissues, in some cases generated by the action of endogenous proteases (Sherman et al. 1978), has also made it difficult to determine which forms are biologically important and how these forms are related to each other. A more detailed understanding of the biochemical pathways involved in receptor-mediated hormone responses will have to await the purification and reconstitution of all important cellular components, including the steroid-occupied and unoccupied receptors, any important chromatin acceptor sites, specific DNA-binding sequences, responsive genes, and other still unknown factors whose presence is required for complete response.

II. Purification of Steroid Receptors

Numerous purification schemes for various steroid receptors have been published. However, real progress has been achieved only in the past few years, primarily as a result of the preparation of suitable steroid affinity adsorbents. Extranuclear estrogen receptors from calf uterus (Sica and Bresciani 1979) and MCF-7 human breast cancer cells (Greene et al. 1980c) have now been purified to near homogeneity in reasonable yields by chromatography through estradiol affinity adsorbents usually in combination with heparin–Sepharose (Molinari et al. 1977). Purification of progesterone receptors from chick oviduct (Kuhn et al. 1975) and human uterus (Smith et al. 1981) by affinity chromatography, as well as by conventional methods, has also been reported. In addition, rat liver extranuclear glucocorticoid receptor has been purified by chromatography through a steroid adsorbent followed by DNA–cellulose (Govindan and Manz 1980). Despite these successes, the composition, sequence, and chemical nature of steroid receptors remains largely a mystery, due in part to a lack of sufficient amounts of purified receptors for detailed analysis and because of problems with proteolytic degradation of receptors during isolation and purification.

III. Antibodies to Steroid Receptors

Virtually all the information about the nature and function of steroid hormone receptors, including estrogen receptors (estrophilins), has come from experiments in which the radioactive hormone serves as a marker for the receptor to which it binds. Although many questions about the structure and function of steroid receptors can be, and have been, answered by using labeled ligands as tags for receptors, a number of the unresolved issues might best be answered with alternative methods of detecting and measuring the receptor proteins. In particular, questions about the synthesis, degradation, and intracellular movement of receptors, as well as questions about the nature of activation and the location of functional domains (e.g., chromatin-binding) on the receptor, or the location of specific receptor sites in the nuclear matrix (Barrack and Coffey 1980) might be answered with the use of specific, high-affinity antibodies that recognize the receptor whether or not it is complexed with hormone. Monoclonal antibodies (Köhler and Milstein 1975) are ideally suited for these purposes because of their purity, potential monospecificity, and high affinity and the ease with which they can be labeled biosynthetically with radioactive amino acids, by iodination with

^{125}I, or by conjugation with enzymes such as horseradish peroxidase. Although several laboratories have now reported the preparation of polyclonal antibodies to various steroid receptor proteins, the only monoclonal antibodies to a steroid receptor reported in the literature are from the Ben May Laboratory (Greene et al. 1980a,b,c).

Despite the fact that polyclonal antibodies generated in rabbits and a goat against estradiol–receptor complexes of calf uterus have proved useful for the study of receptor structure and function (Greene et al. 1977, 1979), as well as for purification and immunoassay of receptor, much of the work has been limited by the heterogeneity of antibodies present in animal sera; some of these antibodies recognize proteins other than estrophilin. To overcome these limitations, we have prepared monoclonal antibodies to calf uterine estrophilin and to MCF-7 human breast cancer estrogen receptor. As expected, the high affinity, monospecificity, and abundance of these monoclonal antibodies have permitted us to use them for the development of specific and sensitive immunoassays for extracted human estrogen receptor, for the development of an immunocytochemical assay for primate and nonprimate estrophilin in several target tissues, for analysis of proteolytic fragments of receptor digests on nitrocellulose blots of SDS gels, for investigating sequence similarities and differences among various mammalian and nonmammalian estrophilins, and for establishing the distinction between estrogen receptors and other classes of steroid hormones (Greene and Jensen, 1982). It is clear that monoclonal antibodies will prove to be powerful probes for investigating the structure, composition, and cellular interaction of all classes of steroid hormone receptors.

IV. Purification of MCF-7 Estrogen Receptor

The use of affinity chromatography for the purification of steroid receptors has generally been limited by resistance of the bound receptor to elution under conditions compatible with its stability, as well as by enzymatic or chemical cleavage of ester and amide groups in the spacer arms that link steroids to the supporting matrix. For estrogen receptors we, as well as others, solved the elution problem by including chaotropic salts such as sodium thiocyanate with or without dimethyl formamide (Musto et al. 1977) in the eluting medium with estradiol to facilitate release of the receptor protein. The problem of adsorbent stability was solved in our work by using a thioether bridge to link estradiol to Sepharose 6B (Fig. 1). As a result, we have established a general purification protocol for estrogen receptor that is simple and reproducible and gives a good yield of highly purified receptor protein as the steroid receptor complex (E*R). The estradiol- and heparin–Sepharose columns used in this scheme both have high capacities for receptor, the former being 5–10 nmol of receptor per milliliter of adsorbent and the latter being about 1 nmol of E*R per milliliter of adsorbent.

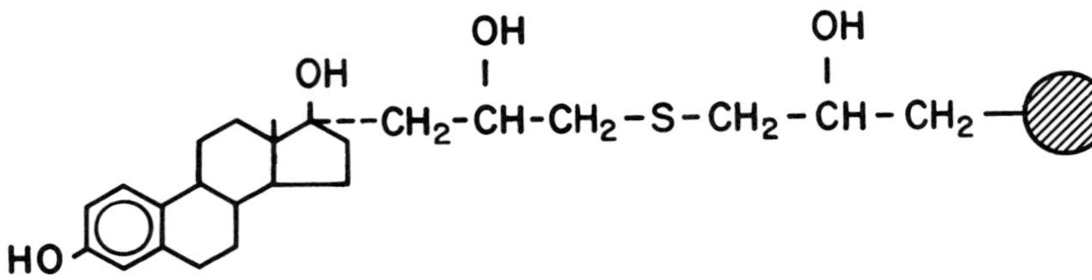

Fig. 1. Structure of the estradiol affinity adsorbent. The shaded circle represents the Sephadex 6B matrix.

A typical purification sequence is summarized in Table 1 for extranuclear estrophilin from MCF-7 cells. When our estradiol affinity adsorbent is used in combination with heparin–Sepharose, the overall recovery of receptor is 30–45% and the purity ranges from 55% to greater than 90% of the specific radioactivity expected for one molecule of [³H]estradiol bound to a 4 S protein of M_r 65,000. We have isolated as much as 5 nmol (325 μg) of receptor in a single experiment. It is possible to achieve up to an 800-fold purification of receptor by steroid affinity chromatography alone if the column is washed extensively before elution of the receptor with [³H]estradiol.

Some of the properties of highly purified MCF-7 estrogen receptor are summarized in Table 2. Unlike the unpurified cytosol E*R, which sediments as an 8–9 S complex in low salt gradients, the purified receptor has lost its ability to aggregate and sediments as a 4.5 S complex in gradients containing either 10 mM KCl or 400 mM KCl. However, if purified E*R is mixed with receptor-depleted MCF-7 cytosol, a 7–8 S E*R complex is observed (unpublished results), indicating that the factors or factor responsible for the formation of 8 S receptor complex are removed during purification. An apparent molecular weight of 120,000 for purified E*R in the presence of 0.4 M KCl determined by HPLC and sedimentation data suggests the formation of a dimer. When highly purified receptor was analyzed by SDS gel electrophoresis, one major band, M_r 65,000, was observed when the gel was stained with silver (Merril et al. 1981). As shown in Table 1, the purified E*R contained more than 59% of the specific radioactivity expected for pure steroid–recep-

Table 1. Purification of MCF-7 Estrogen Receptor

Step	Protein (mg)	Total receptor[a] (nmol)	Specific activity (pmol/mg)	Yield (%)	Purity[b] (%)	Purification factor
1. Cytosol	3979	5.61	1.41	100	—	1
2. Affinity eluate	N.D.	4.58	N.D.	82	—	—
3. G25 eluate	26.6	4.03	152	72	1	108
4. Heparin–Sepharose eluate	0.26	2.33	9058	42	>59	6424

[a] Determined by specific binding to Controlled-Pore glass beads.
[b] Assuming one E* bound to a protein of M_r 65,000 (determined by SDS gel electrophoresis).

Table 2. Properties of Purified MCF-7 Estrogen Receptor

	Native*	Dissociated†
Sedimentation coefficients(s)	4.5 S	4.5 S
Stokes radius‡ (a)	—	57.4 Å
M_r (s & a)	—	120,000
M_r (SDS)	—	65,000
# Hormone sites	—	1

* Buffers containing 10 mM KCl.
† Buffers containing 400 mM KCl.
‡ Determined by gel exclusion high-pressure liquid chromatography.

tor complex consisting of one estradiol bound to a protein molecule of M_r 65,000. With the exception of the calf-specific antibodies, the purified MCF-7 E*R reacts with all tested antiestrophilin antibodies.

V. Antibodies to Estrophilin

Our experience with the preparation of polyclonal and monoclonal antibodies to calf and human estrogen receptors indicates the feasibility of preparing such antibodies with a relatively small amount of material and with receptor that is only partially purified. Rabbits, a goat, and rats (Greene et al. 1980a,b) immunized with nuclear estrogen receptor from calf uterus (10–20% pure; 20–100 μg E*R per injection) produced antibodies that recognized all tested mammalian and nonmammalian estrogen receptors as well as some antibodies that were species specific. Interaction of these antibodies with radioactive estrogen–receptor complexes (E*R) was detected and characterized by sucrose density gradient centrifugation, double antibody precipitation, gel filtration, and adsorption to protein A–Sepharose. In all cases these antibodies have been specific for estrogen receptors, showing no tendency to react with androgen or progesterone receptors. Similar cross-reacting antibodies have been produced by others in animals immunized with estrogen receptor from calf uterus (Radanyi et al. 1979), rat mammary tumor (Al-Nuami et al. 1979), human myometrium (Coffer et al. 1980), and human breast cancer (Raam et al. 1981).

More recently, we have generated polyclonal and monoclonal rat antibodies to estrophilin by immunizing male Lewis rats with partially purified extranuclear estrogen receptor from MCF-7 cells. Receptor eluted from our estradiol affinity adsorbent (~5–10% pure) proved to be immunogenic in a male Lewis rat when injected by the method of Vaitukaitis et al. (Vaitukaitis et al. 1971). Hybridization of splenic lymphocytes from the immunized rat with mouse myeloma cells (P3-X63-Ag8 and Sp 2/0-Ag14) yielded, after cloning by limiting dilution, three hybridoma cell lines, each secreting a unique idiotype of antiestrophilin antibody that recognizes a distinct region of the receptor molecule. The preparation of monoclonal rat antibodies to human estrophilin has been repeated twice by us and once by Larry Miller at

Table 3. Cross-Reactivity of Monoclonal Antibodies to Human Estrophilin

Antibody	Breast cancer		Uterus				Oviduct
	MCF-7	Human	Human	Monkey	Calf	Rat	Hen
Rat serum	+	+	+	+	+	+	+
D58P3μ	+	+	+	+	+	+	−
D75P3γ	+	+	+	+	−	−	−
D547Spγ	+	+	+	+	+	+	−
F88Spγ	+		+		+		+
F344Spμ	+		+		+	+	+
G5Spγ	+						−
G13Spγ	+						

Abbott Laboratories to produce a number of unique monoclonal antibodies that are being used in radiolabeled and unlabeled form to detect, measure, and characterize mammalian estrogen receptors from several species and tissues. Some of the properties of monoclonal antibodies prepared in our laboratory are summarized in Table 3. Of particular significance in this work was our ability to generate pure, monospecific antibodies from a relatively impure (5–10% of theoretical specific activity) preparation of receptor.

VI. Immunochemical Characterization and Assay of Estrophilins

Our experience with the application of immunochemical techniques to the characterization and assay of steroid hormone receptors can best be illustrated by the development of immunoradiometric (Greene and Jensen 1982), immunocolorimetric, and immunocytochemical (Nadji 1981) assays for estrogen receptors in breast cancers and other human tissues. This work is being done in collaboration with Larry Miller and Chris Nolan at Abbott Laboratories and with Mehrdad Nadji at the University of Miami. It was the availability of pure, high-affinity, monospecific antibodies that made these assays possible. Two monoclonal rat antibodies (D547Spγ and D75P3γ, both IgG) that bind additively to different regions of the receptor molecule were used to develop a simple immunoradiometric (IRMA) or immunocolorimetric (ICMA) assay. Polystyrene beads coated with the D547Spγ antibody are incubated with extracts of breast tumors or other tissues to immobilize any estrogen receptor present in the extract. Radiolabeled (IRMA) or peroxidase-conjugated (ICMA) D75P3γ is then used to measure the amount of receptor bound to the bead. When tumor cytosols from 18 human breast cancers were analyzed for receptor content by the IRMA method and the results were compared with estrophilin content determined by sucrose density gradient centrifugation, the relative ranking of receptor levels by the two methods was virtually the same for all 18 tumors (Fig. 2). The same two antibodies, as well as two additional monoclonal antibodies (H222 and H226

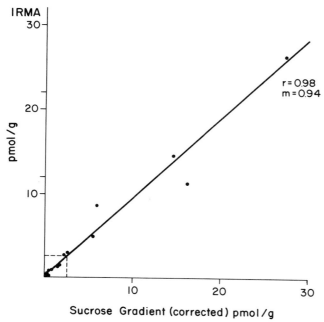

$r = 0.98$
$m = 0.94$

Sucrose Gradient (corrected) pmol/g

Fig. 2. The estrophilin contents of 18 human breast cancer cytosols as determined by the IRMA technique, compared with the results obtained by sucrose gradient ultracentrifugation. In the sedimentation procedure, using 0.5 nM tritiated estradiol which does not saturate all the receptor, the results have been corrected to total binding capacity, under which conditions the distinction between receptor-rich and receptor-poor tumors in postmenopausal patients occurs at a level of 2.5 pmol/g tumor.

from Abbott Laboratories), have also been used to detect and localize estrophilin, with an immunoperoxidase technique, in frozen or paraffin-embedded sections of human breast tumors, human uterus, and other mammalian reproductive tissues. Only nuclear staining was observed for receptor-containing tumors when fixed frozen sections were analyzed by the immunoperoxidase technique, whereas predominantly cytoplasmic staining has been observed in formalin-fixed, paraffin-embedded sections of such tumors. Although the significance of these different staining patterns is still unresolved, the preliminary results indicate the feasibility of using an immunocytochemical assay for estrophilin in human breast cancers and other reproductive tissues. The immunoradiometric (or ICMA) and immunocytochemical assays are simple and rapid and do not depend on the binding of radioactive estradiol to its receptor. In addition, these assays complement each other in that one gives quantitative information about receptor levels and the other provides information about the intra- and intercellular distribution of estrophilin.

An important conclusion can be drawn from Fig. 2 concerning the total amount of estrogen receptor present in the 18 breast cancer cytosols tested by the IRMA method. A comparison of the results for the IRMA and sucrose gradient techniques ($r = 0.98$; $m = 0.94$) indicates that there is no significant amount of occupied receptor or receptor that is unable to bind steroid present in these cytosols. Although this study is hardly exhaustive, the data suggest that the amount of estrophilin determined by steroid-binding assays in breast cancer cytosols is very close to the total amount of receptor present.

To learn more about the location of various antibody determinants on the receptor molecule and whether some of these determinants might provide information about the steroid-, chromatin- and DNA-binding regions of estrogen receptors, we are using radiolabeled forms ($[^{35}S]$-IgG) of several monoclonal antibodies to detect intact receptor and receptor fragments, obtained by proteolysis of purified receptor, on nitrocellulose blots of SDS gels. At least five of the available radiolabeled antibodies recognize their corresponding determinants on such blots. However, although all five antibodies recognize the intact receptor, ($M_r = 65,000$), they differ in their ability to recognize receptor fragments, thereby providing a means of mapping the topography of the receptor molecule with regard to its functional domains.

For purification of various mammalian estrogen receptors, immunoadsorbents prepared from some of the available monoclonal antibodies have proved useful. The high affinity of these antibodies for primate estrophilin has thus far precluded elution of receptor in a form capable of binding estradiol. However, such immunoadsorbents are able to provide, in a single step, denatured receptor (eluted with SDS) that is almost as pure as receptor purified by steroid chromatography on estradiol–Sepharose and heparin–Sepharose. The nearly quantitative yield of this method makes it a valuable source of receptor for peptide mapping and for determining the amino acid composition and sequence of receptor.

The foregoing studies and observations indicate that the purification of mammalian estrogen receptors and preparation of polyclonal and monoclonal antibodies to these receptors are feasible and practical. Further, the antibodies obtained in this manner can be, and have been, invaluable probes for study of the structure and function of hormone receptors.

Acknowledgments. These investigations were supported by research grants from the American Cancer Society (BC-86) and Abbott Laboratories, by a research grant (CA-02897) and contract (CB-43969) from the National Cancer Institute, and by the Women's Board of the University of Chicago Cancer Research Foundation.

References

Al-Nuami N, Davies P, Griffiths K (1979) Cancer Treat Rep 63: 1147
Barrack ER, Coffey DS (1980) J Biol Chem 255: 7265–7275
Coffer AI, King RJB, Brockas AJ (1980) Biochem Internat 1: 126
Gorski G, Toft D, Shyamala S, Smith D, Notides A (1965) Recent Prog Horm Res 24: 45–80
Govindan MV, Manz B (1980) Eur J Biochem 108: 47–53

Greene GL, Closs LE, DeSombre ER, Jensen EV (1977) Proc Natl Acad Sci USA 74: 3681

Greene GL, Closs LE, Fleming H, DeSombre ER, Jensen EV (1979) J Steroid Biochem 11: 33

Greene GL, Fitch FW, Jensen EV (1980a) Proc Natl Acad Sci USA 77: 151

Greene GL, Closs LE, DeSombre ER, Jensen EV (1980b) J Steroid Biochem 12: 159–167

Greene GL, Nolan C, Engler JP, Jensen EV (1980c) Proc Natl Acad Sci USA 77: 5115–5119

Greene GL, Jensen EV (1982) J Steroid Biochem 16: 353–359

Jensen EV, Suzuki T, Kawashima T, Stumpf WE, Jungblut PW, DeSombre ER (1968) Proc Natl Acad Sci USA 59: 632–638

Köhler G, Milstein C (1975) Nature 256: 495

Kuhn RW, Schrader WT, Smith RG, O'Malley BW (1975) J Biol Chem 250: 4420–4228

Merril CR, Goldman D, Sedman SA, Ebert MH (1981) Science 211: 1437

Molinari AM, Medici N, Monocharmont B, Puca GA (1977) Proc Natl Acad Sci USA 74: 4886–4890

Muldoon TG (1980) Endocr Rev 1: 339–364

Musto NA, Gunsalus GL, Miljkovic M, Bardin CW (1977) Endocr Res Commun 4: 147–157

Nadji M, Morales A, Greene GL, Jensen EV (unpublished results)

Raam S, Peters L, Rafkind I, Putnam E, Longcope C, Cohen JL (1981) Mol Immunol 18: 143–156

Radanyi C, Redeuilh G, Eigenmann E, Lebeau MC, Massol N, Secco C, Baulieu EE, Richard-Foy H (1979) C R Acad Sci Paris, Ser D 288: 255

Sherman MR, Pickering LA, Rollwagen FM, Miller LK (1978) Fed Proc 37: 167–173

Sica V, Bresciani F (1979) Biochemistry 18: 2369

Smith RG, d'Istra M, Van NT (1981) Biochem 20: 5557–5565

Vaitukaitis J, Robbins JB, Nieschlag E, Ross GT (1971) J Clin Endocrinol Metab 33: 988–991

Discussion of the Paper Presented by G. L. Greene:

ROY: You presented some data showing the formation of a binding antibody–receptor complex after addition of one type of monoclonal antibody, and then the formation of a ternary antibody–receptor–antibody complex upon addition of a second monoclonal antibody. Under this situation wouldn't you expect to have an aggregate rather than a ternary complex?

GREENE: That would be true except that we are using an excess of antibody. With a large excess of antibody, I would expect only one combining site to react per receptor.

ROY: Your inability to stain the nuclei of the MCF-7 cells with fluorescent-labeled antireceptor antibody is quite interesting. Did you ever examine thin sections of these cells?

GREENE: Yes. Using immunoperoxidase techniques with paraffin-embedded sections of MCF-7 cells and breast cancers we seem to be having trouble seeing any specific staining in the nucleus. I do not have a good explanation for these results at this time.

CLARK: Have you ever done that in the rat uterus? Injected the animal?

GREENE: We do not have any results to report yet with rat uterus.

SCHRADER: You said that none of the antibodies you have produced yet have reacted with hormone binding site. Is that correct?

GREENE: One of the polyclonal antibodies, the goat antibody, definitely affects the binding of steroid to immune complex, perhaps by altering the conformation of the receptor. In general, though steroid binding is not significantly altered by the various antibodies.

SCHRADER: The nature of the assay that you used presumably would preclude your picking up such an antibody.

GREENE: That is correct.

SCHRADER: You would have to do that by some sort of competition screen. I wonder if you have tried that; now that you have one that is positive at an alosteric spot on the protein, you could go back and rescreen your bank for those antibodies which react with the hormone binding domain. I'm first of all wondering have you done that sort of thing?

GREENE: Actually, Bryon Rosner, who is a postdoctoral fellow in the Ben May Laboratory, has tried to screen clones by incubating hybridoma medium first with unlabeled estrogen receptor and then with steroid. However, it is a very time-consuming experiment and so far we haven't obtained a blocking antibody.

SCHRADER: My second question is: You showed in your immunochemistry of fixed cells, if you used whole cells there have been no questions about whether or not receptors might be at the cell surface; if you used whole cells that have not been fixed, have you tried to do, for example, with the ^{35}S antibody, any experiments to see whether or not such an antibody would bind to whole cells in the absence of estradiol?

GREENE: Yes, we tried to see whether ^{35}S-labeled antibody would interact with MCF-7 cells but we did not detect any specific binding.

MOUDGIL: Could you comment as to what will be the role of a hormone under physiological conditions in view of the fact that all of your studies are done with monoclonal antibodies in the absence of the steroid.

GREENE: I did not mean to imply that the hormone was not important, or necessary, for receptor-mediated responses in vivo, but rather that we now have probes for the receptor molecule that do not depend on the binding of radioactive steroid to its receptor. With our library of polyclonal and monoclonal antibodies we hope to answer questions about the synthesis, degradation, and intracellular movement of receptors in target cells, as well as questions about the nature of activation (transformation) and the location of functional domains on the receptor molecule. Some of these questions might best be answered with the use of specific, high-affinity antibodies for receptor that can be radioactively labeled or detected by indirect immunoperoxidase techniques. In this way, we can study the receptor, whether or not it is complexed with hormone.

MOUDGIL: Do you think there are special problems in raising monoclonal antibodies against other steroid receptors such as androgen receptor or progesterone receptor?

GREENE: That is a very difficult question to answer; namely, whether one antigen is immunogenic and another one is not. It does appear as though the mammalian steroid hormone receptors, in hindsight obviously, are immunogenic. First came antibodies to estrogen receptor, followed shortly thereafter by antibodies to glucocorticoid receptor and now, as you said, Milgrom has prepared antibodies to rabbit progesterone receptor. No induced antibodies to chick progesterone receptor have been reported. Also, there are no antibodies to androgen receptor yet, although I know people are working on the problem.

Discussants: G.L. GREENE, A.K. ROY, J.H. CLARK, W. SCHRADER, and V.K. MOUDGIL.

Chapter 13

A Comparison of Central and Peripheral Estrogen Targets

ERNEST J. PECK, JR. AND KATRINA L. KELNER

Estrogens act on a variety of target tissues. In addition to the rat uterus, vagina, pituitary, and hypothalamus (Clark and Peck 1979; Kelner et al. 1980), estrogens are also reported to affect such unlikely tissues as adipose tissue (Gray et al. 1981; Gray and Wade 1980), cerebral cortex (MacLusky et al. 1979), pancreas (Rosenthal and Sandberg 1978), and skin (Punnonen et al. 1980). To investigators studying estrogen action in cell lines, uterus, and vagina, estrogens act as a trophic factor. However, a more catholic view recognizes growth as *one* of a multitude of ways that estrogen can express itself in its many targets. The pituitary and hypothalamus clearly do *not* show hypertrophy on estrogen stimulation, although they do exhibit distinct metabolic changes. Each estrogen target responds in an individual way to estrogen; LHRH content increases in the hypothalamus (Kalra 1976); pituitary cells respond with action potentials and/or changes in membrane properties (Dufy et al. 1979); and corpora lutea show dependence on estrogen for progesterone synthesis (Richards 1974).

The current mechanism of steroid hormone action is based primarily on extensive investigations in the rat uterus and chick oviduct. This approach has advanced our understanding of molecular processes involved in estrogen's action and produced a generalized model of steroid hormone action. However, questions remain in the field that cannot be answered successfully in this manner. One of these was posed succinctly by Notides (1970): "What is the molecular basis for the differential physiological effects of the estrogens in each of the target tissues?" This cannot be answered by single-tissue studies and must necessarily be approached comparatively. By using the current concepts of steroid action as a framework, important steps may be examined in tissues with different physiological responses to steroids. In this manner one should gain insight into biochemical distinctions between target tissues that may be responsible for their varied responses.

Our laboratory has examined estrogen action in the rat hypothalamus and pituitary with this problem in mind. In the following, we outline the results of this comparative study and discuss related data from other estrogen target tissues. In this manner, we hope to illustrate the diversity of intracellular organization in various estrogen target tissues.

I. The Receptor

An early useful definition of an estrogen target tissue required that the organ in question accumulate radioactively labeled estradiol above background or serum levels (Jensen and Jacobson 1962). This radioactivity was subsequently visualized by autoradiography at the light microscope level as silver grains concentrated over the nuclei of target cells (Stumpf 1969). The biochemical correlate was shown to be the existence of high-affinity binding proteins for estrogens in cytosolic and/or nuclear compartments of target cells (Shyamala and Gorski 1967).

In 1972 an exchange assay was reported for the in vitro determination of estrogen-binding sites in rat uterine nuclei (Anderson et al. 1972). Since that time, the basic concept of this assay has been applied with modifications to many tissues. We have adapted this assay for measurement of nuclear receptor in brain tissue (Kelner et al. 1980). Measurement of steroid receptors in neural tissue is fraught with technical difficulties because of the small amount of receptor and the high lipid content of brain. The latter results in high levels of nonspecific binding of [^{3}H]steroid to crude nuclear fractions. To lower nonspecific binding, we have utilized a simple chromatin isolation method that separates the extensive membranous fractions of brain from nuclear material. We have further reduced nonspecific binding by including a nonionic detergent, Tween-80, in the wash buffer. Precipitation of the nuclear pellet with protamine sulfate after [^{3}H]steroid exchange ensures measurement of all receptor in the assay tube, preventing loss receptor because of dissociation from chromatin during exchange and washing (unpublished observation).

Using this procedure and a similar one for the measurement of cytoplasmic receptor, estrogen receptors in target tissues have been characterized with respect to equilibrium binding parameters, i.e., dissociation constant and number of binding sites. A common theme emerges. The dissociation constant of the cytoplasmic estrogen receptor, as determined by saturation analysis, is reported to be approximately $0.1–1.0 \times 10^{-9}$ M in all systems studied. This range is well within the variability expected because of different laboratories and techniques and because of contamination with endogenous competitors (e.g., unlabeled estrogens). The hypothalamus and pituitary are no exception (Kelner and Peck 1981). Both hypothalamic and pituitary cytoplasmic receptors have dissociation constants of 0.2×10^{-9} M, while nuclear receptors have *apparent* dissociation constants of 0.5×10^{-9} M. This difference in dissociation constants between nuclear and cytoplasmic receptors is due to dilution of the radiospecific activity of [^{3}H]steroid by endogenous estradiol present in the nuclear preparation. These values are similar to those found in other target tissues (Clark and Peck 1979).

Unlike the dissociation constant, the number of receptor sites varies greatly from target to target. The receptor content of the hypothalamus, pituitary, and uterus vary from each other by an order of magnitude, although the number of receptor molecules per cell may remain constant. In addition, recent evidence suggests that the number of binding sites for a given target may be altered under different physiological conditions. Treatment of rats with the dopamine, β-hydroxylase inhibitor, U-14624, or with the α-adrenergic receptor blocker, phenoxybenzamine, will decrease the number of progesterone receptor sites in the hypothalamus (Nock et al. 1981). Estrogens increase the number of progesterone binding sites in uterus, pituitary, and hypothalamus and thus increase the responsivity of these tissues to progesterone (Corvol et al. 1972; Hsueh et al. 1975). Progesterone, on the other hand, decreases responsivity of the uterus to estrogen via a decreased production of estrogen receptor (Clark et al. 1977).

The constancy of receptor affinity for estrogens from tissue to tissue suggests that, unlike catecholamine and insulin receptor systems (Williams and Lefkovitz 1977), responsivity to estrogen is not regulated by changes in the affinity of the receptor. This suggests that the basic structure of the receptor molecule is identical from tissue to tissue and plays no part in differentiating estrogen's actions. Only via its occupancy or a secondary regulation of its number can the receptor itself participate in the modulation of estrogen action. In fact, the number of available binding sites does serve as a regulatory device, as evidenced by the estrogen–progesterone interactions. Thus the occupancy of receptor systems and, in turn, the response of the system appear to depend on the availability of both free estrogens and free receptor within the cell.

II. Translocation and Retention

The current model for steroid action includes a process that is termed translocation, the key to the two-step model (Jensen et al. 1968). After estrogen binds to a soluble cytoplasmic receptor, the receptor–steroid complex is thought to move or translocate from the cytoplasm to the nucleus of the cell. The motive force for this process is simply the greater affinity that receptor–steroid complexes have relative to unoccupied receptors for acceptor sites on chromatin and/or DNA in the nucleus (Clark and Gorski 1969; Williams and Gorski 1971). Thus, the partitioning of receptor between cytoplasmic and nuclear compartments changes with the occupation of receptor by ligand. An occupied receptor is more likely to reside in the nucleus.

Do all estrogen receptors have this capacity to translocate? The evidence indicates that most do. Both hypothalamic and pituitary estrogen receptors will, in fact, translocate when an ovariectomized rat is injected with 5 ug of estradiol or estriol in saline. This is illustrated in Fig. 1. In fact, most estrogen targets thus far examined possess a cytoplasmic receptor that has the capacity to translocate to the nucleus.

Although translocation seems a common mechanism in all targets, the extent of translocation for a given stimulus may differ from target to target in vivo. For instance, it appears that steroid concentrating areas in the central nervous system are less sensitive to circulating estradiol than those in the

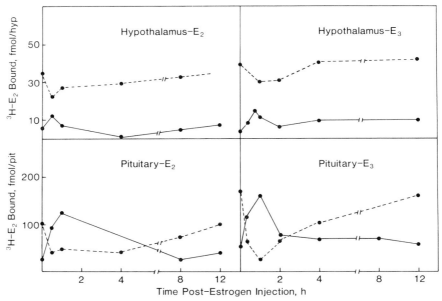

Fig. 1. Simultaneous nuclear accumulation and cytoplasmic depletion of estrogen receptor after estradiol and estriol injection. Ovariectomized rats were injected with 5 μg estradiol or estriol in saline and killed at various times after injection. Cytoplasmic and nuclear fractions of hypothalamus and pituitary were isolated and assayed for [^{3}H]estradiol binding by saturation analysis. Data obtained from the nuclear fractions are indicated by the solid lines; data obtained from the cytoplasmic fractions are shown by the dotted lines. Each point represents data obtained from a pool of six animals.

periphery (i.e., the uterus and the pituitary). This is obvious in Fig. 1 and is further illustrated by Fig. 2. In Fig. 1 animals received 5 μg of either estradiol or estriol and nuclear and cytoplasmic receptors were determined as a function of time after injection. Note that for either steroid a significantly greater fraction of total receptor was translocated to pituitary nuclei than to hypothalamic nuclei. Figure 2 substantiated this observation in an independent investigation, which measured nuclear and cytoplasmic receptors 1 h after injection in hypothalamus, pituitary, and uterus. The solid bars in the left panel of Fig. 2 show that a 5-μg injection of estradiol will drive 70–80% of uterine and pituitary receptor to the nucleus by 1 h after injection, while driving only 20% of hypothalmic receptor to the nuclear fraction at this time. Estriol (stippled bars) is also much more effective in causing nuclear accumulation of receptor in uterus and pituitary than in hypothalamus. The right-hand panel illustrates the simultaneous depletion of cytoplasmic receptor in these tissues. This depletion is much more extensive in uterus and pituitary than in hypothalamus, as expected from the stoichiometry of cytoplasmic to nuclear translocation. Thus, though the ability of receptors from various target tissues to translocate appears ubiquitous, the *extent* of translocation in response to a specific estrogen stimulus varies considerably between estro-

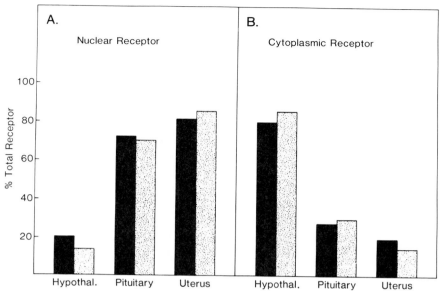

Fig. 2. Differential translocation of estrogen receptor in target tissues. Six ovariectomized rats were injected with 5 μg estradiol or estriol and killed after 1 h. Nuclear and cytoplasmic fractions were isolated from hypothalamus, pituitary, and uterus. Saturation analysis of [^{3}H]estradiol binding yielded an estimate of the number of receptive sites in each fraction. Excess diethylstilbestrol was used to correct for nonspecific binding. Panel A: Solid bars, estradiol; stippled bards, estriol. Panel B shows the corresponding levels of cytoplasmic receptor in these tissues.

gen targets. The mechanism of this phenomenon is as yet unknown, but may relate either to the translocation mechanism itself or to the availability of *free* steroid within a given target tissue. With regard to the latter, it is important to note that both pituitary and uterus have protein-permeable vascular beds, while tight junctions between endothelial cells of brain capillaries exclude protein from extracellular spaces. Thus, steroid bound to serum proteins is excluded from extracellular space in the brain. This exclusion reduces the total steroid in extracellular space and could result in less free steroid as well.

The time course of retention of receptor in nuclei of the target cells also varies between target tissues. As an example, at l h after estrogen injection a substantial fraction of the hypothalamic receptor initially present in the cytoplasmic fraction will be found in the nuclear compartment. However, by 4–8 h the level of receptor in the nuclear compartment has returned to control values. When these time courses of nuclear retention are compared to those of the uterus, a remarkable difference is noted. In uterus, a single injection of estradiol to an ovariectomized animal will result in very high levels of receptor present transiently in the nucleus. Small but significant amounts of receptor remain in the nucleus for up to 24 h. This ''long-term retention'' of receptor in the uterus is thought to mediate the growth responses exhibited by the uterus long after the injection of estradiol. In the hypothalamus and pituitary, no long-term nuclear retention of estrogen receptor is observed. This illustrates that "long-term retention" is a phenomenon restricted to *some* estrogen target tissues, perhaps only those showing estrogen-induced growth.

III. Salt-Inextractable Complexes

The biochemical basis for the long-term retention of nuclear estrogen receptor complexes is thought to be the presence or absence of a specific class of acceptor sites on chromatin. This class of acceptor sites is characterized by its ability to bind receptor–steroid complexes under high salt conditions. These receptor–acceptor complexes are termed "salt inextractable." Thus, at long times after estrogen injection (12–24 h), virtually all of the receptor retained in uterine nuclei is accounted for by salt-inextractable complexes that exist bound to chromatin and nuclear matrix. In light of the absence of long-term retention of receptor in hypothalamic nuclear material, we have compared hypothalamic and uterine nuclei for the presence of salt-inextractable complexes. Crude chromatin was prepared in bulk from hypothalami and uteri of rats injected 1 or 6 h previously with estradiol. The chromatin was extracted with concentrations of KCl up to 0.6 M and then assayed for the presence of specific estrogen binding. We found no evidence for the presence of any estrogen binding after extraction of the nuclear material with salt concentrations of 0.4 M or higher while KCl-inextractable sites were present in uterine nuclei. These data agree with a report by Roy and McEwen (1977) that at both 2 and 6 h after [^{3}H]estradiol injection, over 90% of hypothalamic and pituitary nuclear radioactivity could be extracted with 0.4 M KCl, while only 70–80% could be extracted from uterine nuclei at 6 h after injection. These data suggest that the presence or absence of specific classes of acceptor sites in different target cells may determine the retention time of receptors in the nucleus and thus regulate different biochemical responses to a given hormonal signal.

IV. Type II Sites

Other components of the estrogen-binding system that have been described recently for the rat uterus include a second and third estrogen-binding species, termed the cytoplasmic and nuclear type II estrogen-binding sites. These are discussed in some detail in Chapter 14 of this volume. The functions of these secondary, low-affinity binding species remain unknown. Both species increase significantly in number following estrogen treatment. Nuclear type II estrogen receptors are considered as intermediaries in estrogen's control of uterine growth (Markaverich and Clark 1979; Markaverich et al. 1981).

Nuclear and cytoplasmic type II receptive sites are detected in their respective subcellular fractions as lower-affinity estrogen-binding species. A Scatchard plot of a saturation analysis of [^{3}H]estradiol binding that extends over the range $0.05–100 \times 10^{-9}\,M$ reveals the presence of two estrogen-binding species in both fractions. Uterine cytosol shows a typical two-site binding curve, indicating the presence of 0.1 and 40 nM estrogen-binding components, the classic type I and the cytosolic type II binding sites, respectively. A Scatchard analysis of [^{3}H]estradiol binding to uterine nuclear material taken from a hyperestrogenized animal shows a marked nonlinearity. A discontinuity or concavity in the curve for the lower-affinity site suggests a positive cooperativity in nuclear type II sites. Similar analysis of cytoplasmic and nuclear material isolated from hypothalami or pituitaries of hyperestrogenized rats indicates no evidence of a second estrogen binding component. Figure 3 illustrates Scatchard analyses of binding data for nuclear material of hypothalamus, pituitary, and uterus. Using a modification of the method of Rosenthal (1967), hypothalamus and pituitary prove to contain a single binding class. The Scatchard for uterine data cannot be fit by this procedure. We have also used a nonlinear curve-fitting program, LIGAND (Munson and Rodbard 1980) to attempt a more definitive resolution of the nuclear binding data (Kelner and Peck 1981). As suggested by the Scatchard plots, the data for hypothalamus and pituitary are best approximated by a computer-generated model describing a single high-affinity binding site. The uterine data do prove to contain multiple species.

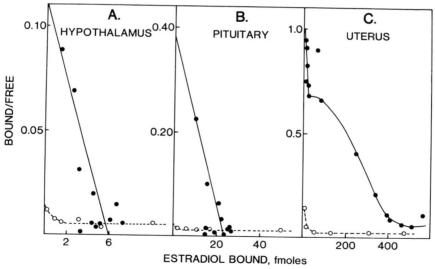

Fig. 3. Scatchard analysis of [^{3}H]estradiol binding to chromatin from hyperestrogenized tissues. Extensive saturation analysis (0.05–40 nM) was performed on crude chromatin isolated from hypothalamus, pituitary, and uterus of hyperestrogenized rats (10 μg estradiol/day for 3 days). Values plotted include specific (closed circles) and nonspecific (open circles) binding as defined by DES competition.

Thus there is evidence for a single estrogen-binding component in both cytosol and nuclear fractions of hypothalamus and pituitary. This is in contrast to the situation in the uterus (see Fig. 3 and Chapter 14), where there are clearly multiple binding sites. This heterogeneity of distribution of type II binding sites strengthens the suggestion that type II estrogen receptors may be involved with the mediation of a limited class of responses. Those tissues that, for instance, hypertrophy in response to estrogen, contain type II sites while others, such as the hypothalamus, do not.

V. Intracellular Responses

Intracellular mediators of estrogen action are undoubtedly numerous. We are aware of only a few. Among these are the three nuclear RNA polymerases responsible for the synthesis of new species and/or increased levels of RNA in estrogen-stimulated cells. For estrogen treatment to result in the increased synthesis of RNAs, the appropriate polymerase must be stimulated. Estrogen treatment is known to cause both transient and sustained increases in RNA polymerase II activity as well as the prolonged activation of RNA polymerase I in the rat uterus (Hardin et al. 1976). We have measured endogenous nuclear RNA polymerase I and II activities in nuclei isolated from estrogen-treated hypothalami (Kelner et al. 1980). The time course and nature of polymerase activation in the hypothalamus is very different from that seen in uterus. Although there is the short-term stimulation of RNA polymerase II at 1 h after treatment, there is no long-term activation of either RNA polymerase I or II. Thus with this parameter as well, there is a distinct heterogeneity between estrogen target tissues. The adult hypothalamus does not grow in response to estrogen and, as expected, does not respond to estrogen injection with the sustained activation of polymerases.

Another enzymatic consequence of estrogen treatment in both uterus and chick oviduct is a large increase in ornithine decarboxylase (ODC) activity (Kaye et al. 1971; Cohen et al. 1970; Hernandez et al. 1973). This enzyme is responsible for the synthesis of the polyamines. Although the function of this enzyme and its product is not as clearly delineated as that for the RNA polymerases, it seems certain that polyamines are involved in nucleic acid condensation. Many reports have suggested that activation of ODC occurs during a tissue's growth, whether during development or when stimulated by exogeneous agents (Raina and Janne 1975). We were therefore curious whether ODC was activated in the hypothalamus after estrogen treatment, since ODC stimulation after estrogen treatment in rat uterus is pronounced. Hypothalamic cytosol was incubated with varying concentrations of L-[1-^{14}C]ornithine for 90 min at 37°C and the incubation was terminated with acid. Subsequent trapping of liberated $^{14}CO_2$ on filter papers saturated with hydroxide of hyamine yielded a measure of ODC activity. No elevation of ODC activity was observed in the hypothalamus of the adult rat after estrogen treatment (unpublished observation). In fact, in agreement with previous reports for chick nervous system (Teng and Tent 1978), there was a slight (30%) depression in the activity of ODC 1 h after a 5-ug injection of estradiol. Thus again, disparate biochemical responses to estrogen have been observed in different estrogen target tissues and the key to these differences appears to be whether or not the tissue grows in response to estrogens.

VI. Terminal Responses

The estrogen responses discussed earlier are all intimately involved with the most central of the cell's biochemical reactions. It is also of interest to examine terminal responses, those which are involved with specialized functions of a particular cell type. A discussion of some of these opened this chapter. A partial listing of those responses that have been investigated in the hypothalamus, pituitary, and uterus—remarkable for its diversity and heterogeneity—is given in Table 1. While some enzymatic responses are seen in all tissues, others observed in the uterus after estrogen treatment are not seen in the hypothalamus and vice versa.

One series of terminal responses, unique to hypothalamus and pituitary, that has been explored in our laboratory is the regulation of hypothalamic LHRH and serum LH. To describe neuroendocrine responses in ovariectomized, estrogen-treated rats, we measured the effect of estrogen implantation (silastic capsules with 0, 10, and 500 μg and 5 mg estradiol) on the content of LHRH in the anterior hypothalamus and medial basal hypothalamus as well as on serum LH. No significant change in LHRH content of the anterior hypothalamic-preoptic region was seen with increasing hormone does 3 days after implantation. The values expressed as pg/hypothalamic equivalent $\pm$ SEM are as follows: Control, 741.4 $\pm$ 24.7; 10-μg implant, 646.3 $\pm$ 102.0; 500-μg implant, 636.5 $\pm$ 99.6; 5-mg implant, 803.6 $\pm$ 236.6. After implantation with estradiol for 3 days, serum LH levels and LHRH content of neurosecretory terminals (synaptosomes isolated from the medial basal hypothalamus) were measured over the course of an afternoon. The top panel of Fig. 4 illustrates that increasing serum estradiol levels caused both quantitative and temporal alterations in the median eminence nerve terminal fraction. At noon, an increase in terminal content of LHRH is observed with increasing hormone dose, while a drop in LHRH levels is observed at 1800 h. Both effects increase in magnitude with increasing estradiol dose. A particularly large drop in LHRH levels is seen at 1800 h with the 5-mg implant. The lower panel of Fig. 4 shows the levels of serum LH from these same animals. The inhibitory or negative feedback effects of estradiol on LH levels are seen at noon. With increasing estradiol dose, the serum LH levels decrease. The induction of the LH surge, that is, the positive feedback effects of estradiol on serum LH, occur late in the afternoon. With increasing estradiol dose, the size of the LH peak at or near lights off increases in size and duration.

Table 1. Exemplary Estrogen-Induced Responses in Hypothalamus, Pituitary, and Uterus

Hypothalamus	Pituitary	Uterus
Incorporation of [^{3}H]lysine into protein (Litteria and Thorner 1974)	Incorporation of alanine and glucosamine into LH (Liu and Jackson 1977)	Net protein synthesis (Hamilton 1963)
Glucose-6-phosphate dehydrogenase activity (Luine et al. 1975)	Glucose-6-phosphate dehydrogenase activity (Luine et al. 1975)	Glucose-6-phosphate dehydrogenase activity (Szego and Roberts 1953)
Progesterone receptor levels (Kato and Onouchi 1977)	Progesterone receptor levels (Kato and Onouchi 1977)	Progesterone receptor levels (Clark and Peck 1979)
Incorporation of [^{3}H]UTP into RNA (Kelner et al. 1980)	[^{3}H]uridine incorporation into RNA (Stone et al. 1977)	[^{3}H]uridine incorporation into RNA (Means and Hamilton 1966)
LHRH content in the medial basal hypothalamus (Kalra 1976)	Sensitivity to LHRH (Vilchez-Martinez et al. 1974)	DNA synthesis (Kaye et al. 1972)
Reuptake of dopamine, serotonin, and norepinephrine in the anterior hypothalamus (Cardinali and Gomez 1977)	Sensitivity to TRH (DeLean et al. 1977)	Water retention (Astwood 1939) Number of α-adrenergic receptors (Roberts et al. 1977; Lefkowitz and Williams 1977)
Spontaneous activity in the preoptic area and medial basal hypothalamus (Bueno and Pfaff 1976)	Input resistance and action potentials in pituitary cell line (Dufy et al. 1979)	
Mating behavior (Green et al. 1970)		

Comparison of the top and bottom panels of Fig. 4 shows that the drop in LHRH levels in the neurosecretory terminal fraction is coincident with the rise in serum LH concentrations at lights off. Thus, these changes in LHRH content of the medial basal hypothalamus reciprocate the LH surge seen upon estradiol treatment. This observation is consistent with our previous report (Kerdelhue and Peck 1981) that there is a circadian rhythm to the releasibility of LHRH from hypothalamic synaptosomes in vitro. That is, the release of LHRH by depolarizing influences is considerably enhanced in the evening, coincident with the LH surge, when compared to earlier times in the day.

To test whether these neuroendocrine responses are differentially sensitive to estrogen and thus might be related to negative and positive feedback,

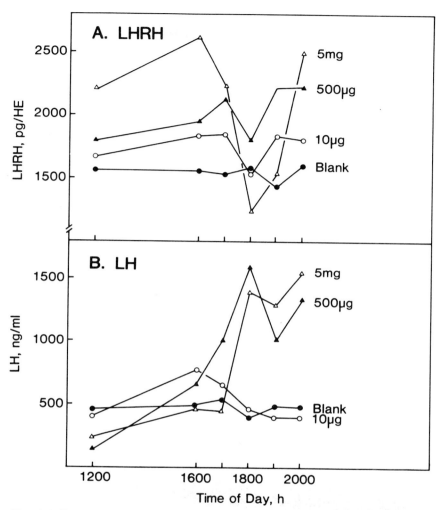

Fig. 4. Effect of estrogen on hypothalamic LHRH content and serum LH. Groups of six ovariectomized rats were implanted with silastic capsules containing 0, 10 μg, 500 μg, or 5 mg estradiol. Hypothalamic P_2 content of LHRH and serum LH levels were determined on day 4 at several times. Lights out was at 1800 h.

groups of six ovariectomized rats were treated with a series of estradiol implants. The extent of six estrogen-induced responses was measured at each estradiol dose and the results expressed at percent maximum response (Fig. 5). Two uterine responses were monitored—uterine wet weight and uterine dry weight. Quantitation of temporal variations in serum LH provided an index of negative and positive feedback systems. Inhibitory or negative feedback effects of estrogen on luteinizing hormone (LH) were defined by the suppression of *serum LH levels at 1200 h* (experimental/

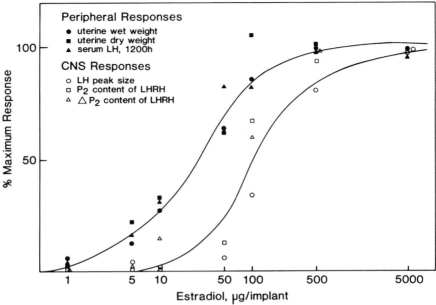

Fig. 5. Dose–response curves of estrogen-induced responses in CNS and peripheral target tissues. Six estrogen-induced responses were measured after implantation of silastic capsules containing various levels of estradiol for 4 days in groups of six ovariectomized rats. Mean errors are the following: uterine wet weight (SD), 17.1 mg (10.9%); uterine dry weight (SD), 3.7 mg (12.4%); 1200 h serum LH or "negative feedback" (SE), 45.2 ng/ml (14.1%); LH peak size or "positive feedback" (SE), 47.7 ng/ml (12.3%); P_2 content of LHRH (SE), 1913 pg/HE (10.8%); P_2 LHRH, not applicable.

control) measured by radioimmunoassay. LH surge induction or positive feedback effects of estrogen on LH were defined as *LH peak size.* This was taken as the increase in serum LH at 1800 h and 2000 h over the 1200 h value. Thus, the average value for LH (1800 h + 2000 h/2) was divided by the 1200 value. *LHRH content of neurosecretory elements* of the medial basal hypothalamus at 2000 h was determined by extraction and radioimmunoassay. The size of the LHRH pulse was defined as the *difference between LHRH content of neurosecretory terminals at 2000 h and 1800 h.* Thus, we measured two indices of hypothalamic response—terminal content of LHRH and the size of the LHRH pulse released from the medial basal hypothalamus at the time of lights out (1800 h). Figure 5 illustrates that these six responses fall into two distinct classes—termed peripheral responses and central nervous system responses. The uterine responses and the negative feedback response of serum LH fall into the peripheral response category. The positive feedback response of serum LH and the two hypothalamic parameters fall into a less sensitive catergory—CNS responses. Thus, three of these six estrogen-induced responses are apparent at low estradiol doses, while three require higher doses for activation.

We have tested the hypothesis that the differential sensitivity between estrogen targets indicated by this dose response curve is mediated by differential estrogen receptor translocation in these target tissues. The experimental results support this hypothesis. Translocation of pituitary and uterine receptors is considerably more sensitive to estradiol than is that of the hypothalamus. Thus, one hour after injection of various estrogens to ovariectomized rats, there is considerably more translocation of receptor in pituitary and uterus than in the hypothalamus (see Figs. 1 and 2). Additionally, after administration of estradiol by implants of various sizes and subsequent measurement of nuclear and cytoplasmic receptor levels, a greater sensitivity of translocation to estradiol is apparent in uterus and pituitary than in hypothalamus. Thus, a higher concentration of serum estradiol is required to translocate hypothalamic estrogen receptor than is required to translocate pituitary receptor.

VII. A Model

And so we return to our initial question. By what mechanism does the estrogen target cell become organized during differentiation to cause its particular array of biochemical and physiological responses and sensitivities to estrogen distinct from those of other target tissues? From the data presented here, we suggest two mechanisms. First is the control of local availability of hormone. The organism seems to have made a broad distinction between central nervous system and peripheral targets that it may differentially control by virtue of these areas' distinct sensitivities to estrogen. The mechanism by which this is accomplished is unknown but probably involves a decreased availability of steroid in the brain when compared to the uterus (see below). Second, we suggest that during the terminal differentiation process, each cell type localizes a unique ensemble of classes of acceptor sites on chromatin. Each of these ensembles controls a separate response or set of responses. Some will be unique to a given cell type; others will be ubiquitous among estrogen targets. Different classes of acceptor allow for transient or short-lived and long-lived nuclear receptor–acceptor complexes. This model is shown schematically in Fig. 6.

In this model we have allowed for either free or restricted access of protein-bound steroid to extracellular spaces (e.g., restricted via nonfenestrated capillaries of the central nervous system) and for the existence of multiple nuclear acceptor sites. In the upper panel is depicted a simple model for the hypothalamus—a tissue that has a low sensitivity to estrogen and that responds only transiently to estrogen. In our view, receptor–acceptor complexes persist in the nuclear compartment of this tissue only as long as intracellular estrogen levels are high enough to maintain occupancy of receptor (i.e., receptor–estrogen complexes can bind acceptors but receptor alone cannot). Thus, a cell endowed only with this acceptor site responds transiently to circulating steroid, with the extent of response being directly proportional to free estrogen in the extracellular space. In the lower panel is

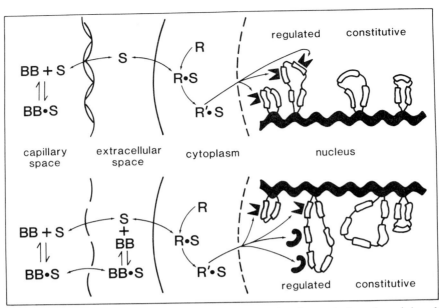

Fig. 6. Steroid control of differential gene expression in central and peripheral target tissues. The upper scheme represents a model of steroid action in a CNS target tissue. Blood-binding proteins (BB) cannot enter extracellular spaces because of the blood–brain barrier. Thus, only free steroid diffusing through the extracellular space and into cytoplasm is available for binding to cytoplasmic receptor. Receptor–steroid complexes (R · S) translocate to the nuclear compartment and bind to *transient* acceptor sites on two batteries of genes, illustrated by the loops of chromatin.

The lower scheme is a model of steroid action in a peripheral target. Complexes of estradiol with blood binders (BB · S), as well as free steroid, can enter extracellular spaces, thus increasing the amount of total steroid available to the cell. The R · S complexes bind to two classes of acceptor on two sets of regulated genes on chromatin. One class of acceptor is identical to that found in the CNS target, i.e., transient. The other is unique to targets that grow in response to estrogen, i.e., are capable of long-term retention of receptor.

depicted a model for the uterus that possesses a protein-permeable vascular bed (thus allowing higher levels of total steroid into the extracellular, non-vascular compartment) and a second class of acceptor sites. These acceptors form complexes with receptor–estrogen complexes and subsequently undergo either conformational changes or covalent alterations such that acceptor–receptor complexes persist *after* intracellular steroid levels are low. This class would explain the existence of salt-inextractable sites and the sustained activation of RNA polymerases I and II in the uterus long after serum estrogens are low.

The model has been developed to fit the data of our comparative study. It allows explanations for differential target tissue sensitivity and differential responses, as well as differences in temporal aspects of response patterns. It also contains considerable speculation that is justified only as far as it allows us to design fruitful experiments for the future.

Acknowledgments. The authors are indebted to Anne-Marie Hedge and Wendy D'Attilio for invaluable technical assistance. This research was supported by grants from the USPHS, HD-08389 and an RCDA, HD-00022, to EJP and from the NSF, predoctoral fellowship NS-SP-17821184 to KLK.

References

Anderson J, Clark JH, and Peck EJ, Jr. (1972) Biochem J 126: 561–567
Astwood EB (1939) Am J Physiol 126: 162–170
Bueno J, Pfaff DW (1976) Brain Research 101: 67–78
Cardinali DP, Gomez E (1977) J Endocrinology 73: 181–182
Clark JH, Gorski J (1969) Biochim Biophys Acta 192: 508–515
Clark JH, Hseuh A, Peck EJ, Jr (1977) Ann NY Acad Sci USA: Biochemical Actions of Progesterone and Progestins, Vol 286, pp 161–179
Clark JH, Peck EJ, Jr (1979) Female Sex Steroids: Receptors and Function. Springer-Verlag, New York
Cohen S, O'Malley BW, Stastny M (1970) Science 170: 336–338
Corvol P, Falk R, Freifeld M, Bardin CW (1972) Endocrinology 90: 1464–1469
DeLean A, Goron M, Kelly PA, Labrie F (1977) Endocrinology 100: 1505–1510
Dufy B, Vincent JP, Fleury H, du Pasquier P, Gourdji D, Tixier-Vidal A (1979) Science 204: 509–511
Gray JM, Dudley SD, Wade GN (1981) Am J Physiol 240: 43–46
Gray JM, Wade GN (1980) Am J Physiol 239: 237–241
Green R, Luttige WG, Whalter RE (1970) Physiol Behav 5: 137–141
Hamilton TH (1963) Proc Natl Acad Sci USA 49: 373–379
Hardin JW, Clark JH, Glasser SR, Peck EJ, Jr (1976) Biochemistry 15: 1370–1374
Hernandez O, Ballesteros LM, Mendez D, Rosado A (1973) Endocrinology 92: 1107–1112
Hseuh AJ, Peck EJ, Jr, Clark JH (1975) Nature 254: 337–339
Jensen EV, Jacobson HI (1962) Recent Prog. Hormone Res. 18: 387–414
Kalra SP (1976) Endocrinology 73: 405–417
Kato J, Onouchi T (1977) Endocrinology 101: 920–928
Kaye AM, Icekson I, Lindner HR (1971) Biochem Biophys Acta 252: 150–159
Kaye AM, Sheratzky D, Lindger HR (1972) Biochim Biophys Acta 261: 475–486
Kelner KL, Miller AL, Peck EJ, Jr (1980) Recept Res 1: 215–237
Kelner KL, Peck EJ, Jr (1981) J Recept Res 2: 47–62
Kerdelhue B, Peck EJ, Jr (1981) Peptides 2: 219–222
Lefkowitz RJ, Williams LT (1977) Proc Natl Acad Sci USA 74: 515–519
Literia M, Thorner MW (1974) J Endocrinology 60: 377–378
Liu TC, Jackson GL (1977) Endocrinology 100: 1294–1302
Luine VN, Khylahavskaya RI, McEwen BS (1975) Brain Res 86: 293–306
MacLusky NJ, Chaptal C, McEwen BS (1979) Brain Res 178: 143–160
Markaverich BM, Clark JH (1979) Endocrinology 105: 1458–1462
Markaverich BM, Upchurch S, and Clark JH (1981) J Steroid Biochem 14: 125–132
Means AR, Hamilton TH (1966) Proc Natl Acad Sci USA 56: 686–693
Munson PJ, Rodbard D (1980) Anal Biochem 107: 220–239
Nock B, Blaustein JD, Feder HH (1981) Brain Res 207: 371–396
Notides AC (1970) Endocrinology 87: 987–992
Punnonen R, Lovgren T, Kouvonen T (1980) J Endocrinol Invest 3: 217–221
Raina JA, Janne J (1975) Med Biol 53: 121–147
Richards JS (1974) Endocrinology 95: 1046–1053
Roberts RM, Insel PA, Goldfein RD, Goldfein A (1977) Clinical Research 25: 465A
Rosenthal H (1967) Anal Biochem 20: 525–532
Rosenthal HE, Sandberg AA (1978) J Steroid Biochem 9: 1133–1139

Roy EJ, McEwen BS (1977) Steroids 30: 657–669
Shyamala G, Gorski J (1967) J Cell Biol 35: 125A–126A
Stone RT, Maurere RA, Gorski J (1977) Biochemistry 16: 4915–4921
Stumpf WE (1969) Endocrinology 85: 31–37
Szego CM, Roberts S (1953) Recent Prog Horm Res 8: 419–469
Teng CS, Teng CT (1978) Biochem J 176: 143–149
Vilchez-Martinez JA, Arimura A, Debeljuk L, Schally AV (1974) Endocrinology 94: 1300–1303
Williams LT, Lefkowitz RJ (1977) J Biol Chem 252: 7207–7213

Discussion of the Paper Presented by E. J. Peck

THOMPSON: Is the hypothalamus really inside the blood–brain barrier?

PECK: It is not completely within the blood–brain barrier, but it certainly is not freely accessible to large-molecular-weight proteins.

THOMPSON: I asked that because I suggest that there is a fourth alternative, based on some studies we have just submitted for publication with Stoney Simons and Louis Mercier. In comparing two different lines of hepatoma cells, we found a similar discontinuity in the induction of tyrosine transaminase, although there is almost identical affinity for glucocorticoid in these two cell lines, and there is obviously no blood–brain barrier to account for the different sensitivity. I think a fourth possibility is that there are simply various intracellular controls and there really is an inherent chromatin difference in sensitivity.

PECK: I would certainly accept that as a fourth alternative.

MEYER: If you took your data and plotted receptor occupancy versus cellular response for the two different tissues, how does pituitary versus hypothalamus appear?

PECK: We've done that with the secondary or what I call "down-the-line" responses. However, it is very difficult to do. If you go back to endogenous RNA polymerase II, a primary response, and you plot nuclear receptor occupancy versus activation of RNA polymerase II, there is a linear correlation with a slope of 1 for the hypothalamus. If you try to do the same thing for uterus, the relationship will not hold because you have a second class of acceptor site there. For the hypothalamus you get a one-to-one correlation between movement of the receptor to the nucleus and the activation of RNA polymerase II. But going down the line and trying to do the same thing with serum levels of LH or even the hypothalamic level of LHRH, there are too many steps in between. We only know that the pituitary is more sensitive than the hypothalamus.

ROY: I could not resist asking this question. You have heard Penti Siteri two weeks ago and in your model there seems to be no room for that kind of intracellular uptake of the steroid bound to serum proteins.

PECK: Actually, I am embarrassed to say, I didn't hear Penti. But I've heard his talk before and I do know his data. It is certainly possible that serum-binding proteins are internalized. His is not the only data to suggest that. Toran-Allerand has also seen the internalization of steroid-binding globulins into cells of the hypothalamus in dispersed culture. This could act as a carrier or as an intracellular buffer to control the differential sensitivity of two classes of cells. What is different about the data of Toran-Allerand is that she found serum globulin in every cell that was not a target cell, i.e., did not have estrogen receptor. As I recall, those cells that had estrogen receptor had no cytoplasmic serum-binding globulin, whereas in Penti's data there is quite a lot of it in cells that have estrogen receptor. So I don't know, there is certainly a possible regulatory mechanism based on serum-binding globulins.

BARDIN: There is another series of experiments from Petra's lab, in which he showed that the TEBG enters androgen-responsive cells, and this might explain an experiment that Robel published a number of years ago for which he never made the explanation clear, that is, cells metabolize steroids differently, depending on the extracellular binding protein that is present outside the cell, and it implied that they got in the cell, although they couldn't demonstrate it at that time. This leads me to ask, "Is there a very low-affinity estrogen binding in plasma or extracellular fluid that can get into the uterus, in other words, what are the data that indicate that type-II protein is not an extracellular protein that is translocated?"

CLARK: Barry Markaverich, who is in the audience, has found type II sites in isolated uterine cells. Other investigators have reported their presence in MCF cells.

PECK: I would add that you can take serum and try to mimic some of the binding curves seen from either cytoplasmic type II or nuclear type II and you cannot measure a type-II binder. It is interesting, though, that whereas the uterus has a "cystoplasmic type II," we cannot measure cytoplasmic type II in the brain either, again suggesting that it is excluded from the nervous system. Either type II is excluded from the brain if it is extracellular or it is not expressed as a gene product in the case of the central nervous system. All of these things, again, make the brain a less sensitive target than the uterus.

GREENE: I don't know if I missed this; the salt-resistant receptor, is this type I or type II?

PECK: Well, the one we are measuring in the brain is type I.

GREENE: And in the uterus?

PECK: In the uterus, in the experiments we originally had done, I am not certain whether we can say; we didn't do it the right way to say for certain.

CLARK: But we did it later on; it is a mixture of both.

GREENE: Because I had the impression that it was type I, but then I wasn't sure whether you were talking about type II.

CLARK: It depends on the hormone treatment, which I will discuss later. It depends on how long the cells have been exposed to hormone, and so there are many stipulations and qualifications, which I will talk about.

GREENE: There is a recent paper also by Puca on extracting receptor from nucleus where he talks about how even 0.25 M thiocyanate will supposedly extract 100% of the nuclear receptor; I don't know what bearing that has exactly on your own observations. It's true that you refer to it as salt resistant rather than necessarily unextractable, but what would you speculate?

MARKAVERICH: In experiments that we've done, nuclear type-II sites do not bind estrogen in the presence of the thiocyanate. So it's unlikely, in the presence of thiocyanate, that you are measuring any type II at all.

GREENE: Then it must be all type-I receptor.

Discussants: E.J. PECK, E.B. THOMPSON, W.J. MEYER, A.K. ROY, C.W. BARDIN, J.H. CLARK, G.L. GREENE and B.M. MARKAVERICH

Chapter 14

Biological Role of Type II Estrogen-Binding Sites and Steroid Hormone Action

B. M. Markaverich and J. H. Clark

I. Introduction

The generally accepted scheme of estrogen action involves the binding of the hormone to cytoplasmic macromolecules, called receptors. The receptor steroid complex undergoes nuclear translocation and binds to nuclear components. The binding of the receptor estrogen complex to nuclear sites is thought to be involved with the stimulation of transcriptional events that are responsible for cell growth and metabolism (for review see Clark and Peck 1979). In addition to these receptor interactions, we have discovered some additional binding sites for estrogenic hormones that appear to be important. The purpose of this article is to summarize our findings and to propose possible functions for these sites.

II. Cytoplasmic Type II Estradiol-Binding Sites

The cytosol from immature rat uteri contains, in addition to the estrogen receptor, a proteinaceous macromolecule, which we have named the cytosol type II estradiol-binding site. These sites are observed when saturation analysis by [^{3}H]estradiol exchange is performed on uterine cytosol obtained from immature rats. Unlike the estrogen receptor, type II sites do not appear to undergo translocation to the nucleus (Eriksson et al. 1978; Clark et al. 1978). That is, an injection of estradiol that causes cytoplasmic depletion and concomitant nuclear accumulation of the estrogen receptor does not deplete type II sites from the cytosol (Fig. 1). Type II sites have an affinity ($K_d \sim 20$ nM) less than that of the estrogen receptor ($K_d \sim 1$ nM) but the number of sites may greatly exceed that of the receptor. Type II sites display stereospecificity for estrogenic compounds and are present in other estrogen target tissues such as the vagina (Clark et al. 1978), mouse (Watson and Clark 1978), and human mammary tumors (Panko et al. 1981).

Although the function of cytoplasmic type II sites is not known, their presence complicates the interpretation of receptor assays. The quantity of these sites varies with many factors and may range from 2 to 10 times the

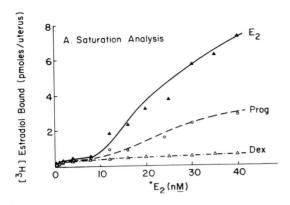

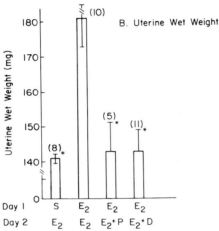

Fig. 1. (A) Saturation analysis of nuclear estrogen-binding sites in rat uterine fractions. The quantity of specifically bound [³H]estradiol (pmoles/uterus) for each estradiol concentration was determined by subtracting the nonspecific binding ([³H]estradiol bound in the presence of 100-fold excess DES) from the total quantity of [³H]estradiol bound. Mature ovariectomized rats were primed with an injection of estradiol (▲), estradiol + progesterone (○), or estradiol + dexamethasone (△). Animals were sacrificed 24 h following the second injection. Estradiol (10 μg), progesterone (2.5 mg), and dexamethasone (5 mg) were injected subcutaneously in 30% ethanol: 0.9% NaCl (v/v). (B) Effects of progesterone and dexamethasone on uterine wet weight. Animals received a priming injection of vehicle (30% ethanol: 0.9% NaCl, v/v) or estradiol (10 μg) on day 1 and a second injection of estradiol, estradiol + progesterone, or estradiol + dexamethasone on day 2 and were sacrificed 24 h later. Hormones were injected exactly as described in legend to part (A). Values represent the mean ± SEM for the number of observations indicated in parentheses. The asterisk indicates a significant difference from animals receiving two daily injections of estradiol ($p < .01$).

quantity of estrogen receptor. The influence of these kinds of variation on the determination of the type I receptor can be significant. As the quantity of type II sites increases, the error introduced in the estimation of the K_d and the number of type I sites progressively increases. This only becomes apparent when saturation analysis is run over a wide range of hormone concentrations. Consequently, assays that are limited to a single concentration of hormone (1–10 nM) will measure both sites and may lead to overestimates of the K_d and numbers of type I sites. These points have been discussed in detail in Clark and Peck (1979).

III. Nuclear Type II Estradiol-Binding Sites

As discussed previously, an injection of estradiol will cause the depletion of type I sites from uterine cytosol. This depletion is accompanied by the accumulation of these sites in the nucleus and represents the well-known cytoplasmic to nuclear translocation phenomenon. Analysis of nuclear fractions for estrogen-binding sites by the [^{3}H]estradiol exchange assay reveals a complex picture that also involves at least two sites (Eriksson et al. 1978; Markaverich and Clark 1979). One conforms to the type I site, which was depleted from the cytosol and is undoubtedly identical to the classically described estrogen receptor. When the quantity of the second site, which we will call nuclear type II, is subtracted from the total quantity of nuclear bound hormone as measured by exchange, one obtains the amount bound to type I. Scatchard analysis of the nuclear type I sites reveals a K_d of 0.60 nM and a maximal number of sites of 0.36 pmol/ml (Markaverich and Clark 1979). These values do not differ significantly from those obtained for the cytosol receptor, which is depleted by estrogen treatment and conforms to the usually accepted properties of the estrogen receptor.

The nuclear type II sites do not appear identical to the cytosol type II sites and display cooperative binding behavior with a Hill coefficient of approximately 2. In addition, these sites bind estrogen specifically, are found in estrogen-sensitive tissues, and may be important in the events involved in estrogen action (Markaverich et al. 1981a).

IV. Effects of Estradiol and Estriol on Nuclear Type II Sites

We have examined the relationship between the stimulation of nuclear type II sites and uterotropic responses by comparing the effects of estradiol and estriol (Markaverich and Clark 1979). Maximal levels of type I and type II sites are reached by 1 h after an injection of estradiol in mature castrate rats. The quantity of type I sites then declines gradually to control levels by 72 h. The quantity of type II sites also declines gradually but is maintained two- to three-fold above controls at 24, 48, and 72 h. Estriol treatment also elevates the quantity of type I sites 1 h after the injection and causes a corresponding increase in uterine wet weight at 4 h. However, only estradiol induced long-term nuclear retention of the type I site (4–6 h), sustained elevated levels of nuclear type II sites (4–48 h), and stimulated true uterine growth (uterine wet weight at 24–48 h). Failure of an injection of estriol to stimulate true uterine growth is correlated with the inability of this hormone to induce long-term (4–6 h) nuclear retention of type I sites or increase the levels of nuclear type II estrogen-binding sites above control levels.

To further examine the relationship between nuclear type II sites and estrogen stimulation of true uterine growth, mature ovariectomized rats were treated with paraffin pellets containing either estradiol or estriol and sacrificed 48 h following hormone administration (Markaverich and Clark 1979). As discussed earlier, under these conditions estriol treatment results in the sustained elevation of nuclear type I sites and the stimulation of true uterine growth in the immature rat (Clark et al. 1799; Martucci and Fishman 1977). If elevated levels of nuclear type II sites are related to estrogen stimulation of true uterine growth (either casually or as a secondary response), then increased quantities of this second nuclear estrogen binding component should be observed in animals treated with an estriol implant. Saturation analysis of nuclear fractions by the [^{3}H]estradiol exchange assay demonstrated that while not as effective as estradiol, administration of estriol by paraffin implant resulted in the sustained elevation of occupied type I sites (0.4 pmol/uterus) and a six- to eight-fold increase in the numbers of nuclear type II sites as compared to the paraffin controls. Elevation of nuclear type II sites also correlated with the ability of estradiol or estriol to stimulate true uterine growth.

These results demonstrate that a positive correlation exists between elevated levels of nuclear type II sites and the stimulation of true uterine growth. This correlation is better than that observed for the classical estrogen receptor (type I site). Type I sites accumulate rapidly in the nucleus after an injection of estradiol; however, they decline to low levels by 24 h. In contrast, the level of type II remains elevated for 24–48 h and true growth of the uterus is observed during this time. An injection of estriol also causes nuclear accumulation of type I, but these sites rapidly disappear from the nucleus and, additionally, neither stimulation of type II sites nor true uterine growth occurs. As discussed earlier, we have shown that a single injection of estradiol stimulated sustained RNA polymerase activity, increased chromatin template activity over long periods of time, and elevated DNA synthesis. A single injection of estriol, in contrast, failed to cause these long-term uterotropic responses (Hardin et al. 1976; Markaverich et al. 1978). We have suggested that these events relate to the ability of estradiol to maintain receptor occupancy in the nucleus for a period of time that is sufficient to stimulate the nuclear mediated events that are obligatory for the production of true growth. One of these obligatory events may be the elevation of type II sites. The failure of an estriol injection to cause true growth results from its inability to maintain type I sites in the nucleus for a sufficient period of time. That type II sites at least attend, if not cause, true growth appears to be the case, since implants of estriol, which sustain occupancy of type I sites, cause the elevation of type II sites and true uterine growth. Thus, estrogen stimulation of true uterine growth correlates with nuclear retention of type I sites and sustained elevation of the level of the nuclear type II estrogen-binding sites.

The precise requirements for estrogenic stimulation of the nuclear type II site remain to be resolved. The ability of estriol when administered by paraffin implant to increase nuclear type II sites and to stimulate true uterine growth suggests that one requirement for the elevation of nuclear type II sites may be sustained nuclear occupancy by receptor–hormone complexes. In addition, the specificity of the interaction between receptor–hormone complexes and nuclear sites that results in the increase in type II sites may also be considered. This conclusion is supported by the observation that while a single injection of either estradiol or estriol resulted in an equivalent accumulation of receptor–hormone complexes at 1–4 h postinjection, only the receptor–estradiol complexes were associated with rapid and sustained elevations of nuclear type II sites between 1 and 48 h after treatment. Whether the increase in nuclear type II sites in estriol-implanted animals was due to long-term nuclear occupancy by receptor–estriol complexes or to saturation of specific nuclear binding sites through a lower affinity interaction with receptor–estriol complexes remains to be established.

In conclusion, these data indicate that two estrogen-binding sites may be involved in the response of the rat uterus to estrogenic hormones. Whereas responses may be mediated through the interaction of estrogen with type I sites (Anderson et al. 1975; Gorski and Gannon 1976; O'Malley and Means 1974), nuclear events associated with true uterine growth (Hardin et al. 1976; Markaverich et al. 1978; Stormshak et al. 1976) may require not only long-term nuclear retention of type I sites, but the sustained elevation of the level of nuclear type II sites.

V. Progesterone and Dexamethasone Antagonism of Type II Sites and Uterine Growth

As discussed earlier, the elevation of nuclear type II sites is closely correlated with the stimulation of true uterine growth, and therefore it is conceivable that these sites might be involved in the mechanism by which estrogens cause uterotropic stimulation. One way to test this hypothesis is to block the stimulation of nuclear type II sites and examine the uterotropic response pattern. Since progesterone and glucocorticoids have been used to block uterotropic responses in various ways (Szego and Roberts 1953; Huggins and Jensen 1955; Lerner 1964), it seemed possible that these hormones could be used for this purpose (Markaverich et al. 1981a).

Mature, ovariectomized rats were given two daily injections of estradiol or a single injection of estradiol on day 1 and an injection of either estradiol plus dexamethasone or estradiol plus progesterone on day 2. All animals were sacrificed 24 h following the second injection. Pretreatment with estradiol (day 1) was to increase the uterine response to progesterone, presumably by increasing the level of progesterone receptor (Milgrom et al. 1973; Leavitt et al. 1974; Walters and Clark 1978). Saturation analysis of specific nuclear binding sites by the [^{3}H]estradiol exchange assay revealed that uterine nuclei from estradiol-treated controls contained approximately 0.2 and 6.0 pmol/uterus of type I and type II sites, respectively (Fig. 1).

Dexamethasone treatment completely blocked the estrogen-stimulated increase in the nuclear type II site and in uterine wet weight ($p < .01$) normally observed 24 h following a second injection of estradiol (Fig. 1 and Markaverich et al. 1981a). Nuclear levels of the type I site were very similar (0.2 pmol/uterus) in the estradiol and estradiol plus dexamethasone treatment groups, suggesting that this antagonist failed to alter nuclear estrogen receptor levels at 24 h. While not as effective as dexamethasone, administration of progesterone to mature ovariectomized rats reduced levels of the nuclear type II site and decreased ($p < .05$) the uterine wet weight response to estradiol but failed to influence nuclear levels of the type I site.

These results suggest that the antagonistic properties of dexamethasone and progesterone on estradiol-induced uterine growth reside in the ability of these compounds to reduce the numbers of nuclear type II sites while not altering nuclear levels of type I estrogen-binding sites. However, these compounds may interfere with nuclear translocation and "processing" of type I sites, thereby reducing the availability of estrogen receptor at 1, 4, and 24 h following dexamethasone or progesterone administration to mature ovariectomized rats (Markaverich et al. 1981a). As shown in Fig. 2, the levels of cytoplasmic type I sites were identical in animals treated with estradiol, estradiol plus dexamethasone, or estradiol plus progesterone at 1 and 4 h postinjection. By 24 h the level of cytoplasmic type I was increased above control (2.0 pmol/uterus) in estradiol (3.6 pmol/uterus) and estradiol plus dexamethasone- (3.0 pmol/uterus) treated animals. The lower level of type I sites in the cytosol of progesterone-treated rats (2.0 pmol/uterus) as compared to the estradiol treatment group (3.6 pmole/uterus) is consistent with previous reports from this laboratory demonstrating that progesterone blocks the estrogen-induced synthesis of cytoplasmic estrogen receptors 8–24 h postinjection (Hseuh et al. 1976). Apparently, dexamethasone treatment does not inhibit this phase of cytoplasmic receptor synthesis. Similarly, the antagonistic effects of dexamethasone and progesterone on nuclear type II sites and uterine growth do not appear to be the result of alterations in nuclear retention patterns of type I sites, since nuclear levels of estrogen receptor were identical at 1, 4, and 24 h following injection of estradiol, estradiol plus dexamethasone or estradiol plus progesterone.

These data suggest that the nature of dexamethasone and progesterone antagonism of uterotropic responses to estradiol is due to an inhibition of the expression of nuclear type II sites rather than an impedance of receptor–nuclear interactions, "processing," and/or cytoplasmic receptor replenish-

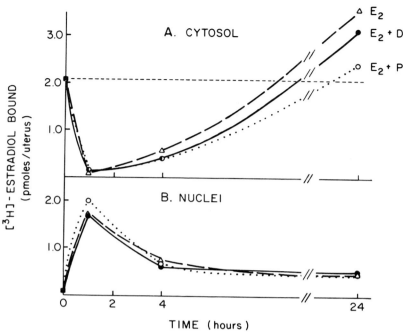

Fig. 2. Temporal effects of progesterone and dexamethasone on cytoplasmic (A) and nuclear (B) levels of nuclear type I sites. Mature ovariectomized rats were treated exactly as described in Fig. 7 and were sacrificed 1, 4, and 24 h following the second injection of estradiol ($\triangle$), estradiol + progesterone ($\bigcirc$), or estradiol + dexamethasone ($\bullet$). Cytoplasmic and nuclear levels of type I sites were estimated at the indicated time by saturation analysis utilizing the [^{3}H]estradiol exchange assay or hydroxylapatite adsorption assay. The values (pmoles/uterus) represent the quantities of type I sites in the cytosol (A) and nuclear (B) fractions corrected for the influence of type II sites by graphic analysis (Clark et al. 1979).

ment. This concept is supported by the observation that a single injection of dexamethasone 24 h prior to estradiol administration inhibited estrogen stimulation of uterine growth and nuclear type II sites even though effects on type I sites do not appear to be involved in this inhibition (Markaverich *et al.* 1981a). Apparently, antagonistic effects of progesterone on nuclear type II sites and uterine growth are dependent on estrogen pretreatment, since progesterone failed to antagonize either of these parameters in the unprimed rat uterus.

In summary, these data demonstrate that dexamethasone and progesterone inhibit the ability of estradiol to elevate nuclear type II estrogen binding sites in the rat uterus, and this is correlated with an antagonism of uterine growth. Since the nuclear binding and cytoplasmic replenishment of type I receptors are normal under these circumstances, we propose that the estrogen-induced elevation of nuclear type II sites may be involved in the mechanism by which estrogen causes uterine growth.

VI. Differential Uterine Cell Stimulation and Nuclear Levels of Nuclear Type I and II Sites

Triphenylethylene drugs, such as Nafoxidine and Clomiphene, stimulate epithelial cells of the uterus while having little effect on the stromal or myometrial components (Clark and Peck 1979). This finding plus the discovery of type II sites prompted us to examine the relationships between these two phenomena.

Mature castrate rats were implanted with estradiol, Nafoxidine or Clomiphene and epithelial, stromal, and myometrial cells were prepared as described in McCormack and Glasser (1980). Saturation analysis demonstrated that these cells contain both type I and type II estrogen-binding sites and that sustained nuclear occupancy by type I sites and the elevation in nuclear type II sites are correlated to a certain degree with the growth of the epithelium that we observed histologically (Markaverich et al. 1981b). Surprisingly, estradiol stimulation of nuclear type II sites in the uterine epithelium was approximately two-fold greater than that obtained with Nafoxidine and Clomiphene, even though histologically the triphenylethylenes were more active in this regard (Fig. 3). Thus, whether or not elevations in this second nuclear binding site for estradiol are directly proportional to the epithelial growth response remains to be established. Likewise, nuclear levels of the type I site were not directly correlated with luminal epithelial

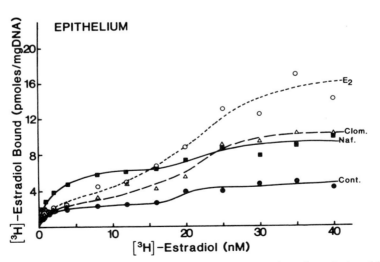

Fig. 3. Saturation analysis of uterine nuclear fractions from isolated luminal epithelial cells by [^{3}H]estradiol exchange. Epithelial cell nuclei obtained from mature ovariectomized rats were incubated with increasing concentrations of [^{3}H]estradiol (0.4–40 nM) ± 100-fold excess diethylstillbestrol (0.04–4.0 μM) at 37°C for 30 min. Specific binding in the nuclear fractions from controls (●), estradiol (○); nafoxidine (■) and clomiphene (△)-implanted (2 mg; 96 h) animals was obtained by subtraction of nonspecific binding (DES-compatible) from the total quantity of [^{3}H]estradiol bound.

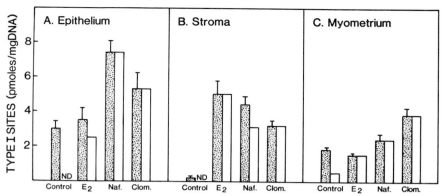

Fig. 4. Nuclear levels of type I sites in the (A) luminal epithelium, (B) stroma, and (C) myometrium of mature ovariectomized rat uterus. The total quantity of type I sites (stippled bars; unoccupied plus occupied) was determined by saturation analysis at 37°C for 30 min (see Section VII). Occupied type I sites (open bars) were determined by subtraction of the numbers of unoccupied type I sites measured by saturation analysis at 4°C for 60 min from the total quantity of estrogen receptors measured at 37°C. Mature ovariectomized rats were implanted with beeswax pellets (controls) containing 2 mg of estradiol (E_2), nafoxidine (Naf.), or clomiphene (Clom.) for 96 h prior to sacrifice.

growth. Both estradiol and Clomiphene treatment stimulated uterine growth to the greatest extent both histologically and on a wet weight basis. Yet these compounds elevated nuclear type I sites only slightly above controls. Conversely, Nafoxidine caused maximal accumulation of type I sites, was equally effective in stimulating epithelial growth, but was the least effective of the three compounds in increasing uterine wet weight. The elevated levels of type I sites in the control group appear to be due to a redistribution artifact resulting from the cell separation technique (McCormack and Glasser 1980). As shown in Fig. 4, these sites are unoccupied by hormone in control animals whereas those from the hormone-treated groups are occupied and assumed to be physiologically active receptor–hormone complexes.

Nuclear levels of both estrogen-binding sites in uterine stromal cells more closely correlated with the uterotropic responses to estradiol and the triphenylethylene derivatives (Fig. 5). Similar but more pronounced effects were observed in cellular preparations of myometrium (Fig. 6). In this tissue estradiol elevated the level of type II sites 30-fold above controls, while Nafoxidine and Clomiphene had little effect on type II sites in the myometrium. This ability of estradiol and the failure of Nafoxidine and Clomiphene to cause elevations in type II sites is highly correlated with their differential capacities to stimulate growth in the myometrium (Clark and Peck 1979; Markaverich *et al.* 1981b). Nuclear levels of type I sites in estradiol-, Clomiphene-, and Nafoxidine-treated rats were not significantly elevated above controls; however, as noted earlier, these sites in the control animals are not occupied by hormone.

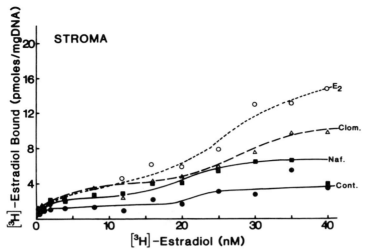

Fig. 5. Saturation analysis for estrogen-binding sites by [³H]estradiol exchange in nuclear fractons from isolated uterine stromal cells. Results are expressed as specific binding determined as described in Fig. 4 for beeswax controls (●), estradiol (○), nafoxidine (■), and clomiphene (△)-implanted rats.

From these studies we conclude that estradiol, Clomiphene, and Nafoxidine cause accumulation of type I sites in the epithelium, stroma, and myometrium of the rat uterus. Likewise, nuclear type II sites are elevated to some degree in all three tissues; however, the ability of estradiol to stimulate these sites in the myometrium greatly exceeds that of Nafoxidine and

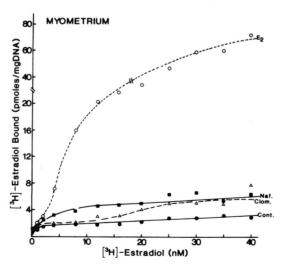

Fig. 6. Saturation analysis for estrogen-binding sites by [³H]estradiol exchange in nuclear fractons from isolated uterine myometrial cells. Results are expressed as specific binding determined as described in Fig. 4 for beeswax controls (●), estradiol (○), nafoxidine (■), and clomiphene (△)-implanted rats.

Clomiphene. This stimulation of nuclear type II sites is correlated with the agonistic properties of estradiol, while the reduced responses observed with Clomiphene and Nafoxidine are correlated with their antagonistic properties.

The differential cellular response to Nafoxidine and Clomiphene has important implications with respect to their proposed activity as "antiestrogens." Compounds such as Clomiphene, Tamoxifen, and Nafoxidine are used clinically to induce ovulation in anovulatory women and in the treatment of breast cancer. The rationale for their use is based on their definition as antiestrogens. However, the differential cellular response to the triphenylethylene derivatives requires that their activity as estrogen antagonists be reevaluated. Since these compounds have mixed agonist–antagonist activities that are related to target cell response, their definition as "antiestrogens" requires careful reassessment.

VII. Influence of Nuclear Type II Sites on Estrogen Receptor Quantitation

As stated earlier, nuclear type II sites can have profound effects on estrogen receptor measurement. For example, uterine myometrium from estradiol-implanted rats (Figs. 6) contains ~30-fold more nuclear type II sites than estrogen receptor when assayed by full saturation analysis. If, however, a single saturating concentration of [^{3}H]-estradiol is used to estimate type I sites (10 nM; Fig. 11), one can see that estrogen receptor (~2 pmol/ml DNA) would be overestimated approximately eight-fold (~16 pmol/mg DNA; 10 nM, Fig. 6) because of the influence of the type II site. Obviously, assays employing higher levels of [^{3}H]-estradiol would be more subject to this error.

Until recently quantitation of estrogen receptor in the presence of type II sites was extremely difficult and could only be achieved by full saturation analysis, as described earlier. However, we have recently described simple methods that can be used to quantitate these two sites separately based upon the susceptibility of the type II sites to reducing agents (Markaverich et al. 1981c). In these experiments uterine nuclei from estradiol-implanted rats were prepared in TE (10 mM Tris; 1.5 mM EDTA) or TED (10 mM Tris; 1.5 mM EDTA; 0.1–1.0 mM dithiothreitol) buffer and assayed for estrogen-binding sites by [^{3}H]estradiol exchange. As can be seen in Fig. 7A, addition of dithiothreitol to the buffer (TED 10 mM Tris; 1.5 mM EDTA; 0.1 mM dithiothreitol) has marked effects on the nuclear binding profile of [^{3}H]estradiol. Saturation analysis performed at 37° × 30′ in the TE buffer permits the measurement of type I and type II sites (Fig. 7A + B), whereas addition of dithiothreitol (TED) to the buffer completely eliminated [^{3}H]estradiol binding to the type II site. Therefore, only the type I site was measured in the presence of dithiothreitol. Comparison of the Scatchard plots for TE (Fig. 7B) and TED (Fig. 7C) buffers demonstrates the dramatic influence the type

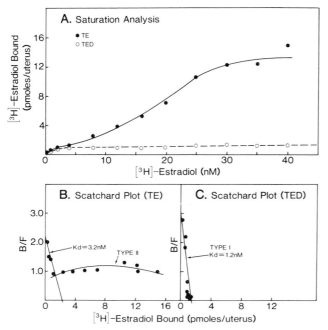

Fig. 7. Dithiothreitol effects on nuclear estrogen-binding sites. Uterine nuclei from estradiol-implanted rats (4 mg × 96 h) were prepared and assayed in TE (10 mM Tris, 1.5 mM EDTA; ●) or TED (10 mM Tris, 1.5 mM EDTA, 0.1 mM dithiothreitol; ○) buffer by saturation analysis (A). Scatchard plots of these data in TE (B) or TED (C) buffer are presented as specific [³H]estradiol binding.

II site has on receptor measurement. Scatchard analysis for type I sites in TE buffer shows that the K_d and the numbers of these sites are overestimated approximately two-fold due to the influence of the type II site as compared to the estimates obtained in the TED buffer (Fig. 7C), which measures only [³H]estradiol binding to the type I site. Identical results are also obtained with 0.1 mM monothiglycerol and iodocetamide (data not shown).

The decreased nuclear binding of [³H]estradiol in the presence of dithiothreitol (Fig. 7) suggests this reducing agent is inhibiting estrogen binding by the type II site. However, it is also possible that exposure to dithiothreitol results in extraction of the type II sites from the nuclear pellet during tissue preparation. To examine this possibility in detail, we performed saturation analysis by [³H]estradiol exchange to measure type II sites in the soluble extract and residual nuclear pellet after incubating nuclei in the presence of DTT. In these experiments, uterine nuclei from estradiol-implanted rats were prepared in TE buffer. The washed nuclear pellet was split into two equal aliquots and incubated in either TE or TED buffer (100 mg fresh tissue equivalent/ml) at 4°C for 140 min. Following incubation, the nuclear suspensions were centrifuged (800g × 10 min) and the nuclear pellets prepared in TE (nonextracted nuclei) and TED (DTT-residual nuclear fraction) buffers were assayed for nuclear estrogen-binding sites by saturation analysis. In addition, the supernatant (TED extract) obtained following centrifugation of the TED nuclear pellet was assayed for DTT-soluble type II sites by the [^{3}H]estradiol exchange–hydroxlapetite adsorption assay exactly as previously described (Markaverich et al. 1981d). The results of these experiments demonstrated that exposure of nuclei to dithiothreitol resulted in virtually a 100% reduction of [^{3}H]estradiol binding to type II sites (Fig. 8: TE, o vs TED res, •), and these sites were not recovered in the soluble nuclear extract (TED ext. []). Therefore, the inability to measure nuclear type II sites in the presence of dithiothreitol most likely results from an inhibition of [^{3}H]estradiol binding to these sites. It is unlikely that the type II site is extracted from uterine nuclei by reducing agents.

In addition, we have shown that the quantitation of nuclear type II sites requires different assay conditions than those routinely utilized for the measurement of estrogen receptor (Markaverich et al. 1981c; Anderson et al. 1972). Maximal quantities of type II sites are measured in TE buffer under

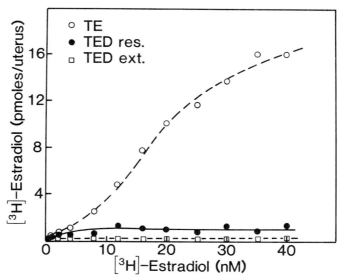

Fig. 8. Effect of dithiothreitol on extraction of nuclear type II sites. Uterine nuclei from estradiol-implanted rats were prepared in TE buffer as described in methods. Following preparation, the washed nuclear pellet was split into two aliquots that were resuspended and incubated in TE or TED buffer at 4°C for 140 min. Following incubation, the nuclear suspensions were centrifuged (800 $g \times 10$ min) to obtain the supernatant (dithiothreitol extracted fraction; TED ext., □) and nuclear pellet (dithiothreitol resistent; TED res., ●), which were assayed for estrogen-binding sites by [^{3}H]estradiol exchange (see methods). In addition, the nuclear pellet prepared in TE buffer (○) was also assayed by [^{3}H]estradiol exchange to quantitate nuclear levels of type I and type II estrogen-binding sites prior to extraction with dithiothreitol.

[^{3}H]estradiol exchange conditions (4°C × 60 min) which do not measure unoccupied estrogen receptor. We also found that 300–500-fold excess DES is an optimum concentration of competitor for measuring the type II site. Lower levels of DES will result in an overestimate of nonspecific binding and an underestimation of specific [^{3}H]estradiol binding to type II sites. These higher concentrations (300–500-fold excess DES) of competitor have no measurable adverse effects on receptor measurement and therefore can be routinely used for measurement of both nuclear estrogen-binding sites.

VIII. Effects of Nuclear Dilution on the Quantitation of Type I and Type II Sites

In a further attempt to optimize assay conditions for quantitating nuclear type I and type II sites, we evaluated the effects of nuclear dilution on [^{3}H]estradiol binding (Markaverich and Clark 1982). In these experiments uterine nuclei from estradiol-injected (10 μg × 30 min; type I) or estradiol-implanted (4 mg × 96 h; type II) rats were prepared and resuspended for assay in final volumes equivalent to 2.5–40 mg fresh tissue/ml of TE buffer. [^{3}H]estradiol binding to nuclear type I and type II sites was evaluated by saturation analysis, and the results are presented in Table 1. These data clearly demonstrate that dilution of uterine nuclei prior to analysis by [^{3}H]estradiol exchange has profound effects on measurable quantities of the nuclear type II site. Although estimates of type I sites (pmol/uterus) were identical at any nuclear dilution, measurable quantities of nuclear type II sites were increased six- to seven-fold by diluting nuclei from 40 mg/ml to 2.5–5.0 mg/ml. These results suggested uterine nuclei contain an endogenous inhibitor that interferes with [^{3}H]estradiol binding to type II sites but has no effect on the estrogen receptor. More recent experiments have demonstrated that this inhibitor is a small molecule (MW ~350) that is present in the cytosol and nuclei of target and nontarget tissues and does not appear to be regulated by

Table 1. Effects of Dilution on the Quantitation of Nuclear Type I and Type II Sites by [^{3}H]Estradiol Exchange[a]

	Estrogen-binding sites (pmol/uterus)	
Nuclear dilution	Type I	Type II
40 mg/ml	1.1	7.0
20 mg/ml	1.1	23.0
10 mg/ml	1.1	33.0
5 mg/ml	—	45.0
2.5 mg/ml	—	47.5

[a] Uterine nuclei from estradiol injected (10 mg × 30 min type I sites) or estradiol implanted (4 mg × 96 h; type II sites) were prepared in TE buffer, diluted to the indicated concentration (mg fresh tissue equivalent/ml) and assayed for type I (37°C × 30 min) or type II (4°C × 60 min) sites by saturation analysis.

estrogen. The purification and characterization of this inhibitor are currently under investigation.

In summary, these studies have shown that type I and type II estrogen binding sites in rat uterine nuclei can be separately quantitated by exchange assays employing a single saturating concentration of [³H]estradiol. Measurement of type I sites is readily accomplished by preparing nuclei in TED buffer (1.0 mM DTT) and incubation (37° × 30′) with 10–15 mM [³H]estradiol ± 100–300-fold excess DES. Nuclear type II sites can be measured in TE buffer with a single concentration of [³H]estradiol (40–80 nM ± 300-fold excess DES) under incubation conditions (4° × 60′) that do not measure occupied estrogen receptors.

In light of the nuclear dilution experiments (Table 1) we must also place additional constraints on the assay conditions used for the measurement of nuclear type II sites. Since uterine nuclei contain an endogenous inhibitor of [³H]estradiol binding to nuclear type II sites, maximal quantities of this site can only be measured when the inhibitor concentration is reduced to very low or nondetectable levels. This would appear to be in the nuclear dilution range of 2.5–5.0 mg fresh tissue equivalents/ml of TE buffer. Concentrations of uterine nuclei in excess of 5-mg equivalents fresh tissue/ml of buffer will result in an underestimate of these sites.

IX. Overview

Our current concept of estrogen action in the rat uterus involves both the binding of the hormone to type I sites and the stimulation or "activation" of nuclear type II sites. The significance of the cytoplasmic type II site is subject only to speculation, since the precise relationships between these sites and uterotropic response is unknown. We proposed earlier (Clark and Peck 1979) that perhaps cytoplasmic type II sites serve as a "sink" for estrogen within target cells and may be involved in the preferential accumulation and/or retention of estrogen. Such a hypothesis does not necessarily obviate the estrogen receptor from such a function, since type I sites bind estradiol with a higher affinity (K_d ~1 nM). However, cytoplasmic type II sites could serve to bind "excess" estrogen under conditions where the comparatively low levels of receptor sites are saturated and/or depleted from the cytosol.

Alternatively, cytoplasmic type II sites may play a protective role in uterine physiology similar to that proposed for α-fetoprotein. Throughout the course of our investigations we have observed that the neonatal (unpublished observations) and the immature (Eriksson et al. 1978) rat uterus contain very high levels of cytoplasmic type II sites relative to the quantities found in estrogen-treated adult-castrate animals (Markaverich and Clark, unpublished). Although type II sites bind estradiol with a much lower affinity (K_d ~30 nM) than the estrogen receptor, their number exceeds receptor levels by at least three- to four-fold [Eriksson et al. 1978]. Therefore, cytoplasmic type II sites could bind significant quantities of estrogen and subsequently influence the availability of hormone for receptor binding. Perhaps cytoplasmic type II sites sequester estrogen in the young animal, subsequently impeding its uterotropic effects, a function that diminishes with the onset of puberty.

The precise role of nuclear type II sites in estrogen action has also defied definition. However, indirect evidence suggests that the stimulation of these sites may be an important step in hormone action. Studies from numerous laboratories have shown positive correlations between nuclear occupancy by type I sites and uterotropic responses (Clark and Peck 1979) and these sites are undoubtedly involved in estrogen action. Likewise, we have shown that nuclear type II sites are only stimulated or "activated" by estrogens under conditions that cause true uterine growth (Markaverich and Clark 1979). In addition, in tissues that do not grow in response to estrogen, like the hypothalamus and pituitary, no type II sites can be detected (Kelner and Peck 1981). Compounds, such as dexamethasone and progesterone, that inhibit the stimulation of true uterine growth block the nuclear increase in type II sites while not interfering with nuclear estrogen receptor binding and retention (Markaverich et al. 1981a). This also appears to be the case with the triphenylethylene derivatives, such as Nafoxidine and Clomiphene. Although these compounds cause the nuclear accumulation of receptors in all uterine cell types, their agonistic properties in the luminal epithelium and stroma are associated with the "activation" of nuclear type II sites. Conversely, their antagonistic properties in the myometrium may result from the inability of Nafoxidine and Clomiphene to stimulate nuclear type II sites, even though the nucleus is continuously occupied by the estrogen receptor (Markaverich et al. 1981b). These results demonstrate there is a good correlation between the level of nuclear type II sites and uterotropic response following the administration of a variety of steroidal and nonsteroidal estrogen agonists and antagonists.

Unfortunately, these experimental results are only correlational and the precise involvement of nuclear type II sites in estrogen action remains speculative. However, these sites appear to be associated with nuclear matrix (Markaverich and Clark, unpublished) and may play a role in DNA replication similar to that proposed for the estrogen and androgen receptor by Coffey and colleagues (Barrack et al. 1977; Barrack and Coffey 1980). Whether or not nuclear type II sites bind only the ligand or the receptor–estrogen complex is unknown. However, nuclear type II sites exhibit remarkable specificity for estrogens (Markaverich et al. 1980) and do not bind the triphenylethylene derivatives (Markaverich et al. 1981). This specificity does not rule out the possibility that the differential uterine cell response to Nafoxidine and Clomiphene is related to the capacity of nuclear type II sites to recognize and bind these compounds. Our competition experiments with Nafoxidine and Clomiphene were done with whole uterine nuclei. Therefore, such experiments do not represent the response of each individual cell type. There is a good possibility that nuclear type II sites in the myometrium do not bind these triphenylethylene derivatives, whereas binding does occur in the epithelium and stroma. Alternatively, the agonistic–antagonistic properties of Nafoxidine and Clomiphene may be related to the metabolism of these compounds in the various uterine cell types. Katzenellenbogen and colleagues have shown that a similar triphenylethylene derivative, CI-628, is converted to a more polar metabolite which is bound in uterine nuclei (Katzenellenbogen et al. 1978). Perhaps the luminal epithelium and stroma are capable of converting the parent compounds to a metabolite that will "activate" nuclear type II sites, whereas the myometrium is deficient in this capability.

Therefore, although the precise relationships between estrogen receptors, nuclear type II sites, and uterine response remain an enigma, studies concerning these interactions should lead to some exciting new concepts regarding hormone action. In addition to the rat uterus, multiple binding sites for estrogens has also been described for mouse (Watson and Clark 1980) and human (Panko et al. 1981) mammary cancer, the chick oviduct (Smith et al. 1979), and the rabbit corpus luteum (Yuh and Keyes 1979). Heterogeneity in hormone-binding sites has also been observed for glucocorticoids (Barlow et al. 1979; Do et al. 1979) and progesterone (Giannopoulos and Munowitz 1980), demonstrating that this is a general phenomenon. Consequently, these secondary hormone-binding sites may be an important component in steroid hormone action.

References

Anderson JN, Clark JH, Peck EJ, Jr. (1972) Biochem J 126: 561–567

Anderson JN, Peck EJ, Jr., Clark JH (1975) Endocrinology 96: 160–167

Barlow JW, Kraft N, Stockigt JR, Funder JW (1979) Endocrinology 105: 1055–1063

Barrack ER, Hawkins EF, Allen SL, Hicks LL, Coffey DS (1977) Biochem Biophys Res Commun 79: 829–836

Barrack ER, Coffey DS (1980) J Biol Chem 255: 7265–7275

Clark JH, Paszko Z, Peck EJ, Jr. (1977) Endocrinology 100: 91–96

Clark JH, Markaverich B, Upchurch S, Eriksson H, Hardin JW, (1979) In Steroid Hormone Receptor Systems. Leavitt, WW, Clark, JH, (eds.) Plenum Press, New York, pp 17–46

Clark JH, Hardin JW, Upchurch S, Eriksson H (1978) J Biol Chem 253: 7630–7634

Clark JH, Peck EJ, Jr (1979) In: Female Sex Steroids: Receptors and Function. Springer-Verlag, Berlin

Eriksson H, Upchurch S, Hardin JW, Peck EJ, Jr, Clark JH (1978) Biochem Biophys Res Commun 81: 1–7

Giannopoulos G, Munowitz P (1980) Abstract #7, 62nd Annual Endocrine Meeting, Washington, DC

Gorski J, Gannon F (1976) Ann Rev Physiol 38: 425–450

Hardin JWH, Clark JH, Glasser SR, Peck EJ, Jr. (1976) Biochemistry 17: 3146–3152

Hseuh AJW, Peck EJ, Jr., Clark JH (1976) Endocrinology 98: 438–444

Huggins C, Jensen EV (1955) J Exptl Med 102: 335–346

Katzenellenbogen BS, Katzenellenbogen, JA, Ferguson ER, Krauthammer N (1978) J Biol Chem 255: 697–707

Kelner KL, Peck EJ, Jr. (1981) J Receptor Res (in press)

Leavitt WW, Toft DO, Strott CA, O'Malley BW (1974) Endocrinology 94: 1041–1053

Lerner JL (1964) Rec Progr Horm Res 20: 435–490
McCormack SA, Glasser SR (1980) Endocrinology 106: 1634–1649
Markaverich BM, Clark JH (1979) Endocrinology 105: 1458–1462
Markaverich BM, Clark JH (1982) J Steroid Biochem (in press)
Markaverich BM, Clark JH, Hardin JW (1978) Biochemistry 17: 3146–3152
Markaverich BM, Upchurch S, Clark JH (1980) In: Perspectives in Steroid Receptor Res, Bresciani, F (ed), Raven Press, New York, pp 143–164
Markaverich BM, Upchurch S, Clark JH (1981a) J Steroid Biochem 14: 125–132
Markaverich BM, Upchurch S, McCormack SA, Glaser SR, Clark JH (1981b) Biol Repr 24: 171–181
Markaverich BM, Williams M, Upchurch S, Clark JH (1981c) Endocrinology 109: 62–69
Markaverich BM, Upchurch S, Clark JH (1981d) J Receptor Res 1: 415–438
Martucci C, Fishman J (1977) Endocrinology 101: 1709–1715
Milgrom E, Thi L, Atger M, Baulieu EE (1973) J Biol Chem 248: 6366–6377
O'Malley BW, Means AR (1974) Science 183: 610–620
Panko WB, Watson CS, Clark JH (1981) J Steroid Biochem (in press)
Stormshak F, Lake R, Wertz N, Gorski J (1976) Endocrinology 99: 1501–1511
Szego CM, Roberts S (1953) Rec Progr Horm Res 8: 419–464
Smith RG, Clarke SG, Zalta E, Taylor RN (1979) J Steroid Biochem 10: 31–38
Walters MR, Clark JH (1978) Endocrinology 98: 601–609
Watson CS, Clark JH (1980) J Recep Res 1: 91–111
Yuh K-C, Keyes PL (1979) Endocrinology 105: 690–696

Discussion of the Paper Presented by J. H. Clark

SHAPIRO: Would you care to give us your thoughts as to what extent the class II binding you see is due to a single entity associating with different cellular components or to actual different proteins?

CLARK: No.

ROY: In your talk you often used the terms "binding site" and "receptor" in an interchangeable manner. Would you care to distinguish them or redefine these terms?

CLARK: The term "receptor site" implies that the binding of an appropriate ligand to this site will elicit a biological response. "Binding site" does not carry this connotation. More specifically, we think of estrogen receptors as macromolecules that manifest the following properties: high-affinity binding of estrogenic compounds, low capacity, i.e., a small number of sites per cell ($\sim$5–20K), and nuclear accumulation via cytoplasmic translocation. Cytoplasmic type II sites also bind estrogenic compounds specifically; however, the number of sites per cell is higher and these sites do not undergo nuclear translocation. Therefore, it is rather easy to distinguish between type I and II sites. Nuclear type II sites generally have a much lower affinity (10–30) for estrogenic compounds and their number/cell varies. They can be readily distinguished because exposure to sulfhydril reagents such as (DTT) will block the binding of [^{3}H]estradiol to type II sites but not type I sites. The link between receptor estrogen binding and the stimulation of nuclear type II binding sites has not been established; however, we feel that such interactions are involved in the biosynthetic responses that culminate in cellular growth.

ROY: The problem is you may be confusing many of us by calling the estrogen receptor the "type I site" and the nonreceptor "type II sites." Can you think of a better term to use. You may want to call the type I as receptor and the type II as Clarkophilin or whatever?

CLARK: Clarkophilin wouldn't go, and obviously Peckophilin would even be worse. Markaverichophilin, that would be too much. We didn't want to give them silly names like A and B, so for now we are left without a suitable name.

GREENE: Physiologically, under in vivo condition the levels of estradiol in the serum are low, and as you said, you need a very-high-affinity binder for them, that is the binder would be the classical estrogen receptor. So how is it that enough estradiol gets to the type II site to do anything, particularly when its affinity is at least an order of magnitude lower than the classical estrogen receptor?

CLARK: The function of the RE complex could be to concentrate estradiol in the nucleus. Such a concentration mechanism may bring intranuclear levels of estradiol well within the operating limits of nuclear type II sites. Another consideration is that estradiol may not be the endogenous ligand. We simply use [^{3}H]estradiol in our assays, and type II sites obviously have some affinity for this ligand, but the affinity for the endogenous ligand could be much higher.

PECK: I'd like to add to that. The interaction of a ligand with a binding site is dependent on two things: the affinity of a given binding site and the number of those sites. If you have 10 times more type II site than type I, then, in fact, occupied type II will exist for a given concentration of steroid in the nucleus. In some of the hyperestrogenized systems that Jim showed, there's easily 10–20 times more type II than type I. Under those circumstances, type II will be occupied, at least theoretically.

GREENE: I can believe that, but that's a manipulated system too.

CLARK: It is very difficult to show in vivo occupancy and perhaps nuclear type II sites are not occupied by estrogen as I just mentioned. However, we do know that they are elevated under totally physiological circumstances.

LIAO: Jim, I just want to make a comment. When Dr. K.M. Anderson was in my lab in 1967, he did exactly the same type of experiment you did, by taking the rat ventral prostate, incubating it with different concentrations of radioactive androgen. He obtained exactly the same type of curve you showed. At low concentrations of androgen the receptor was saturated by the prostate and continued to take up more androgen as the concentrations of androgen in the medium increased.

CLARK: You mean you were not crazy enough to publish it.

LIAO: Well, Dr. Anderson included it in his Ph.D. dissertation (University of Chicago 1969). Another point I wish to make is that, in vivo, only about 5–10% of the nuclear androgen is not extractable by 0.4 M KCl solution. If the nuclear retention of the radioactive receptor complex was carried out in the cell-free system, more than 30% of the nuclear radioactive androgen was in the salt-resistant fraction. The number of these salt-resistant binding sites increased considerably if nuclei were sonicated or heated at 40°C or higher temperatures (S. Fange and S. Liao, J. Biol. Chem., 246, 16–24, 1971). These salt-resistant sites appeared to be very similar to type II nuclear sites you mentioned. Also, the protein responsible for your cytosol type II binding may be similar to α-protein of the rat ventral prostate. α-protein is the major secretory protein that binds phospholipid as well as androgens, cholesterols, and other steroids (see Chapter 16) with low affinity. I wonder if your type II cytosol binding is due to a major low-affinity, high-capacity uterine protein.

CLARK: The cytosol type II site is very specific for estrogenic structures; therefore it doesn't seem likely that it would bind cholesterol.

LANDAU: Do you have a hypothesis as to why the implant of estriol will stimulate type II sites, whereas the injection of the hormone will not—specifically, since the estriol isn't required to have the implanted form to produce the same effect?

CLARK: The hypothesis is a simple one, based upon nuclear retention time; if you keep the hormone around in the case of estriol, it occupies the receptor and keeps it in the nucleus; this prolonged nuclear occupancy is sufficient to stimulate the secondary sites. If you give it by injection it's cleared rapidly and does not stimulate these sites. So it's just a matter of nuclear occupancy, we think.

LANDAU: Do you think it's a nuclear exchange of estriol, that the receptor is staying in there, because estriol is exchanged?

CLARK: No. These animals are exposed only to estriol. They are castrated animals and there is very little estradiol present.

LANDAU: I realize that. I am referring to the prolonged retention of estriol after the implant in the nuclear site; do you think that it is because new receptor complexes are going into the nucleus?

CLARK: Oh, I don't know. We can't evaluate that. I would suspect that it is a kinetic bombardment of those sites, yes. I didn't realize that was your question.

Discussants: D.J. SHAPIRO, J.H. CLARK, A.K. ROY, G.L. GREENE, E.J. PECK, S. LIAO and T. LANDAU

Chapter 15

Hormone Receptors and the Nuclear Matrix

EVELYN R. BARRACK AND DONALD S. COFFEY

I. Introduction

The induction by steroid hormones of a wide variety of biological actions has been visualized to result from the interaction of steroids with specific cytoplasmic receptor proteins and their translocation into the nucleus (see reviews by Gorski and Gannon 1976; Thrall et al. 1978; Jensen 1979; Liao et al. 1979). In spite of the progress that has been made to elucidate various aspects of these events, there remain many unknowns in our understanding of how hormones actually bring about their effects. New concepts and techniques may be needed to provide new insights. In this regard, recent studies of the binding of receptor hormone complexes to specific gene sequences promises to yield fruitful new information (Mulvihill et al. 1982). At present it appears that steroid receptor complexes may bind both to DNA and to nuclear proteins. While insight is being gained on possible DNA sequences to which the steroid receptor complex may bind, there is little agreement on the identification of specific nuclear proteins that may interact with the steroid–receptor complex. These nuclear proteins have been collectively referred to as acceptors, but their exact location and nature have not been fully resolved. We have chosen to focus on nuclear structural proteins as important elements in hormone action, for several reasons. First, there is a growing belief that much of cellular structure and function may be governed by specific arrays of structural proteins termed the cytoskeleton; in the nucleus, a similar concept has been developed, with the skeleton structure being referred to as the nuclear matrix (see reviews by Shaper et al. 1979; Porter and Tucker 1981; Isaacs et al. 1981; Barrack and Coffey 1982). Second, evidence is accumulating to suggest that the nuclear matrix may play a fundamental role in many dynamic aspects of nuclear function, including DNA replication and heterogeneous nuclear RNA synthesis (see Table 1 and reviews by Berezney and Coffey 1976; Shaper et al. 1979; Berezney 1979; Agutter and Richardson 1980; Barrack and Coffey 1982). Since steroid hormones affect these processes in specific target tissues, it is of interest that hormones interact in a specific manner with the nuclear matrix (Barrack et al. 1977, 1979; Barrack and Coffey 1980, 1982).

In the following sections, we briefly describe the nuclear matrix and review our findings on the association of specific steroid-binding sites with this structure. We also discuss our data in the context of other information on nuclear receptors and hormone action.

II. The Nuclear Matrix

The nuclear matrix is an insoluble, skeletal framework of the nucleus. This underlying structure is revealed following extraction from the nucleus of the nuclear membrane phospholipid and the chromatin. This is generally achieved by a series of sequential extractions with a nonionic detergent (Triton X-100), brief digestion with DNase I, a hypotonic buffer containing very low concentrations of divalent cations (0.2 mM MgCl$_2$), and hypertonic salt buffer (e.g., 2 M NaCl). The resulting residual structure, the nuclear matrix, resembles the nucleus in size and shape and consists of a peripheral lamina that forms a continuous structure surrounding the nuclear sphere and represents a residual component of the nuclear envelope; residual nuclear pore complexes that are embedded in the lamina; an internal fibrogranular protein- and RNA-containing network; and residual nucleoli. The proteins of the matrix represent only about 10% of the total nuclear protein mass and comprise a subset of nonhistone proteins. The matrix proteins are primarily, though not exclusively, in the molecular weight range of 40,000 to 75,000. A great deal of effort is being made in a number of different laboratories to elucidate the nature of these proteins, how they are organized and interact with each other, and their specific intranuclear localization. There is much interest in elucidating their fate during mitosis and whether they form the core scaffolding of the metaphase chromosomes (Detke and Keller 1982; Bekers et al. 1981; Gerace and Blobel 1980).

The nuclear matrix is a universal feature of eukaryotic nuclei; such residual structures have now been isolated from a wide variety of mammalian and nonmammalian sources. Different investigators have referred to the nuclear matrix in various ways, including nuclear framework, nuclear skeleton, nuclear scaffold, nuclear ghost, nuclear cage, or chromatin-depleted nucleus. All these terms refer to the same structure. There have also been numerous modifications of the original procedure described for isolating the nuclear matrix, and it is important to recognize that many factors can affect the nature of the final product (e.g., see Kaufmann et al. 1981). Details of the isolation and characterization of the nuclear matrix have been reviewed recently (Barrack and Coffey 1982). Because the nuclear matrix has been isolated or visualized directly following relatively drastic manipulations in vitro (e.g., 2 M NaCl), the question of the existence in vivo of a nuclear matrix structure can be raised. In addition, the insoluble nature of the matrix proteins raises questions concerning denaturation artifacts. These issues have prompted the use of alternative methods for elucidating the nature and existence of the nuclear matrix in situ (Berezney and Coffey 1976, 1977;

Berezney 1979; Ghosh et al. 1978; Brasch and Sinclair 1978; Goldfischer et al. 1981).

We have concentrated our efforts on determining if the nuclear matrix is associated with important biological properties. Evidence has accumulated over the past several years that the nuclear matrix is not simply a static structure, but rather is a residual scaffolding system that has dynamic properties and is intimately associated with such fundamental nuclear processes as DNA organization, DNA replication, heterogeneous nuclear RNA (hnRNA) synthesis and processing, and hormone action (see Table 1). These findings suggest that many important nuclear events occur not in solution but rather in association with insoluble structural components, firmly bound to the nuclear matrix. The nuclear matrix appears to provide ordering and organization of complex processes heretofore visualized as soluble systems within the nucleoplasm. For example, the proposal that the DNA of the eukaryotic nucleus is organized in the form of supercoiled loops that are anchored at their bases to the nuclear matrix (see Table 1) and that during DNA replication these loops of DNA are reeled through fixed replication sites on the nuclear matrix, where the DNA is replicated (Pardoll et al. 1980), provides an explanation for the mechanism by which an enormous length of DNA must be ordered spatially during replication such that the daughter strands remain untangled yet coupled in a precise fashion for later entry into mitosis. In addition, as indicated in Table 1, the demonstration that newly synthesized hnRNA and its processing intermediates are associated with the insoluble nuclear matrix and that certain actively transcribed genes are preferentially associated with the nuclear matrix has done much to further our understanding of the role of structural organization in cellular function. That the nuclear matrix appears to be a major site of steroid hormone receptor binding in the nucleus is consistent with the role of steroid hormones in stimulating hnRNA synthesis, which itself appears to occur in association with the nuclear matrix.

Table 1. Biological Functions Associated with the Nuclear Matrix

1. *The nuclear matrix contains structural elements of the pore complex, lamina, internal network, and nucleolus* (Berezney and Coffey 1974, 1976, 1977; Comings and Okada 1976; Wunderlich and Herlan 1977; Hodge et al. 1977; Fisher et al. 1982; also see review by Barrack and Coffey 1982).
2. *DNA binding proteins in the nuclear matrix*
   a. DNA is tightly attached to the nuclear matrix (Berezney and Coffey 1975, 1976, 1977; Pardoll et al. 1980).
   b. Tenacious binding of DNA to specific nuclear matrix proteins—Bowen et al. 1980; Razin et al. 1981.
   c. Mouse liver nuclear matrix proteins have a high affinity for DNA and show a preference for binding single-stranded DNA, AT-rich DNA and poly(dT) (Comings and Wallack 1978).
3. *Role in DNA organization*
   a. DNA is organized in the nucleus in the form of supercoiled loops, each containing 60,000–100,000 bp per loop. These loops are anchored at their

Table 1 (continued)

bases to the nuclear matrix (Vogelstein et al. 1980; Georgiev et al. 1978; Cook et al. 1976; Razin et al. 1979).

    b.   In bovine kidney cells, 60% of the tightly attached matrix-associated DNA is tandemly repeated satellite DNA (Matsumoto 1981).

    c.   Repeated DNA sequences are associated with the nuclear matrix in a specific, nonrandom fashion (Small et al. 1982).

4. *Role in DNA replication*

    a.   Newly replicated DNA is associated preferentially with the nuclear matrix (Berezney and Coffey 1975, 1976; Dvorkin and Vanyushin 1978; Dijkwel et al. 1979; Pardoll et al. 1980; Vogelstein et al. 1980; McCready et al. 1980; Hunt and Vogelstein 1981; Berezney and Buchholtz 1981a, b).

    b.   The nuclear matrix contains fixed sites of DNA replication (Pardoll et al. 1980; Vogelstein et al. 1980; Hunt and Vogelstein 1981; Berezney and Buchholtz 1981a).

    c.   DNA polymerase α is tightly bound to the nuclear matrix of actively replicating liver but not of normal liver (Smith and Berezney 1980).

5. *Enrichment of certain actively transcribed genes with the nuclear matrix*

    a.   Ribosomal RNA genes are enriched at least six-fold in the residual DNA associated with liver nuclear matrix (Pardoll and Vogelstein 1980).

    b.   SV40 sequences are enriched four- to seven-fold in nuclear matrix-associated DNA of SV40 transformed 3T3 cells (Nelkin et al. 1980).

    c.   The transcriptionally active ovalbumin gene is preferentially associated with the nuclear matrix in hen oviduct, whereas the inactive β-globin gene is not enriched on the oviduct nuclear matrix (Robinson et al. 1982).

6. *Association of hnRNA with the nuclear matrix*

    a.   RNP particles are a component of the internal network of the nuclear matrix (Berezney and Coffey 1974, 1976, 1977; Miller et al. 1978a; Berezney 1979, 1980; Steele and Busch 1966; Pogo 1981; Maundrell et al. 1981).

    b.   Essentially all of the hnRNA and snRNA are associated exclusively with the nuclear matrix (Miller et al. 1978a, b; Herman et al. 1978; Long et al. 1979; van Eekelen and van Venrooij 1981; Jackson et al. 1981; Mariman et al. 1982).

    c.   Globin RNA, containing introns, is tightly associated with the nuclear matrix of chicken erythroblasts (Ross 1980). Precursor mRNAs for ovomucoid and ovalbumin are on the nuclear matrix of hen oviduct (Ciejek et al. 1982).

    d.   hnRNA is attached to the nuclear matrix *via* two of the major hnRNP proteins (van Eekelen and van Venrooij 1981).

    e.   RNA is synthesized at fixed transcription complexes on the nuclear matrix (Jackson et al. 1981).

    f.   Nascent RNA is attached at the 5' cap, and perhaps also at the 3' end, to the nuclear matrix (Jackson et al. 1981).

    g.   Processing of hnRNA to mRNA occurs in association with the nuclear matrix (Mariman et al. 1982).

7. *Interaction with viruses*

    a.   Certain SV40 specific proteins are associated exclusively with the nuclear matrix of infected HeLa cells (Deppert 1978).

    b.   Significant amounts of polyoma T-antigen and intact viral genomes are associated with the nuclear matrix of lytically infected 3T6 cells; the nuclear matrix is implicated as a site of viral DNA replication (Buckler-White et al. 1980).

Table 1 (continued)

8. *Phosphorylation of matrix and lamina proteins*
   a. Increased phosphorylation of specific rat liver nuclear matrix proteins occurs just prior to the onset of DNA synthesis in regenerating rat liver (Allen et al. 1977).
   b. Specific proteins of clam nuclear lamina–pore complex fraction are phosphorylated by endogenous nuclear envelope-associated enzymes (Maul and Avdalović 1980).
   c. Protein phosphokinase activity in the pore complex–lamina fraction of rat liver phosphorylates endogenous pore complex–lamina proteins (Steer et al. 1980; Lam and Kasper 1979).
   d. The three major nuclear lamina proteins of the matrix are phosphoproteins and are phosphorylated to a greater extent prior to or during mitosis than in interphase. Phosphorylation of the major lamina proteins appears to be involved in the reversible depolymerization of the lamina during cell division (Gerace and Blobel 1980).
   e. Nuclear matrix proteins of sea urchin embryos are phosphorylated by endogenous matrix kinase activity in vitro; more phosphorylation occurs in the blastula than in the pluteus stage (Sevaljević et al. 1981).

9. *Reversible expansion/contraction of the matrix*
   The nuclear matrix isolated from *Tetrahymena* macronuclei reversibly contracts when Ca^{2+} and Mg^{2+} concentration is decreased to 5 mM or increased to 125 mM (Wunderlich and Herlan 1977).

10. *Modulation of nuclear membrane lipid fluidity by internal components of the nuclear matrix*
    Lipid fluidity is higher in isolated nuclear membrane ghosts than in intact nuclei (Giese and Wunderlich 1980; Wunderlich et al. 1978).

11. *Contains binding sites for steroid hormones, EGF[a], lectins, and polyribonucleotides*
    a. Specific, high-affinity, tissue- and steroid-specific receptors for androgens and estrogens are associated with the nuclear matrix of target tissues (rat uterus, chicken liver, rat prostate) (Barrack et al. 1977, 1979; Barrack and Coffey 1980, 1982; Agutter and Birchall 1979).
    b. Lentil binding sites are found on the nuclear matrix of sea urchin embryos (Sevaljević et al. 1981).
    c. WGA binds to the internal structure and conA binds to the periphery of HeLa cell nuclear scaffolds (Hozier and Furcht 1980).
    d. Poly(A) binds to the nuclear lamina (McDonald and Agutter 1980).
    e. Binding sites for EGF are found in chromatin-depleted nuclei (Johnson et al. 1980).

12. *Preferential binding of carcinogens to the nuclear matrix*
    a. Benzo[*a*]pyrene binds to the nuclear matrix of rat lung, rat liver, and thymocytes (Hemminki and Vainio 1979; Blazsek et al. 1979; Ueyama et al. 1981).
    b. Retinol markedly inhibits binding of benzo[*a*]pyrene to DNA and protein of the rat liver nuclear matrix but has little or no effect on binding of benzo[*a*]pyrene to bulk DNA and protein in chromatin (Nomi et al. 1981).

13. *Heat shock proteins become associated with the nuclear matrix*
    Heat shock proteins may be structural components of the nucleus, serving to protect the cell from effects of heat shock (Levinger and Varshavsky 1981; Sinibaldi and Morris 1981).

[a] EGF, epidermal growth factor; WGA, wheat germ agglutinin; conA, concanavalin A.

III. The Binding of Steroids to the Nuclear Matrix

A. Background

When we began our studies to investigate the role of the nuclear matrix in steroid hormone action, there already existed evidence for nuclear steroid receptors being associated with salt-resistant nuclear components, although these salt-insoluble fractions had not be characterized or identified. Although nuclear steroid receptors have most often been characterized as being extractable from nuclei with 0.3–0.6 M KCl or NaCl (Jensen et al. 1968; Puca and Bresciani 1968; Bruchovsky and Wilson 1968), in many different target tissue systems a significant amount of steroid-binding activity has been noted to remain resistant to solubilization by salt (Fang et al. 1969; DeHertogh et al. 1973; Lebeau et al. 1973; Mester and Baulieu 1975; Klyzsejko-Stefanowicz et al. 1976; Nyberg and Wang 1976; Clark and Peck 1976; Barrack et al. 1977, 1979; Honma et al. 1977; Boesel et al. 1977; Davies et al. 1977; Ruh and Baudendistel 1977; Snow et al. 1978; Gschwendt and Schneider 1978; Danzo and Eller 1978; Kaufman et al. 1978; Wang 1978; Franceschi and Kim 1979; Sanborn et al. 1979; Sato et al. 1979; Barrack and Coffey 1980, 1982; Cidlowski and Munck 1980; Tsai et al. 1980; Brown et al. 1981).

The work of several investigators has provided evidence that salt-resistant nuclear steroid receptors represent a physiologically meaningful compartment of nuclear receptors. For example, DeHertogh et al. (1973) proposed, on the basis of in vivo infusion studies of [^{3}H]estradiol labeling of rat uterine nuclei, that the salt-resistant nuclear fraction represented the ultimate site of hormone localization in the nucleus. Clark and Peck (1976) later demonstrated that estrogen-induced long-term growth of the rat uterus, which involves both hypertrophy and hyperplasia, depends on the specific interaction of estradiol–receptor complexes with nuclear salt-resistant binding components. Using an in vitro [^{3}H]estradiol exchange assay (Anderson et al. 1972) to quantitate the amount of hormone that had become bound to these sites as a result of in vivo processes of hormone–receptor complex translocation, Clark and Peck (1976) observed that the number of salt-resistant nuclear estradiol receptors was identical with the number of receptors required for maximal uterine growth, and they proposed that these binding sites represent specific nuclear acceptor sites. Similar observations were made by Ruh and Baudendistel (1977), who in addition observed that treatment of immature rats with estradiol results in the appearance of both salt-extractable and salt-resistant nuclear estradiol receptors in the uterus, whereas treatment with antiestrogens, which stimulate only limited uterine growth, results in the appearance of only salt-extractable uterine nuclear receptors (see also Jordan et al. 1977). Ruh and Baudendistel (1978) concluded that the salt-resistant nuclear estradiol binding sites may be involved primarily in events that result in the replenishment and processing of receptors, an event required for a continued growth response to estrogen stimulation.

Nuclear salt-resistant steroid hormone receptors have also been implicated in the mechanism of induction by estrogens of Leydig cell tumors in mice, since in strains that are susceptible to tumor induction, the Leydig cells contain salt-resistant nuclear receptors, whereas Leydig cells of resistant mice contain only salt-extractable nuclear estradiol receptors (Sato et al. 1979). The presence of nuclear salt-resistant receptors for glucocorticoids in dexamethasone-sensitive mouse myeloid leukemic cells but not in certain clones of dexamethasone-resistant cells (Honma et al. 1977) further suggests that nuclear salt-resistant steroid hormone receptors may be involved in the responsiveness of normal and neoplastic cells to specific steroid hormones.

Other evidence that nuclear salt-resistant hormone-binding sites play an important role in the mechanism of action of hormones has come from yet a different approach. In cell-free binding experiments, high-affinity, tissue-specific nuclear acceptor sites for isolated steroid hormone–receptor complexes have been described in residual fractions of chromatin that resist dissociation by 0.5–2 M NaCl or 4–5 M GuHCl. These salt-resistant acceptor sites have been characterized for androgen receptors in the rat prostate (Wang 1978), testis (Klyzsejko-Stefanowicz et al. 1976), and Sertoli cell (Tsai et al. 1980) and for progesterone receptors in the chick oviduct (Thrall et al. 1978).

Thus, by three different experimental approaches—labeling in vivo, exchange in vitro, and cell-free reconstitution—nuclear salt-resistant hormone binding sites have been implicated to be a significant intranuclear site of hormone action. Any yet, while some investigators have questioned the true nature of the nuclear matrix (see earlier discussion), others have expressed doubt on the existence of salt-resistant nuclear receptors. However, many of these discrepant interpretations can be explained on the basis of methodological differences (for further details see Barrack et al. 1977; Barrack and Coffey 1982).

B. Properties of Steroid Hormone-Binding Sites on the Nuclear Matrix

We have identified and characterized specific sex steroid binding sites associated with the nuclear matrix of both estrogen- and androgen-responsive tissues. The properties of these specific binding sites for estradiol on the nuclear matrix of rat uterus and chicken liver, and for dihydrotestosterone (DHT) on the nuclear matrix of rat ventral prostate, have all the characteristics of bonafide steroid hormone receptors. The binding of these steroids to the nuclear matrix is saturable, high affinity (K_d ~1 nM), and heat and Pronase sensitive. These binding sites exhibit steroid specificity, show tissue specificity, and occur only in response to an appropriate hormonal stimulus. More detailed information may be found in previous publications (Barrack et al. 1977, 1979; Barrack and Coffey 1980, 1982).

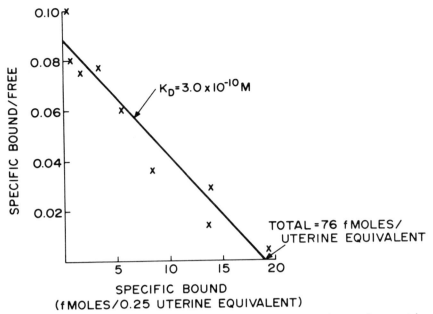

Fig. 1. Specific, high-affinity binding of estradiol to rat uterine nuclear matrix. One hour after the injection of 0.1 μg unlabeled estradiol to immature female rats, uterine nuclei were isolated and extracted to yield the nuclear matrix (see Barrack et al. 1977). The final nuclear matrix pellet was washed in 10 mM Tris, pH 7.4, 1.5 mM EDTA; estradiol binding was measured in vitro by saturation analysis in the presence of increasing concentrations of [^{3}H]estradiol (70 pM to 14 nM) without (total binding) or with a 100-fold excess of unlabeled estradiol (nonspecific binding). Free steroid was removed by washing; specific binding was calculated by subtracting nonspecific from total binding, and plotted by the method of Scatchard (1949).

Scatchard plots are shown of the specific, high-affinity binding of [^{3}H]estradiol to the isolated nuclear matrix fractions of uteri from estrogenized rats (Fig. 1) and of livers from estrogen-treated chicks (Fig. 2) and egg-laying hens (Fig. 3). Figure 4 demonstrates the in vitro binding of [^{3}H]DHT to the nuclear matrix of ventral prostates from intact adult male rats.

The steroid specificities of these sites are characteristic of estrogen and androgen receptors. The specific binding of [^{3}H]estradiol to the nuclear matrix of estrogen target tissues such as the avian liver is inhibited by unlabeled estrogens but not by androgens, progestins, or cortisol (Fig. 5). Similarly, the binding of [^{3}H]DHT to the nuclear matrix of androgen target tissues (rat ventral prostate) is inhibited by androgens, but 100–1000-fold less effectively by other classes of steroids (Fig. 6).

The appearance of specific steroid-binding sites with the nuclear matrix occurs in response to an appropriate hormonal stimulus, not indiscriminately. Thus, for example, the liver of egg-laying hens, in response to high blood levels of estrogen, synthesizes vitellogenin (the precursor of the major egg yolk proteins); the liver nuclear matrix of hens binds about 42 fmol estradiol/100 μg starting nuclear DNA equivalents. In contrast, the liver nuclear matrix of untreated roosters, which do not produce yolk proteins, contains only one-eighth as many specific binding sites for estradiol (5.5 fmol/100 μg nuclear DNA equivalents) as that of the laying hen (Fig. 7). However, the administration of pharmacological doses of estrogen to roosters or immature chicks results in a stimulation of vitellogenin mRNA synthesis (Deeley et al. 1977) and a marked increase (12-fold) in the number of nuclear matrix-associated specific estradiol binding sites (65 fmol/100 μg nuclear DNA equivalents; Fig. 8).

Similarly, in the ventral prostate, the growth and functions of which are androgen-dependent, the presence of DHT receptors on the nuclear matrix is specifically associated with androgen stimulation of this gland. These binding sites are present in the prostate nuclear matrix of intact adult male rats; but following withdrawal of androgen (castration), there is a rapid loss (within 24 h) of these binding sites from the nuclear matrix that precedes the

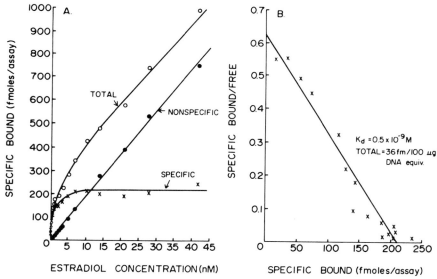

Fig. 2. Binding of estradiol to liver nuclear matrix of DES-treated chicks. Immature chicks were treated with diethylstilbestrol (25 mg/kg) and killed 20 h later. Livers were perfused with 0.9% NaCl, and purified nuclei were isolated by centrifugation through 2.2 M sucrose, 10 mM Tris, pH 7.4, 5 mM MgCl$_2$, as described by Berezney and Coffey (1977). The nuclear matrix was isolated from these purified liver nuclei, and estradiol binding was quantitated by an in vitro assay (see Barrack and Coffey 1980). Each assay tube contained 165 μg nuclear matrix protein that was derived from an amount of nuclei equivalent to 580 μg of starting nuclear DNA. (A) Total binding (O——O) was measured in the presence of [^{3}H]estradiol alone (70 pM to 42 nM). Nonspecific binding (●——●) was measured in the presence of [^{3}H]estradiol plus unlabeled estradiol (0.7 μM). Specific binding (×——×) is the difference between total and nonspecific binding. (B) Scatchard plot of specific binding (×——×) in A. Total binding capacity of the nuclear matrix was 210 fmol/assay. [From Barrack and Coffey (1980).]

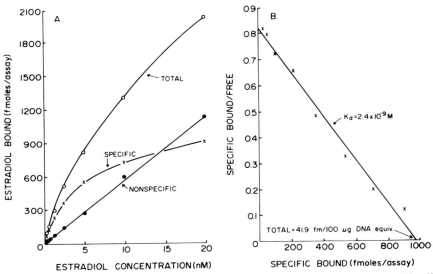

Fig. 3. Binding of estradiol to liver nuclear matrix of egg-laying hens. The liver nuclear matrix was isolated from egg-laying hens and assayed in vitro for specific [³H]estradiol binding activity, as described in the legend to Fig. 2.

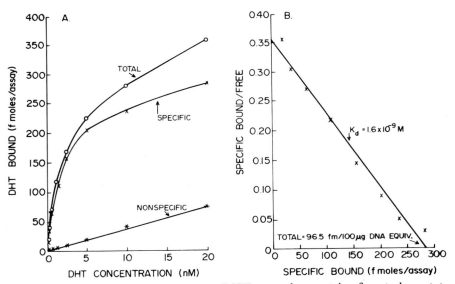

Fig. 4. Binding of dihydrotestosterone (DHT) to nuclear matrix of ventral prostates of intact adult rats. The nuclear matrix was isolated from ventral prostate nuclei (Barrack and Coffey 1980), and [³H]DHT binding to the nuclear matrix was assayed in vitro. Each assay tube contained nuclear matrix derived from nuclei equivalent to 295 μg nuclear DNA. Phenylmethylsulfonyl fluoride (1 mM) was added at all steps. (A) O——O, total binding in the presence of [³H]DHT only. *——* nonspecific binding in the presence of [³H]DHT with unlabeled DHT. ×——×, specific binding represents the difference between total and nonspecific binding. (B) Scatchard plot of specific binding (×——×) in A. [From Barrack and Coffey (1980).]

Fig. 5. Steroid specific inhibition of estradiol binding to chick liver nuclear matrix. Liver nuclear matrix, isolated from the liver nuclei of DES-treated chicks, was incubated in vitro in the presence of [^{3}H]estradiol (7 nM) alone (total binding) or with increasing concentrations of unlabeled steroids (7 nM to 7 μM; non-displaceable binding). Specific binding of [^{3}H]estradiol measured in the presence and absence of a 100-fold excess of unlabeled estradiol (0.7 μM) was set equal to 100%. DHT, 5α-dihydrotestosterone, DES, diethylstilbestrol. [From Barrack and Coffey (1980).]

Fig. 6. Steroid-specific inhibition of DHT binding to rat ventral prostate nuclear matrix. Ventral prostate nuclear matrix from intact adult rats was incubated in vitro in the presence of [^{3}H]DHT (20 nM) with or without increasing concentrations of unlabeled steroids (10 nM to 20 μM). Specific binding of [^{3}H]DHT measured in the presence and absence of a 100-fold excess of unlabeled DHT (2 μM) was set equal to 100%. R 1881 (methyltrienolone) is a potent synthetic androgen. [From Barrack and Coffey (1980).]

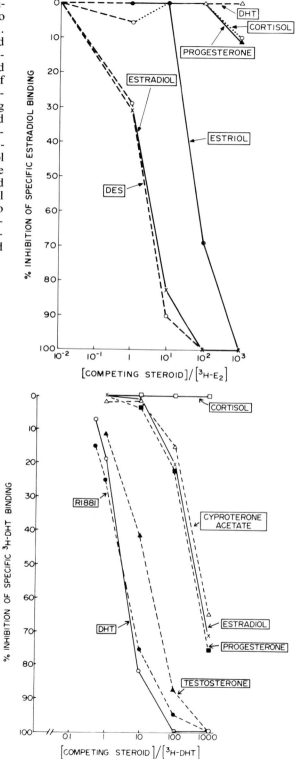

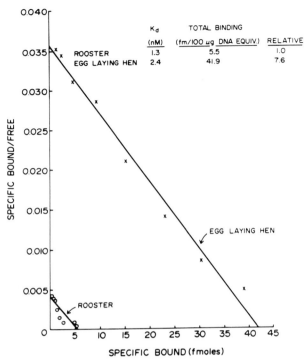

	K_d	TOTAL BINDING	
	(nM)	(fm/100 µg DNA EQUIV.)	RELATIVE
ROOSTER	1.3	5.5	1.0
EGG LAYING HEN	2.4	41.9	7.6

Fig. 7. Specific binding of estradiol to liver nuclear matrix of adult roosters vs. egg-laying hens. Saturation analyses were carried out in the presence of 0.15–20 nM [^{3}H]estradiol alone or together with 1 μM unlabeled estradiol. Specific binding results have been normalized to fmol per 100 μg starting nuclear DNA equivalents [From Barrack and Coffey (1980).]

involution of the gland (Fig. 9). Administration of androgens, but not of estrogens, restores these sites to normal levels within 1 h (Barrack and Coffey 1980). In order to quantitate androgen receptors on the prostate nuclear matrix, the inclusion of phenylmethylsulfonyl fluoride in the isolation buffers is essential. The degradation of binding sites and decreased recoveries of nuclear matrix spheres and protein in the absence of protease inhibitor are probably due to the high concentration of degradative enzymes normally present in the prostate.

The association of steroid receptors with the nuclear matrix is tissue specific. Following treatment of the immature female rat with a physiological dose of estrogen (0.1 μg) that is sufficient to induce maximal uterine growth, specific estradiol-binding sites are found associated with the uterine nuclear matrix, but not with the liver nuclear matrix of these animals (Barrack et al. 1977). This latter observation is consistent with the lack of responsiveness of the rat liver to these small doses of estrogen (Aten et al. 1978). Recently, Agutter and Birchall (1979) have confirmed that the rat uterine nuclear matrix contains specific estradiol-binding sites and have shown in addition that

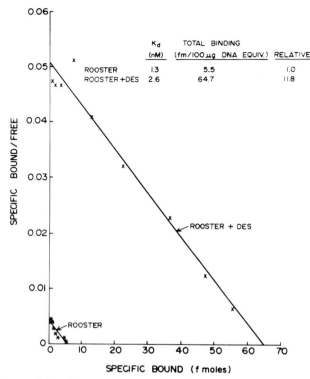

$Fig.\ 8.$ Specific binding of estradiol to liver nuclear matrix of untreated (*——*) vs. estrogenized (×——×) roosters. Saturation analyses were carried out as described in legend to Fig. 7 [From Barrack and Coffey (1980).]

the rat lung nuclear matrix contains none. From all of the preceding considerations, therefore, we conclude that the appearance of nuclear matrix-associated steroid-binding sites correlates with the stimulation of a biological response.

Clark and his colleagues have recently demonstrated that rat uterine nuclei contain a class of low-affinity (type II, K_d ~30 nM) specific estradiol-binding sites (Clark et al. 1980), in addition to the well-known high-affinity (type I, K_d ~1 nM) estradiol receptors. We have looked for low-affinity, type II binding of steroid to the nuclear matrix of avian liver and rat prostate by carrying out saturation analyses in the presence of concentrations of labeled steroid ranging from 0.1 nM to 42 or 80 nM but find only a single class of specific, high-affinity (K_d ~1 nM) steroid-binding sites.

A major proportion (50–70%) of the total steroid-binding capacity of nuclei is localized to the nuclear matrix. In the avian liver, for example, the recovery of estradiol binding sites with the nuclear matrix constitutes 61% of the total number of specific estradiol receptors in unextracted nuclei (Table 2). In contrast, only 7% of the total nuclear protein and 2% of the DNA are recovered with the nuclear matrix; thus, per unit amount of protein, there is

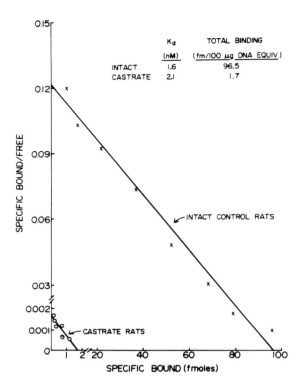

	K_d	TOTAL BINDING
	(nM)	(fm/100 μg DNA EQUIV.)
INTACT	1.6	96.5
CASTRATE	2.1	1.7

Fig. 9. Specific binding of DHT to ventral prostate nuclear matrix of intact and castrate rats. Specific binding of DHT to the nuclear matrix of intact adult rat ventral prostates ($\times$——$\times$) was obtained from the data of Fig. 4B. Specific binding of DHT to the ventral prostate nuclear matrix of adult rats 20 h after castration (O——O) was determined by saturation and Scatchard analysis. Specific binding data have been normalized to fmol per 100 μg starting nuclear DNA equivalents [From Barrack and Coffey (1980).]

a five-fold enrichment of nuclear estradiol receptors on the matrix compared to total nuclei.

In the rat prostate (Table 3A) a similar percentage (67%) of the total number of nuclear DHT binding sites is recovered with the nuclear matrix, which contains only 10–15% of the total nuclear protein and 1–2% of the DNA. The remainder of the nuclear androgen receptors is easily extracted by 2 M NaCl. On a protein basis, the specific activity of the DHT binding sites on the nuclear matrix is 4-fold higher than that of intact nuclei and 10-

Table 2. Comparison of Specific Estradiol Binding in Chick Liver Nuclei and Nuclear Matrix[a]

	Recovery (% of total nuclear)		Specific binding of estradiol (fmol)		
	Protein	DNA	Per 100 μg protein in fraction	Per 100 μg starting DNA equivalents	K_d (nM)
Total nuclei	100	100	24.2(1.0)	59.2(1.0)	0.9
Nuclear matrix	7	2	127.3(5.3)	36.2(0.61)	0.5

[a] Liver nuclei and nuclear matrix were isolated from estrogenized chicks. Specific estradiol-binding capacities and dissociation constants were derived from Scatchard analyses (see Barrack and Coffey 1980 for experimental details). Numbers in parentheses are values relative to total nuclei.

Table 3. Distribution and Enrichment of DHT Binding Sites in Subfractions of Prostate Nuclei and Nuclear Matrix[a]

Pretreatment of nuclei[b]	Distribution			Specific Activity		
	Nuclear matrix	Salt extract	Total	Nuclear matrix	Salt extract	Total
	(fmol DHT/100 μg starting DNA equivalents)			(fmol DHT/100 μg protein)		
A. DNase (control)	101 ± 8	49 ± 14	150 ± 17	194 ± 48	19 ± 6	49 ± 6
B. DNase + RNase + DTT	17 ± 4	136 ± 3	153 ± 2	52 ± 20	44 ± 6	45 ± 3

[a] The isolation of nuclei and nuclear matrix from prostates of intact rats, and the measurement of specific DHT binding, are described in detail in Barrack and Coffey 1980.
[b] Pretreatments of the nuclei were followed by extraction with 2 M NaCl to obtain the insoluble nuclear matrix (experiment A) or pore complex-lamina (experiment B), and the soluble salt extract.

fold higher than that of the salt extract (Table 3-A). It is not likely that the nuclear matrix-associated steroid-binding sites result from exposure of sites normally masked in the intact nucleus, since binding in the soluble extracts and in the insoluble nuclear matrix accounts for essentially all of the binding that is observed in unextracted nuclei.

The nuclear matrix, which we prepare by extracting nuclei with Triton X-100, DNase I, and 2 M NaCl (Barrack and Coffey 1980), consists of a peripheral lamina, residual nucleolus, and an internal ribonucleoprotein network. To determine whether the hormone-binding sites are distributed uniformly throughout the matrix structure or are enriched in a specific morphological component of the matrix, we have taken advantage of the observation that when Triton X-100-washed prostate nuclei are treated with DNase I and RNase A in the presence of 1 mM dithiothreitol (DTT) and then extracted with 2 M NaCl, matrix spheres are obtained that are devoid of internal structure and contain only 60% as much protein as control matrix preparations (Table 3B).

By comparing the distribution and specific activity of the hormone-binding sites associated with intact vs. empty matrix spheres, we have been able to show that the matrix-associated steroid-binding sites appear to be enriched on internal matrix structures (Barrack and Coffey 1980). Binding sites are not localized exclusively to the lamina, nor are they distributed uniformly on all matrix components. As shown in Table 3B, the recovery of specific DHT-binding sites with empty prostate matrix spheres (consisting only of a peripheral pore complex-lamina) represents only 17% of the DHT-binding activity associated with intact nuclear matrix structures that contain internal network material (17 vs. 101 fmol/100 μg starting nuclear DNA equivalents). If the binding sites had been distributed uniformly on all matrix components, then the specific activity of binding in the residual fraction would have been unchanged. We find, however, that the specific activity of the DHT binding sites that remain associated with these empty matrix spheres (52 fmol/100 μg protein) is only 27% of that of intact nuclear matrix structures (194 fmol/100

μg protein). The DHT binding sites that were originally associated with the intact nuclear matrix are recovered in the salt extract, the specific activity of which is thereby increased 2.3-fold (19 vs. 44 fmol/100 μg salt-extractable protein). The specific activity of the matrix-associated binding sites that become salt extractable as a result of the additional pretreatment with RNase and DTT (i.e., the difference between the amount of salt-resistant matrix binding and protein in experiments A vs. B) is calculated to be about 400 fmol/100 μg of protein extracted from the matrix; assuming that these DHT binding sites had been associated with the internal ribonucleoprotein network of the nuclear matrix, then the specific activity of these internal sites is almost eight-fold greater than that of the sites remaining with the peripheral lamina fraction (52 fmol/100 μg protein). A similar localization of matrix-associated estradiol-binding sites is observed in hen liver (Barrack and Coffey 1982). In addition, Agutter and Birchall (1979) find that whereas the nuclear matrix of rat uterine nuclei contains estradiol-binding sites, the pore complex-lamina fraction, isolated by a different method, contains virtually none of these sites.

Nuclease treatments alone solubilize no binding activity. DTT treatment alone also does not extract steroid-binding sites; it only renders them capable of being solubilized by subsequent or simultaneous treatment with salt. Thus, even if DTT is added only to the tissue homogenizing buffer, extraction of the isolated nuclei with NaCl will result in the solubilization of most of the nuclear receptors. Indeed, this may explain why some investigators do not find significant amounts of salt-resistant binding (Traish et al. 1977; Chamness et al. 1978).

We have considered the possibility that the conditions that result in solubilization of the internal ribonucleoprotein network may also extract minor components of the lamina and pore complexes and that these components contain the hormone receptors. Other approaches will be required to confirm the precise localization of these hormone receptors within a specific component of the nuclear matrix. However, with regard to the apparent localization of the majority of the matrix-associated hormone-binding sites to the internal ribonucleoprotein network, it is interesting to note that Liao et al. (1973) have reported the ability of prostate cytosol DHT–receptor complexes to bind to isolated nuclear RNP particles that were resistant to solubilization from nuclei by 1 M KCl and DNase I but could be released by deoxycholate treatment. That these RNP-associated binding sites (Liao et al. 1973) may have been a component of the internal RNA-protein network that is observed in the isolated nuclear matrix (Herman et al. 1978; Miller et al. 1978a; Faiferman and Pogo 1975) is supported by the observation of Miller et al. (1978a) that the RNP complexes of this network are highly susceptible to disruption by deoxycholate. In addition, electron autoradiographic studies have demonstrated that the DNA replication sites (Pardoll et al. 1980) and the newly labeled heterogeneous nuclear RNA (Fakan et al. 1976; Herman et al. 1978) are associated with the internal ribonucleoprotein network structure and indicate the potential biological importance of this component of the nuclear matrix.

C. Considerations of Steroid–Receptor–Acceptor Interactions

An important aspect of the nature of these specific steroid hormone-binding sites associated with the nuclear matrix concerns the question of whether these sites represent a native steroid–receptor–acceptor interaction.

1. Occupancy of Matrix-Associated Hormone-Binding Sites. The first point relates to whether these sites are occupied by steroid in vivo. The measurement of steroid-binding sites on the nuclear matrix involves in vitro incubation of the isolated nuclear matrix fraction with labeled steroid. The underlying assumption of this approach is that under equilibrium conditions in vitro the labeled steroid exchanges with unlabeled steroid that had become bound to these sites as a result of the in vivo processes of steroid hormone–receptor complex translocation following administration of the hormone (Anderson et al. 1972). However, the state of occupancy of these binding sites in the nuclear matrix is not known. Thus, indirect binding studies such as those we have used do not distinguish between the binding of the labeled steroid to receptors previously occupied by endogenous unlabeled steroid (i.e., exchange) and direct binding to unoccupied sites. Unoccupied binding sites might simply represent receptor from which steroid has dissociated during the time it takes to isolate the nuclear matrix or a form of unoccupied receptor that is bound to the nuclear matrix in vivo without steroid attached.

2. Receptor Interactions with the Nuclear Matrix. As to whether labeled steroid binds to the nuclear matrix via receptor proteins that had been translocated to the nucleus remains to be established directly. Although the equilibrium dissociation constant for the interaction of steroids with the nuclear matrix is similar to that of the cytoplasmic and nuclear salt-extractable steroid receptors, this is consistent with the hypothesis, but does not prove, that the binding site on the matrix is the same receptor moiety. In this regard, however, soluble macromolecule-associated estradiol binding, which on sucrose density gradient analysis had a sedimentation coefficient of 7 S, somewhat larger than that of the nuclear salt-extractable estradiol receptor (Snow et al. 1978), has been obtained by extracting the chicken liver nuclear matrix with 2 M LiCl and 4 M urea (Hardin and Clark 1979). In addition, when the nuclear matrix-associated androgen-binding sites of the rat ventral prostate are rendered salt-soluble following NaCl extraction of nuclei that had been digested with DNase I plus RNase A and dithiothreitol (see Table 3B), sucrose density gradient analysis reveals that the bound steroid is macromolecule-associated (E.R. Barrack, unpublished observations).

Additional evidence that (steroid)–receptor complexes become associated with the nuclear matrix following their translocation from the cytoplasm is the observation that in the hormone-withdrawn state (roosters or castrate rats) there are virtually no high-affinity, specific steroid-binding sites in the nucleus, whereas shortly after appropriate hormonal treatment (within less than an hour in the androgen-treated castrate rat) significant levels of specific receptor activity are found associated with the nuclear matrix (Barrack and Coffey 1980). These data are consistent with the types of observations that form the basis for the well-established model of steroid hormone-mediated translocation of cytoplasmic receptors into the nucleus (Gorski and Gannon 1976; Jensen 1979; Liao et al. 1979).

Of course, the possibility that these matrix-associated steroid-binding sites do not reflect the native state, but rather merely result from adventitious adsorption of receptors, must always be considered. The adsorption of uncharged cytosol receptors to the nuclear matrix, however, can be ruled out, since in the prostate of the castrate there are large amounts of androgen receptor in the cytosol (Bruchovsky and Wilson 1968; Fang et al. 1969; and E. R. Barrack, unpublished observations), but under these conditions only very few binding sites are found associated with the matrix (Fig. 9). Nevertheless, although high-salt conditions are often used to eliminate nonspecific adsorption, the possibility cannot be ruled out entirely that during the isolation of nuclei and nuclear matrix some activated cytoplasmic steroid–receptor complexes or salt-extractable nuclear receptors become associated with the nuclear matrix in a nonphysiological interaction that remains resistant to disruption and solubilization by high ionic strength.

On the other hand, however, since the precise mechanism by which steroid hormone–receptor complexes act in the nucleus to modulate gene expression by affecting the synthesis of premessenger heterogeneous nuclear RNA is still unresolved, the possibility that salt-resistant interactions of receptors with the nuclear matrix in fact reflect a meaningful association should not be discounted (see also Clark and Peck 1976; Honma et al. 1977; Ruh and Baudendistel 1977, 1978; Sato et al., 1979). Moreover, the overwhelming accumulation of data that implicate a functional role of the nuclear matrix in replication, transcription, and processing lend added credence that the association of hormone receptors with the nuclear matrix represents a meaningful interaction.

3. Considerations of Salt-Extractable and Salt-Resistant Nuclear Receptors. The apparent existence of two forms of steroid–receptor–acceptor complexes in the nucleus, one that is easily disrupted by high-salt conditions and the other that is resistant, raises some interesting questions. For example, one possibility is that the acceptors are the same but that the cytoplasm-derived nuclear receptors themselves are heterogeneous. Though there may be little evidence to support this notion, the methods that are currently used for characterizing receptors may not be sufficiently sensitive to detect such heterogeneity. A second possibility is that the receptors are homogeneous, but the acceptor sites for salt-extractable and salt-resistant receptors are different. It is not known whether the interactions of receptors with these acceptor sites (electrostatic vs. hydrophobic) might have different biological consequences. For example, Clark et al. (1976) have suggested that the salt-resistant form may represent a transformation of the salt-extractable receptor, whereas Ruh and Baudendistel (1978) have proposed that nuclear salt-extractable receptor–acceptor interactions are involved in the induction of both DNA and RNA synthesis and that the salt-resistant receptor–acceptor interactions are involved in receptor processing. Both groups of investigators, nevertheless, attribute special biological significance to the salt-resistant receptor–acceptor interaction.

Additionally, one must keep in mind that salt-extractable and salt-resistant receptors are operationally defined and may not reflect in a precise manner the in vivo interaction of receptors with the nucleus. For example, alterations in the conditions for extracting nuclear receptors, such as the inclusion of reducing agents, can alter the relative proportions of nuclear receptor that are salt extractable and salt resistant (Barrack and Coffey 1982). Therefore, the extent to which, for example, oxidation of protein sulfhydryl groups may be involved in the in vivo interaction of receptors with the nucleus remains to be elucidated.

Another example of operational considerations is that the inclusion of inhibitors of proteolysis can affect the amount of nuclear receptor found to be salt-resistant (Barrack and Coffey 1982). Proteolysis can also increase the proportion of nuclear steroid receptors that are solubilized by buffer and/or 0.4 M KCl (Katzenellenbogen et al. 1980). This raises the intriguing question of whether under standard conditions the appearance of salt-extractable nuclear receptors might arise from endogenous proteolytic activity.

Problems associated with the characterization of subnuclear hormone receptors may also arise from the use of sodium molybdate, which has been shown to stabilize many classes of steroid hormone receptors (Leach et al. 1979; Toft and Nishigori 1979; Noma et al 1980; Miller et al. 1981). Several investigators have adopted this methodology but have included sodium molybdate in the homogenization buffer (Anderson et al. 1980; Gaubert et al. 1980; Hawkins et al. 1981). It is important to point out that when sodium molybdate is added directly to the homogenization buffer, prostate nuclear androgen receptor levels are dramatically reduced while cytosol receptor levels increase, suggesting that there is a redistribution of a large proportion of the nuclear receptors to the cytosol fraction (Trachtenberg et al. 1981; Barrack et al. 1983). Other experiments have shown that sodium molybdate extracts glucocorticoid receptors from isolated liver nuclei (Murakami and Moudgil 1981). All of these examples therefore emphasize the difficulty of defining subcellular localization.

D. Relationship to Chromatin-Associated Receptors

The role of hormone-binding components of chromatin, and subfractions thereof, in the mechanism of steroid hormone action has been investigated extensively (see review by Thrall et al. 1978). Unfortunately, however, chromatin is operationally defined and a variety of different methods are used to prepare and subfractionate chromatin. Hence individual preparations may represent different structural and functional components of the nucleus. In addition, chromatin preparations often contain fragments of the nuclear envelope (Jackson 1976; Smith and von Holt 1981) and of the nuclear matrix (Berezney 1979). Thus the concept of "chromatin" is undergoing a redefinition in terms of modern concepts of the nucleus (Busch 1978).

Since all the DNA in the nucleus is organized into supercoiled loop domains that are tightly anchored to the nuclear matrix (see Table 1), it should not be surprising to find that isolated chromatin contains components of the nuclear matrix, particularly if chromatin is prepared by simply washing nuclei with a low-ionic-strength buffer to extract the soluble nuclear proteins (Spelsberg et al. 1971; Thrall and Spelsberg 1980) or by sonicating nuclei to disrupt nuclear (and nuclear matrix) structure (e.g., Pederson 1974; de Boer et al. 1978; Rennie 1979; Berezney 1980; van Eekelen and van Venrooij 1981). In this regard it is important to note that nuclear acceptor sites for steroid hormone–receptor complexes have been found in chromatin fractions that resist solubilization by both low and high ionic strength (Klyzsejko-Stefanowicz et al. 1976; Wang 1978; Thrall et al. 1978; Tsai et al. 1980) and in chromatin subfractions obtained by sonicating nuclei (de Boer et al. 1978).

Localization of hormone receptor binding sites has also been probed by DNase I or micrococcal nuclease digestion of nuclei or chromatin. Many of the receptors that are solubilized by nuclease digestion are associated with nucleosomes (Senior and Frankel 1978; Scott and Frankel 1980) or internucleosomal DNA (Rennie 1979). On the other hand, Schoenberg and Clark (1981) find no such association. Only 5–70% of nuclear steroid receptors are released by nuclease digestion (Senior and Frankel 1978; Alberga et al. 1979; Scott and Frankel 1980; Schoenberg and Clark 1981), and some investigators find that these receptors can be dissociated from nuclease-sensitive sites by 0.4 M KCl or 0.6 M NaCl (Senior and Frankel 1978; Rennie 1979). These nuclease-sensitive hormone-binding sites therefore appear to be distinct from the DNase I-resistant sites associated with the nuclear matrix. On the other hand, however, it is important to recognize that micrococcal nuclease will attack not only DNA, but also RNA, releasing both nucleosomes and RNP particles from nuclei or chromatin (Walker et al. 1980). Therefore, the possibility exists that some of these nuclease-sensitive receptors might be derived from the internal RNA–protein network of the nuclear matrix.

Template-active regions of chromatin are known to be preferentially sensitive to nuclease digestion (Garel and Axel 1976; Weintraub and Groudine 1976; Bloom and Anderson 1978), yet hormone receptors have been found in association with both template-active and inactive fractions (Alberga et al. 1979; Scott and Frankel 1980; Davies et al. 1980). In MCF-7 cells, estradiol–receptor complexes were enriched in the transcriptionally active fraction of chromatin (Scott and Frankel 1980). In contrast, the template-inactive fraction of the ventral prostate appears to contain about three times as many acceptor sites as the transcriptionally active chromatin (Davies et al. 1980). These apparent discrepancies remain to be resolved.

It is important to comment on the observations, on the one hand, that transcriptionally active genes in nuclei and chromatin are preferentially sensitive to nuclease digestion, with the recent findings, on the other hand, that certain actively transcribed genes are enriched in the regions of DNA that are tightly attached to the nuclease-resistant nuclear matrix (Pardoll and Vogelstein 1980; Nelkin et al. 1980; Robinson et al. 1982). Although these two sets of data *appear* contradictory, they are not. It is precisely because of the former observations that experiments to investigate the spatial organization of genes along the loops of DNA must be carried out by first extracting all histones from the nuclei with 2 M NaCl (Nelkin et al. 1980; Robinson et al. 1982). With the histones removed, DNase I makes random cuts in the DNA, with no preference for active or inactive genes (Nelkin et al. 1980; Robinson et al. 1982). Following progressive cleavage of the DNA loops by controlled DNase I digestion, the DNA that remains associated with the nuclear matrix is believed to represent sequences at or near the base of the loops that are anchored to the matrix (Nelkin et al. 1980; Robinson et al. 1982).

Therefore, the ability of nuclease digestion to release transcriptionally active regions from chromatin belies the fact that their native localization appears to be in association with a discrete, nuclease-resistant structure, the nuclear matrix. Whether a similar situation applies to the preferential release of some nuclear steroid hormone–receptor complexes by nucleases or salt remains to be determined.

IV. Potential Role of Cellular Structural Components in Hormone Action

Cells contain extensive and elaborate three-dimensional skeletal networks that form integral structural components of the plasma membrane, the cytoplasm, and the nucleus. If these matrix systems were interconnected and could undergo dynamic shifts in structure and conformation, such as by polymerization, depolymerization, cross-linking, biochemical modifications, or contractile movements, then one might visualize how a change in one matrix component could transmit and couple these changes to the rest of the system. Perhaps by such a mechanism externally applied signals might be transmitted along this communications network from one part of the cell to another. Alternatively, these interconnecting networks might play a role in the transport of informational molecules to their effector sites. A third function of these matrix systems might be to act as "solid state catalysts" or organizational support structures that could facilitate the interactions of molecules. Since cellular skeleton networks are composed of a number of different types of macromolecules, one could imagine a system in which all three types of interactions might be feasible. The intrinsic appeal of such an organization is the capacity for vectorial chemomechanical coupling as a means of signal transduction. [For additional discussion of these concepts, see review by Isaacs et al (1981).]

One of the basic questions to be resolved in endocrinology is how hormonal signals are transmitted to the nucleus to alter the structure and function of DNA. To date, the working hypothesis of steroid hormone action has been that steroids freely diffuse into the cell, where they become bound to specific intracellular soluble cytoplasmic proteins, which then, by mechanisms not fully understood, enter the nucleus and bind with high-affinity to specific acceptor sites on the chromatin. Given the recent discoveries about the cytoskeleton and the organization of DNA relative to the nuclear matrix and the organization and synthesis of hnRNA relative to the matrix, the nuclear matrix represents a conceptually attractive candidate for the site in the nucleus where steroid hormones might act to regulate specific gene transcription.

Hormones affect not only gene regulation but also structure. Such changes in cellular structure include nuclear swelling (Cavazos and Melampy 1954; Ritter 1969; Chung and Coffey 1971), and rapid and extensive changes in the organization of large areas of chromatin (Tew et al. 1980; Vic et al. 1980). Given the possibility of interacting matrix systems throughout the cell, it is tempting to speculate that these structural changes may involve dynamic changes in the nuclear matrix following the interaction of steroid hormone–receptor complexes with the nuclear matrix.

Whether hormones might also affect other cellular structural components as part of their mechanism of action is not known, but a number of recent noteworthy observations deserve consideration in this regard. For example, there is evidence that there are specific estradiol receptors on the plasma membrane of target cells (Szego and Pietras 1981; Nenci et al. 1981) and that these membrane receptors undergo a temperature- and hormone-dependent capping process (Nenci et al. 1981). In addition, Puca et al. (1981) have reported the presence of specific, high-affinity binding sites for estradiol–receptor complexes on the plasma membrane cytoskeleton of erythrocytes. Some investigators now believe that most of the soluble proteins that comprise the cytosol may not in fact exist in a soluble state in the intact cell but rather become solubilized by the buffers and salts of homogenizing media from their association with the cytoskeleton or microtrabecular lattice that appears to extend throughout the cytoplasm (Porter and Tucker 1981). We thus note with interest the report by Szego and Pietras (1981) that membrane-associated estradiol receptors can be rendered soluble, depending on the method of cell disruption. It is therefore conceivable that the soluble steroid receptors that have been studied previously have been solubilized by cell disruption and fractionation techniques, and might have been bound in situ to structural components. These matrix systems may represent such structural components for the binding of hormone–receptor complexes. Further progress in resolving what the steroid receptor does in the nucleus to regulate DNA function must await a clearer understanding of the organization of the nucleus.

Acknowledgments. We would like to thank Miss Ruth Middleton for help in preparing this manuscript. This work was supported by Grant AM 22000, USPHS, National Institute of Arthritis, Metabolism, and Digestive Diseases.

References

Agutter PS, Birchall K (1979) Exp Cell Res 124: 453–460
Agutter PS, Richardson JCW (1980) J Cell Sci 44: 395–435
Alberga A, Tran A, Baulieu EE (1979) Nucleic Acids Res 7: 2031–2044
Allen SL, Berezney R, Coffey DS (1977) Biochem Biophys Res Commun 75: 111–116
Anderson J, Clark JH, Peck EJ Jr (1972) Biochem J 126: 561–567
Anderson KM, Phelan J, Marogil M, Hendrickson C, Economou S (1980) Steroids 35: 273–280
Aten RF, Weinberger MJ, Eisenfeld AJ (1978) Endocrinology 102: 433–442
Barrack ER, Coffey DS (1980) J Biol Chem 255: 7265–7275
Barrack ER, Coffey DS (1982) Recent Prog Horm Res 38: 133–195
Barrack ER, Hawkins EF, Allen SL, Hicks LL, Coffey DS (1977) Biochem Biophys Res Commun 79: 829–836
Barrack ER, Hawkins EF, Coffey DS (1979) In: Leavitt WW and Clark JH (eds) Steroid Hormone Receptor Systems. Plenum Press, New York, pp 47–70
Barrack ER, Bujnovszky P, Walsh PC (1983) Cancer Res (in press)
Bekers AGM, Gijzen HJ, Taalman RDFM, Wanka F (1981) J Ultrastruct Res 75: 352–362
Berezney R (1979) In: Busch H (ed) The Cell Nucleus. Academic Press, New York, Vol. 7, pp 413–456
Berezney R (1980) J Cell Biol 85: 641–650
Berezney R, Buchholtz LA (1981a) Exp Cell Res 132: 1–13
Berezney R, Buchholtz LA (1981b) Biochem 20: 4995–5002
Berezney R, Coffey DS (1974) Biochem Biophys Res Commun 60: 1410–1417
Berezney R, Coffey DS (1975) Science 189: 291–293
Berezney R, Coffey DS (1976) Adv Enz Reg 14: 63–100
Berezney R, Coffey DS (1977) J Cell Biol 73: 616–637
Blazsek I, Vaukhonen M, Hemminki K (1979) Res Commun Chem Path Pharmacol 23: 611–626
Bloom KS, Anderson JN (1978) Cell 15: 141–150
Boesel RW, Klipper RW, Shain SA (1977) Endocr Res Commun 4: 71–84
Bowen B, Steinberg J, Laemmli UK, Weintraub H (1980) Nucleic Acids Res 8: 1–20
Brasch K, Sinclair GD (1978) Virchows Arch B Cell Pathol 27: 193–204
Brown TR, Rothwell SW, Migeon CJ (1981) J Steroid Biochem 14: 1013–1022
Bruchovsky N, Wilson JD (1968) J Biol Chem 243: 2012–2021
Buckler-White AJ, Humphrey GW, Pigiet V (1980) Cell 22: 37–46
Busch H (1978) In: Busch H (ed) The Cell Nucleus. Academic Press, New York, Vol. 6, pp xxiii–xxvii
Cavazos LF, Melampy RM (1954) Endocrinology 54: 644–648
Chamness GC, Zava DT, McGuire WL (1978) In: Stein G, Stein J, Kleinsmith LJ (eds) Methods in Cell Biology. Academic Press, New York, 17: 325–333
Chung LWK, Coffey DS (1971) Biochim Biophys Acta 247: 570–583
Cidlowski JA, Munck A (1980) J Steroid Biochem 13: 105–112
Ciejek EM, Nordstrom JL, Tsai M-J, O'Malley BW (1982) Biochemistry 21: 4945–4953
Clark JH, Peck EJ Jr (1976) Nature 260: 635–637
Clark JH, Eriksson HA, Hardin JW (1976) J Steroid Biochem 7: 1039–1043
Clark JH, Markaverich B, Upchurch S, Eriksson H, Hardin JW, Peck EJ Jr (1980) Recent Prog Horm Res 36: 89–134

Comings DE, Okada TA (1976) Exp Cell Res 103: 341–360
Comings DE, Wallack AS (1978) J Cell Sci 34: 233–246
Cook PR, Brazell IA, Jost E (1976) J Cell Sci 22: 303–324
Danzo BJ, Eller BC (1978) J Steroid Biochem 9: 477–483
Davies P, Thomas P, Griffiths K (1977) J Endocrinol 74: 393–404
Davies P, Thomas P, Borthwick NM, Giles MG (1980) J Endocrinol 87: 225–240
de Boer W, deVries J, Mulder E, van der Molen HJ (1978) Nucleic Acids Res 5: 87–103
Deeley RG, Gordon JI, Burns ATH, Mullinix KP, BinaStein M, Goldberger RF (1977) J Biol Chem 252: 8310–8319
DeHertogh R, Ekka E, Vanderheyden I, Hoet JJ (1973) J Steroid Biochem 4: 313–320
Deppert W (1978) J Virol 26: 165–178
Detke S, Keller JM (1982) J Biol Chem 257: 3905–3911
Dijkwel PA, Mullenders LHF, Wanka F (1979) Nucleic Acids Res 6: 219–230
Dvorkin VM, Vanyushin BF (1978) Biochem (USSR) 43: 1297–1301
Faiferman I, Pogo AO (1975) Biochemistry 14: 3808–3816
Fakan S, Puvion E, Spohr G (1976) Exp Cell Res 99: 155–164
Fang S, Anderson KM, Liao S (1969) J Biol Chem 244: 6584–6595
Fisher PA, Berrios M, Blobel G (1982) J Cell Biol 92: 674–686
Franceschi RT, Kim KH (1979) J Biol Chem 254: 3637–3646
Garel A, Axel R (1976) Proc Natl Acad Sci USA 73: 3966–3970
Gaubert CM, Tremblay RR, Dubé JY (1980) J Steroid Biochem 13: 931–937
Georgiev GP, Nedospasov SA, Bakayev VV (1978) In: Busch H (ed) The Cell Nucleus. Academic Press, New York, 6: 3–34
Gerace L, Blobel G (1980) Cell 19: 277–287
Ghosh S, Paweletz N, Ghosh I (1978) Exp Cell Res 111: 363–371
Giese G, Wunderlich F (1980) J Biol Chem 255:1716–1721
Goldfischer S, Kress Y, Coltoff-Schiller B, Berman J (1981) J Histochem Cytochem 29: 1105–1111
Gorski J, Gannon F (1976) Ann Rev Physiol 38: 425–450
Gschwendt M, Schneider W (1978) Eur J Biochem 91: 139–149
Hardin JW, Clark JH (1979) J Cell Biol 83: 252a
Hawkins EF, Lieskovsky G, Markland FS Jr (1981) J Clin Endocrinol Metab 53: 456–458
Hemminki K, Vainio H (1979) Cancer Lett 6: 167–173
Herman R, Weymouth L, Penman S (1978) J Cell Biol 78: 663–674
Hodge LD, Mancini P, Davis FM, Heywood P (1977) J Cell Biol 72: 194–208
Honma Y, Kasukabe T, Okabe J, Hozumi M (1977) J Cell Physiol 93: 227–236
Hozier J, Furcht LT (1980) Cell Biol Intl Rep 4: 1091–1099
Hunt BF, Vogelstein B (1981) Nucleic Acids Res 9: 349–363
Isaacs JT, Barrack ER, Isaacs WB, Coffey DS (1981) In: Murphy GP, Sandberg AA, Karr JP (eds) The Prostatic Cell: Structure and Function. Alan R. Liss, Inc., New York, Part A, pp 1–24
Jackson RC (1976) Biochemistry 15: 5652–5656
Jackson DA, McCready SJ, Cook PR (1981) Nature 292: 552–555
Jensen EV (1979) Pharmacol Rev 30: 477–491
Jensen EV, Suzuki T, Kawashima T, Stumpf WE, Jungblut PW, DeSombre ER (1968) Proc Natl Acad Sci USA 59: 632–638
Johnson LK, Vlodavsky I, Baxter, JD, Gospodarowicz, D (1980) Nature 287: 340–343
Jordan VC, Dix CJ, Rowsby L, Prestwich G (1977) Mol Cell Endocrinol 7: 177–192
Katzenellenbogen BS, Lan NC, Rutledge SK (1980) J Steroid Biochem 13: 113–122
Kaufman M, Pinsky L, Kubski A, Straisfeld C, Dobrenis K, Shiroky J, Chan T, MacGibbon B (1978) J Clin Endocrinol Metab 47: 738–745

Kaufmann SH, Coffey DS, Shaper JH (1981) Exp Cell Res 132: 105–123

Klyzsejko-Stefanowicz L, Chiu J-F, Tsai Y-H, Hnilica LS (1976) Proc Natl Acad Sci USA 73: 1954–1958

Lam KS, Kasper CB (1979) Biochemistry 18: 307–311

Leach KL, Dahmer MK, Hamond ND, Sando JJ, Pratt WB (1979) J Biol Chem 254: 11884–11890

Lebeau MC, Massol N, Baulieu EE (1973) Eur J Biochem 36: 294–300

Levinger L, Varshavsky A (1981) J Cell Biol 90: 793–796

Liao S, Liang T, Tymoczko JK (1973) Nature New Biol 241: 211–213

Liao S, Mezzetti G, Chen C (1979) In: Busch H (ed) The Cell Nucleus. Academic Press, New York, 7: 201–227

Long BH, Huang C-Y, Pogo AO (1979) Cell 18: 1079–1090

Mariman ECM, van Eekelen CAG, Reinders RJ, Berns AJM, van Venrooij WJ (1982) J Mol Biol 154: 103–119

Matsumoto LH (1981) Nature 294: 481–482

Maul GG, Avdalović N (1980) Exp Cell Res 130: 229–240

Maundrell K, Maxwell ES, Puvion E, Scherrer K (1981) 136: 435–445

McCready SJ, Godwin J, Mason DW, Brazell IA, Cook PR (1980) J Cell Sci 46: 365–386

McDonald JR, Agutter PS (1980) FEBS Lett 116: 145–148

Mester J, Baulieu EE (1975) Biochem J 146: 617–623

Miller TE, Huang C-Y, Pogo AO (1978a) J Cell Biol 76: 675–691

Miller TE, Huang C-Y, Pogo AO (1978b) J Cell Biol 76: 692–704

Miller LK, Tuazon FB, Niu E-M, Sherman MR (1981) Endocrinology 108: 1369–1378

Mulvihill ER, LePennec J-P, Chambon P (1982) Cell 24: 621–632

Murakami N, Moudgil VK (1981) Biochem J 198: 447–455

Nelkin BD, Pardoll DM, Vogelstein B (1980) Nucleic Acids Res 8: 5623–5634

Nenci I, Marchetti E, Marzola A, Fabris G (1981) J Steroid Biochem 14: 1139–1146

Noma K, Nakao K, Sato B, Nishizawa Y, Matsumoto K, Yamamura Y (1980) Endocrinology 107: 1205–1211

Nomi S, Matsuura T, Ueyama H, Ueda K (1981) J Nutr Sci Vitaminol 27: 33–41

Nyberg LM, Wang TY (1976) J Steroid Biochem 7: 267–273

Pardoll DM, Vogelstein B (1980) Exp Cell Res 128: 466–470

Pardoll DM, Vogelstein B, Coffey DS (1980) Cell 19: 527–536

Pederson T (1974) J Mol Biol 83: 163–183

Pogo AO (1981) In: Busch H (ed) The Cell Nucleus. Academic Press, New York, Vol 8, pp 331–367

Porter KR, Tucker JB (1981) Sci Am 244: 56–67

Puca GA, Bresciani F (1968) Nature 218: 967–969

Puca GA, Nola E, Molinari AM, Armetta I, Sica V (1981) J Steroid Biochem 15: 307–312

Razin SV, Mantieva VL, Georgiev GP (1979) Nucleic Acids Res 7: 1713–1735

Razin SV, Chernokhvostov VV, Roodyn AV, Zbarsky IB, Georgiev GP (1981) Cell 27: 65–73

Rennie PS (1979) J Biol Chem 254: 3947–3952

Ritter C (1969) Endocrinology 84: 844–854

Robinson SI, Nelkin BD, Vogelstein B (1982) Cell 28: 99–106

Ross DA (1980) Fed Proc 39: 2196

Ruh TS, Baudendistel LJ (1977) Endocrinology 100: 420–426

Ruh TS, Baudendistel LJ (1978) Endocrinology 102: 1838–1846

Sanborn BM, Steinberger A, Tcholakian RK (1979) Steroids 34: 401–412

Sato B, Spomer W, Huseby RA, Samuels LT (1979) Endocrinology 104: 822–831

Scatchard G (1949) Ann NY Acad Sci 51: 660–672

Schoenberg DR, Clark JH (1981) Biochem J 196: 423–432

Scott RW, Frankel FR (1980) Proc Natl Acad Sci USA 77: 1291–1295
Senior MB, Frankel FR (1978) Cell 14: 857–863
Sevaljević L, Poznanović G, Petrović M, Krtolica K (1981) Biochem Intl 2: 77–84
Shaper JH, Pardoll DM, Kaufmann SH, Barrack ER, Vogelstein B, Coffey DS (1979) Adv Enz Reg 17: 213–248
Sinibaldi RM, Morris PW (1981) J Biol Chem 256: 10735–10738
Small D, Nelkin B, Vogelstein B (1982) Proc Natl Acad Sci USA 79: 5911–5915
Smith HC, Berezney R (1980) Biochem Biophys Res Commun 97: 1541–1547
Smith P, von Holt C (1981) Biochemistry 20: 2900–2908
Snow LD, Eriksson H, Hardin JW, Chan L, Jackson RL, Clark JH, Means AR (1978) J Steroid Biochem 9: 1017–1026
Spelsberg TC, Steggles AW, O'Malley BW (1971) J Biol Chem 246: 4188–4197
Steele WJ, Busch H (1966) Biochim Biophys Acta 129: 54–67
Steer RC, Goueli SA, Wilson MJ, Ahmed K (1980) Biochem Biophys Res Commun 92: 919–925
Szego CM, Pietras RJ (1981) In: Litwack G (ed) Biochemical Actions of Hormones. Academic Press, New York, Vol 8, pp 307–463
Tew KD, Schein PS, Lindner DJ, Wang AL, Smulson ME (1980) Cancer Res 40: 3697–3703
Thrall CL, Spelsberg TC (1980) Biochemistry 19: 4130–4138
Thrall CL, Webster RA, Spelsberg TC (1978) In: Busch H (ed) The cell nucleus. Academic Press, New York, Vol 6, pp 461–529
Toft D, Nishigori H (1979) J Steroid Biochem 11: 413–416
Trachtenberg J, Hicks LL, Walsh PC (1981) Invest Urol 18: 349–354
Traish A, Müller RE, Wotiz HH (1977) J Biol Chem 252: 6823–6830
Tsai Y-H, Sanborn BM, Steinberger A, Steinberger E (1980) J Steroid Biochem 13: 711–718
Ueyama H, Matsuura T, Nomi S, Nakayasu H, Ueda K (1981) Life Sci 29: 655–661
van Eekelen CAG, van Venrooij WJ (1981) J Cell Biol 88: 554–563
Vic P, Garcia M, Humeau C, Rochefort H (1980) Mol Cell Endocrinol 19: 79–92
Vogelstein B, Pardoll DM, Coffey DS (1980) Cell 22: 79–85
Walker BW, Lothstein L, Baker CL, LeStourgeon WM (1980) Nucleic Acids Res 8: 3639–3657
Wang TY (1978) Biochim Biophys Acta 518: 81–88
Weintraub H, Groudine M (1976) Science 193: 848–856
Wunderlich F, Herlan G (1977) J Cell Biol 73: 271–278
Wunderlich F, Giese G, Bucherer C (1978) J Cell Biol 79: 479–490

Discussion of the Paper Presented by D.S. Coffey

GREENE: I'd like to ask a question about the various types of steroid-binding sites and either Don or Jim can answer it. In nuclear matrix you have looked at type I receptor, at salt-resistant sites, and at type II binders. So far, what you are looking at appears to me to be type I receptor; the number of sites and the K_d appear to be consistent with this idea. What would be the role of type II sites in all of this and has this entered into your thinking?

COFFEY: We have been talking only about high affinity (type I) receptors. We have looked for lower affinity (type II) sites in the nuclear matrix of the chick liver and rat prostate, but found none. We had never looked for type II sites on the uterine nuclear matrix, because when we started this work in 1977, type II sites had not yet been reported by Jim Clark. Now from the work of Clark and his colleagues, it looks like type II estradiol binding sites in uterine nuclei are important in the mechanism of action, and that they too are associated with the nuclear matrix (Clark and Markaverich *in* The Nuclear Envelope and the Nuclear Matrix, p. 259 (G.G. Maul, ed.), Alan R. Liss, N.Y., 1982).

LIAO: Could I ask you, Don, have you taken nuclear matrix and mixed it with DHT–receptor complex or estrogen–receptor complex?

COFFEY: We have not carried out those experiments.

LIAO: Many of your numbers, they are by exchange assay?

COFFEY: Yes, all of this is by exchange assay.

LIAO: With exchange assay, sometimes you see the things you don't see in vivo. Is that right?

COFFEY: I'm sure that this must be true. It's hard to believe that you could do an exchange assay and isolate something exactly like in vivo. I think it has meaning, at least it seems to have some relation with some biological function. That is all I can say.

THOMPSON: It seems to me that the critical question that everybody's wondering about is whether or not these are true "estrogen receptor sites" that you are measuring in the matrix. And the two kinds of experiments that I can think of that would address the question would be those in which you try to characterize the sites' physical properties, as has been done fairly extensively for the estrogen receptor type I, and those using antiserum developed against the estrogen receptor to examine the matrix. Have you done such studies? Another approach would be to look at the kinetics of cytoplasmic to nuclear movement of type I sites, to see if you can define the kinetics of accumulation of these sites on the matrix. Have you done those experiments?

COFFEY: We have not carried out any physical studies of the matrix-associated steroid receptors, one reason being that they are so tightly associated with the nuclear matrix, and not easily solubilized. The second point is that we do not have any antibodies to estrogen or androgen receptors. On the other hand, though, we have looked at the kinetics of disappearance of the androgen receptor from the prostate nuclear matrix following castration and its reappearance following androgen readministration. After castration, androgen receptors disappear from the nuclear matrix and cytosol receptor levels increase. Within less than an hour after injecting DHT into castrate rats, the number of receptors on the nuclear matrix is restored to pre-castration levels. All these data are consistent with a hormone-mediated translocation of receptors to the nuclear matrix.

ROY: Don, do you have any evidence that this kind of nuclear matrix exists in a living cell?

COFFEY: Many people have noted that certain aspects of the isolated nuclear matrix closely resemble the appearance of interchromatinic and non-chromatinic structures seen in intact nuclei and whole tissue (Berezney, Coffey 1976; Berezney 1979; Brasch, Sinclair 1978; Ghosh et al. 1978; Brasch 1982, *Exp. Cell Res. 140*: 161–171); these studies were done using thin sections and conventional EM techniques. Others have visualized by EM an underlying nuclear network using osmium-potassium ferrocyanide fixation of tissue (Goldfischer et al. 1981), or using whole mounts of Triton-treated cell monolayers (Capco et al. 1982, *Cell 29*: 847–858). All of these studies support the notion that the nuclear matrix exists in vivo, though its exact structure in vivo is not known.

ROY: What I am implying is that this matrix structure that you see, could that be due to precipitation when you fix a cell and the cell dies. Is there any way of looking at a living cell with some kind of probe and looking through polarized light or that sort of thing and see whether they have this nuclear matrix like the cytoskeleton?

COFFEY: Probably the most convincing approach would be to use antibodies made against nuclear matrix proteins and to be able to decorate a skeleton structure in the nucleus of an intact cell. This type of approach has been highly successful in convincingly demonstrating the presence of a cytoskeleton. Unfortunately, we do not yet have any antibodies that will light up a nuclear skeleton, but I know that many labs are working towards this goal. Yet even with this technique the cells would have to be fixed.

O'MALLEY: I have two questions, which are (1) since you are the expert on the matrix, do you think the matrix is an entity itself? I think you have support for your concept, because this year on the Cell Biology program, there is a section on the nuclear matrix. But many people as you know, feel that it is really just membrane containing a lot of precipitated, aggregated nuclear protein, and that it is not really a structural entity.

COFFEY: Obviously, we've been fighting this war with reviewers and with granting agencies for a long time, and we had to make a decision, Bert, very early in the game. If someone says that something is a denatured product, then no matter what you do to get it out, or how quickly you get it out, they will say it denatured in a microsecond, so you are up against an endless argument. To get around that we decided that it wasn't important unless it had biologically interesting properties, and if it has biologically interesting properties, then it deserves further study. So you can see why we concentrated on DNA synthesis and hormone receptors.

O'MALLEY: My second question relates to whether phosphorylation plays a role in matrix or association–dissociation reactions among matrix proteins, DNA and RNA.

COFFEY: The lamina proteins become phosphorylated just prior to mitosis and disperse into the cytoplasm during mitosis (Gerace, Blobel 1980).

O'MALLEY: What about matrix-bound RNA (or DNA)?

COFFEY: Well, during mitosis, of course, the DNA is condensing down to form the chromosomes, and there is a scaffolding to the chromosome and we and others would like to believe that some of these proteins in the matrix are part of the scaffolding. We don't think those DNA loops ever leave the matrix.

Discussants: G.L. GREENE, D.S. COFFEY, S. LIAO, E.B. THOMPSON, A.K. ROY and B.W. O'MALLEY

Chapter 16

Alpha-Protein: A Marker for Androgen Action in the Rat Ventral Prostate

SHEILA M. JUDGE, ALAN G. SALTZMAN, AND
SHUTSUNG LIAO

A common feature of the interaction between steroid hormones and their target tissues is the induction of specific protein synthesis in response to steroid hormone administration. Some well-established examples of this phenomenon include estrogen induction of ovalbumin synthesis in the chick oviduct, glucocorticoid induction of tyrosine aminotransferase in hepatoma cells, androgen induction of α_{2u} globulin synthesis in the liver, and synthesis of uteroglobin by the rabbit uterus in response to progesterone. Isolation and characterization of these steroid hormone-induced proteins and their corresponding mRNAs is crucial to the study of steroid hormone regulation of gene expression.

The isolation of an androgen-induced protein from the rat ventral prostate has led to a new model system for studying androgen action in the prostate. This "α-protein," so-called to distinguish it from the prostatic androgen receptor that was referred to as "β-protein" at that time (Fang and Liao 1971), has been isolated from both prostatic fluid and from cytosol preparations. Investigators working in this area have referred to this protein by a variety of names, some of which reflect the propensity of this protein for binding androgens and other steroids. For example, Lea et al. (1979) referred to this protein as "prostatein"; "prostatic binding protein" (PBP) is the term preferred by Heyns and DeMoor (1977) and Fosgren et al. (1979) designated it "estramustine binding protein." Isolation of a similar steroid-binding protein has also been described by Ichii (1975). Despite distinct differences in estimations of the size of the protein and its constituents and variations in reports of its steroid-binding affinities, the proteins described by these investigators appear to be identical to α-protein.

I. Structure

The studies by Heyns et al. (1978) and by us (Chen et al. 1979, 1982) have shown that α-protein has a molecular weight of about 50,000 and can be dissociated by sodium dodecyl sulfate into two different subunits (A and B, Fig. 1). In the presence of β-mercaptoethanol or dithiothreitol, subunit A dissociates into components I and III and subunit B dissociates into compo-

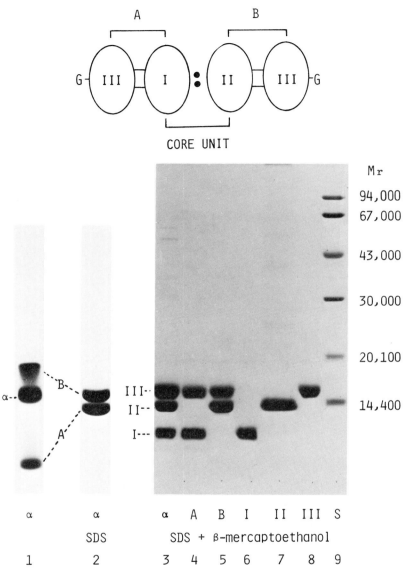

Fig. 1. Polyacrylamide gel electrophoretic patterns of purified subunits and components of α-protein. Gels 1 and 2 were of α-protein analyzed in the absence and presence, respectively, of SDS (samples were not treated with β-mercaptoethanol). Gels 3–9 contained SDS and samples were treated with SDS and β-mercaptoethanol. Molecular weight standards (gel S) were α-lactalbumin ($M_r = 14,400$), trypsin inhibitor ($M_r = 20,100$), carbonic anhydrate ($M_r = 30,000$), ovalbumin ($M_r = 43,000$), albumin ($M_r = 67,000$) and phosphorylase b ($M_r = 94,000$). Protein migrated from the top of the gels toward anode at the bottom of the gels. A diagrammatic representation of the arrangement of the protein components and subunits is also shown. Carbohydrate (G) is attached to component III.

Table 1. Apparent Molecular Weights, Carbohydrate Contents and Isoelectric Points of α-Protein, Prostatein, and their Components

Protein	$M_r \times 10^{-3}$	Carbohydrate content	PI
α^a	50	16.07	ND^c
A^a	24	11.8	4.8
B^a	24	8.0	5.6
I^a	10	Not detectable	4.7
II^a	14	Not detectable	ND^c
III^a	15	31	5.7
Prostatein^b	46	ND	4.8

^{a} Chen et al. (1982).
^{b} Lea et al. (1979).
^{c} Not determined.

nents II and III (Table 1). Since component III isolated from subunit A shares a common amino acid sequence and identical electromobility and immunochemical properties with component III isolated from subunit B, it appears that these peptides are identical.

Of the three peptide components of α-protein only component III contains carbohydrate (Table 2). The amino acid sequence of components I and III have been determined by us (Liao et al. 1982) and by Peeters et al. (1981). They contain 88 (M_r: 10,191) and 77 (M_r: 8653) amino acids, respectively (Fig. 2). Component III contains approximately 20% carbohydrate by weight. Only a partial amino acid sequence has been determined for component II at this time. The available data indicate that components I and II show considerable homology in their amino acid sequences (Heyns et al. 1981).

How the two subunits (A and B) interact with each other is not clear. During the development of a protocol for purifying the protein we found that component II would associate with component I throughout various chromatographic processes if SDS was omitted; components I and II therefore may form a core unit while maintaining individual disulfide linkages to component III (Fig. 1).

In the presence of $ZnCl_2$ α-protein forms dimers and tetramers with molecular weights of about 90,000 and 180,000. Freshly isolated α-protein con-

Table 2. Carbohydrate Content of α-Protein and Component III

Sugar	nmol Sugar/nmol protein^a		
	IIIA^b	α-Protein	IIIB^c
Fucose	3.16	6.56	2.86
Galactose	2.87	4.83	2.53
Mannose	11.08	22.5	10.79
Glucose	0.3	0.62	0.1
Glucosamine	5.46	10.6	4.51

^{a} Determined by GLC.
^{b} Isolated from subunit A.
^{c} Isolated from subunit B.

Component I

NH$_2$-Ser-Gln-Ile-Cys-Glu-Leu-Val-Ala-His-Glu^{10}-Thr-Ile-Ser-Phe-Leu-Met-Lys-
Ser-Glu-Glu^{20}-Glu-Leu-Lys-Lys-Glu-Leu-Glu-Met-Tyr-Ans^{30}-Ala-Pro-Pro-Ala-
Ala-Val-Glu-Ala-Lys-Leu^{40}-Glu-Val-Lys-Arg-Cys-Val-Asp-Gln-Met-Ser^{50}-Asp-
Gly-Asp-Arg-Leu-Val-Val-Ala-Glu-Thr^{60} Leu-Val-Tyr-Ile-Phe-Leu-Glu-Cys-
Gly-Val^{70}-Lys-Gln-Trp-Val-Glu-Thr-Tyr-Tyr-Pro-Glu-Ile^{80} Asp-Phe-Tyr-Tyr-Asp-
Met-Asn^{80}OH

Component III

Ser-Gly-Ser-Gly-Cys-Ser-Ile-Leu-Asp-Glu^{10}-Val-Ile-Arg-Gly-Thr-Ile-Asn(CHO)-
Ser-Thr-Val^{20}-Thr-Keu-His-Asp-Tyr-Met-Lys-Leu-Val-Lys^{30}-Pro-Tyr-Val-Gln-
Asp-His-Phe-Thr-Glu-Lys^{40}-Ala-Val-Lys-Gln-Phe-Lys-Gln-Cys-Phe-Leu^{50}-Asp-
Gln-Thr-Asp-Lys-Thr-Leu-Glu-Asn-Val^{60}-Gly-Val-Met-Glu-Ala-Ile-Phe-Asn-
Ser-Glu-Ser-Chs^{20}-Gln-Gln-Pro-Ser

Fig. 2. Primary structure of component III and component I. [Component I: Liao et al. (1982). Component III: Peeters et al. (1981).]

tains phospholipid and 0.7–1.0 mol of cholesterol per mole of protein. The cholesterol-bound α-protein sediments as a discrete 3.5 S component during sucrose gradient centrifugation (Chen et al. 1982).

Heyns et al. (1981) have found that a small proline-rich protein (M_r: 5200) is associated with α-protein isolated from the rat ventral prostate that dissociates from α-protein under conditions of low pH.

II. Properties

α-Protein freshly isolated from the prostate binds less than 0.2 mol of radioactive androgen per mole of protein. If the purified protein is delipidated by treatment with acetone, the steroid-binding capacity of the protein is significantly increased (cf. Heyns and DeMoor 1977). At steroid concentrations greater than 5 μM, about 1 mol of 5α-dihydrotestosterone binds to 1 mol of protein. The association constants (K_a) of delipidated α-protein for various steroids are: androstenedione, $1.15 \times 10^6 \, M^{-1}$; pregnenolone, $1.1 \times 10^6 \, M^{-1}$; 5$\alpha$-dihydrotestosterone, $0.87 \times 10^6 \, M^{-1}$ (Chen et al. 1982). Fosgren et al. (1979) have shown that α-protein binds estramustine with a K_a of 5.9×10^7 M^{-1}. We found that cholesterol is the most avidly bound natural ligand with a K_a of approximately $10^7 \, M^{-1}$. Cortisol does not bind to this protein.

On a molar basis the B subunit of α-protein appears to bind steroid nearly as well as intact α-protein, whereas steroid binding by subunit A is feeble. Of the three peptide components of α-protein only component II shows significant steroid-binding activity, but the insolubility of this peptide in aqueous solution in the absence of detergents precludes more detailed studies.

α-Protein does not bind radioactive spermine at pH 7.5; however, at pH 8.7 α-protein or a mixture of A and B will bind spermine and sediment as a 3 S complex. Besides spermine this protein also binds cadaverine, spermidine, and putrescine at the higher pH without notable specificity. In contrast, the androgen-sensitive spermine-binding protein (SBP) of the rat ventral prostate is highly specific toward spermine and does not bind other di- or poly amines well (Liang et al. 1977; Mezzetti et al. 1979). The presence of 1 μM 5α-dihydrotestosterone does not affect spermine binding by α-protein. Spermine at 1 mM also has no effect on the steroid-binding activity of the protein.

III. Regulation of the mRNA for α-Protein by Androgens

Using a cDNA hybridization technique, Parker and Mainwaring (1977) found that poly(A)-containing RNA from prostates of normal animals consists of several abundance classes; the most abundant RNA sequence class, which appears to code for the major prostatic polypeptides, was markedly reduced 3 days after castration (Parker and Scrace 1978).

Heyns et al. (1978) were the first to report that castration of rats markedly reduces the amount of prostatic binding protein (PBP or α-protein) mRNA in the rat ventral prostate. They were able to measure PBP mRNA levels by translating prostatic RNA in vitro, then precipitating radioactively labeled PBP from the translation products using a monospecific antiserum to PBP. These authors concluded that the mRNA for PBP decreases significantly within 4–7 days after castration and that this decrease could be reversed by treating castrated rats for 3 days with androgens.

The mRNAs corresponding to the three components of α-protein have been isolated and enriched by affinity chromatography on oligo(dT)-cellulose and by gradient centrifugation (Peeters et al. 1980; Wilson et al. 1981). Parker and coworkers (1980) have cloned the DNAs complementary to these RNAs in a bacterial plasmid, which has enabled them to isolate pure DNA probes specific for the α-protein RNAs. By measuring the extent of hybridization of prostatic RNAs with these specific cDNA probes it was discovered that the level of the mRNAs that code for α-protein increases three-fold within 2 h after testosterone administration to castrated rats (Parker et al. 1980; Hiremath et al. 1981a). Hiremath and Wang (1981) showed that castration reduces to 50% of normal levels these mRNA species in the prostate in about 18 h.

IV. Biological Aspects

α-Protein was first identified as a protein that binds steroids and that can inhibit the binding of radioactive DHT–receptor complex to prostate nuclei (Fang and Liao 1971). This protein promotes the release of the androgen–receptor complex already attached to chromatin through a temperature-dependent process in vitro (Shyr and Liao 1978). These effects of α-protein are not due to irreversible destruction of the receptor complex or to damage to the nuclear binding site. As shown in Fig. 3 only subunit A is active. Component I is at least five-fold more active than subunit A, whereas components II and III are inactive. Since component I does not bind DHT or other steroids, the steroid-binding activity of α-protein is probably not related to the dissociation of the receptor complex from chromatin (Chen et al. 1982). In addition, α-protein and component I inhibit the binding and promote the release of DHT–receptor complex (or estradiol–receptor complex) from DNA or DNA–cellulose.

There is no direct evidence that α-protein or its components act as regulators in vivo of prostatic cell function. Our observations, however, provide

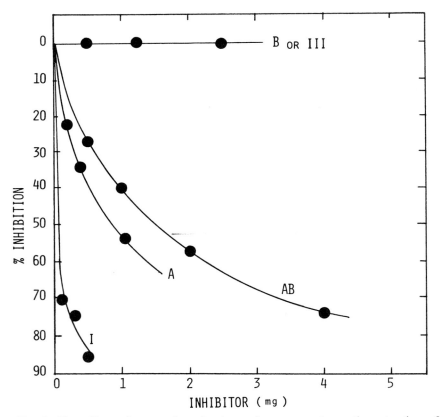

Fig. 3. The effect of α-protein subunits and components on the retention of the radioactive 5α-dihydrotestosterone–receptor complex by prostate cell nuclei. The nuclear retention assay was performed in the presence of various amounts of α-protein preparations that had been dialyzed extensively against medium ET. The results were compared with the amount of the radioactive-receptor complex retained in the absence of α-protein preparations and plotted as percent inhibition of retention on the ordinate. The protein fractions tested were α-protein (AB), subunits A and B, and components I and III at the concentrations shown on the abscissa. [From Chen et al. (1982).]

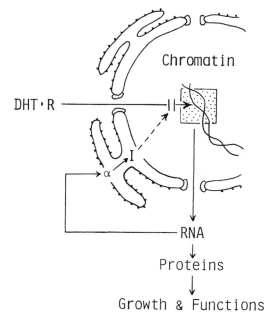

Fig. 4. A hypothetical model for cellular organelle-linked control of chromatin binding of the steroid–receptor complex and RNA synthesis. Steroid hormone withdrawal results in the disappearance of the steroid–receptor complexes from the functional nuclear sites (or chromatin), the reduction of cellular synthetic activity, and atrophy of some cytoplasmic organelle structures. Resupply of the steroid hormone promotes interaction between steroid–receptor complexes and functional sites in the nucleus and increases specific RNA synthesis. Steroid hormone treatment also stimulates the development of the endoplasmic reticulum structure, which is the site of synthesis of secretory protein (such as α-protein). When the intracellular concentration of these proteins regains maximal levels, a component of α-protein (such as component I) may enter the nucleus and inhibit the interaction of the receptor complex with the nuclear sites, thereby reducing the amount of specific RNA transcription. A similar 'feedback' process may apply to the actions of steroid hormones at nongenomic sites (cf. Shyr and Liao 1978; Chen et al. 1979).

the first model Fig. 4 for the study of protein-mediated control of the interaction of steroid–receptor complexes with nuclear chromatin. It is not inconceivable to consider that an ultimate protein product of androgen action could exert feedback control at the genomic level by promoting the release of steroid–receptor complexes from their site of action (Shyr and Liao 1978; Chen et al. 1979).

The secretion of α-protein into prostatic fluid and its presence in the seminal fluid and copulatory plug of rats (Lea et al. 1979; Borgstrom et al. 1981) has led us to speculate that α-protein may play an as yet undiscovered role in male fertility. These observations, coupled with our discovery of the cholesterol-binding properties of α-protein, have suggested to us that this protein might affect the phenomenon of sperm capacitation by altering the lipid composition of sperm cells. The importance of sperm lipid composition to capacitation, in particular the ratio of cholesterol of phospholipid, has already been described (David 1981). Through its potential to act as a cholesterol donor α-protein may prevent premature sperm capacitation; that is, it may act as a decapacitation factor (DF). Such a "decapacitation factor" has been previously described as a lipid-binding glycoprotein that is associated with the membrane vesicle components of seminal plasma; however there are many conflicting reports regarding the nature of DF and isolation of DF has proved to be difficult.

The rat ventral prostate is just one of several model systems for studying androgen action. As will be discussed in the following chapters in greater detail, important discoveries regarding the mechanism of androgen action have also been made using the mouse kidney and rat liver as model systems. The mouse kidney model and the development of several variant mouse kidney cell lines have played an important part in our understanding of the phenomenon of androgen insensitivity. The use of the liver model system for studying androgen action has other advantages; in particular it allows us to examine the interaction of androgens with other hormones, as described in Chapter 19.

The ability to describe the mechanism of androgen action is particularly important when it involves treatment of androgen-dependent disease processes, such as benign prostatic hyperplasia. The development of new 5α-reductase inhibitors (as discussed in Chapter 19) has shown that androgen action at the level of the target cell may be modulated therapeutically at the enzyme level.

References

Borgstrom E, Pousette A, Bjork P, Hogberg B, Carlstrom K, Sundelin B, Gustafsson J-A (1981) The Prostate 2: 425–432
Chen C, Hiipakka RA, Liao S (1979) J Steroid Biochem 11: 401–405
Chen C, Schilling K, Hiipakka RA, Huang I-Yih, Liao S (1982) J Biol Chem 257: 116–121
Davis BK (1981) Proc Natl Acad Sci USA 78: 7560–7564
Fang S, Liao S (1971) J Biol Chem 246: 16–24
Forsgren B, Bjork P, Carlstrom K, Gustafsson J, Pousette A, Hogberg B (1979) Proc Natl Acad Sci USA 76: 3149–3153
Heyns W, DeMoor P (1977) Eur J Biochem 78: 221–230
Heyns W, Peeters B, Mous J (1977) Biochem Biophys Res Commun 77: 1492–1499
Heyns W, Peeters B, Mous J, Rombauts W, DeMoor P (1978) Eur J Biochem 77: 181–186
Heyns W, Peeters B, Mous J, Bossyns D, Rombauts W, DeMoor P (1981) Prog Clin Biol Res 75A: 339
Hiremath ST, Wang TY (1981) Biochemistry 20: 6672–6676
Hiremath ST, Mpanias OD, Wang TY (1981) Exp Cell Res 134: 193–200
Ichii (1975) Endocrinol Jpn 5: 433
Lea OA, Petrusz P, French FS (1979) J Biol Chem 254: 6196–6202
Liao S, Chen C, Huang I-Yih (1982) J Biol Chem 257: 122–125
Parker MG, Mainwaring WIP (1977) Cell 12: 401–407

Parker MG, Scrace GT (1978) Eur J Biochem 85: 399–406
Parker MG, Scrace GT, Mainwaring WIP (1978) Biochem J 170: 115–121
Parker MG, White R, Williams JG (1980) J Biol Chem 255: 6996–7001
Peeters BL, Mous JM, Rombauts WA, Heyns WJ (1980) J Biol Chem 255: 7017–7023
Peeters B, Rombauts W, Mous J, Heyns W (1981) Eur J Biochem 115: 115–121
Shyr C-I, Liao S (1978) Proc Natl Acad Sci USA 75: 5969–5973
Wilson EM, Viskochil DH, Bartlett RJ, Lea OA, Noyes CM, Petrus P, Stafford DW, French FS (1981) Progr Clin Biol Res 75A: 351

Chapter 17

How Changes in Cytosol and Nuclear Androgen Receptors Relate to the Testosterone Responses: Studies with New Exchange Assays

OLLI A. JÄNNE, VELI V. ISOMAA, ANTTI E.I. PAJUNEN, WILLIAM W. WRIGHT, AND C. WAYNE BARDIN

I. Introduction

Testosterone and other androgenic steroids facilitate synthesis of several renal proteins in the mouse (reviewed in Bardin and Catterall, 1981). The postulate that androgen receptors participate in this regulation has been supported by experiments on androgen-insensitive (Tfm/Y) animals (Bardin et al. 1973; Bullock and Bardin 1974; Attardi and Ohno 1974), by studies on the uptake of radioactive testosterone by renal nuclei in vivo and in vitro (Bullock and Bardin 1975; Jänne et al. 1976; Bardin and Catterall 1981), and by the demonstration that the binding affinity of androgens to cytoplasmic androgen receptors relates to their biological potency in mouse kidney (Brown et al. 1979). However, the lability of renal androgen receptors has until recently hampered acquisition of data correlating changes in nuclear and cytosol androgen receptor concentrations to subsequent biological responses. The cytosol androgen receptor assays from mouse kidney were recently improved in our laboratory by taking advantage of the stabilizing effect of sodium molybdate on the unoccupied androgen receptors (Wright et al. 1981). In subsequent studies, a reliable method was devised for measurement of nuclear androgen receptors in this tissue (Isomaa et al. 1982). Combined use of these assay techniques allowed us, for the first time, to relate androgen receptor concentrations in vivo to the expression of biological androgen activity such as induction of renal β-glucuronidase and ornithine decarboxylase enzymes (Pajunen et al. 1982). The present chapter summarizes the main methodological aspects and experimental findings in these latter studies.

II. Stabilization of Cytosol Androgen Receptors by Sodium Molybdate

Sodium molybdate has been shown to prevent degradation in vitro of glucocorticoid, progesterone, estrogen, and androgen receptors (Nielsen et al. 1977; Toft and Nishigori 1979; McBlain and Shyamala 1980; Anderson et al. 1980; Gaubert et al. 1980; Noma et al. 1980). In the case of mouse renal androgen receptors, inclusion of 5–20 mM sodium molybdate in the homogenization and assay buffers increased measurable cytosol receptor content by 50–100% (Fig. 1), from about 20 to 40 fmol/mg protein (Wright et al. 1981). Sodium molybdate did not influence to an appreciable extent the apparent equilibrium dissociation constant (K_D) for the methyltrienolone–receptor interaction, nor did it significantly alter the apparent association and dissociation rate constants for this interaction (Wright et al. 1981).

Sodium molybdate stabilized the unoccupied androgen receptor over a wide range of temperatures. For example, in the presence of molybdate, the half-times of receptor inactivation at 4°C and 35°C were 204 h and 0.75 h,

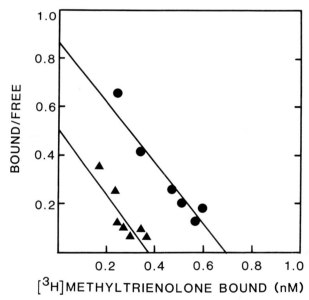

Fig. 1. Effect of 20 mM sodium molybdate on the measurable androgen receptor content in the mouse renal cytosol. Cytosol was prepared in TEGD buffer [50 mM Tris/HCl, 0.1 mM Na$_2$EDTA, 5 mM dithiothreitol, 10% (v/v) glycerol, pH 7.4 at 20°C] in the presence (●) or absence (▲) of sodium molybdate. Portions of the cytosol samples were incubated with different concentrations of [^{3}H]methyltrienolone with and without a 1000-fold molar excess of radioinert testosterone for 24 h at 4°C, after which bound and free steroids were separated by filtration through DE-81 discs and specific binding data plotted according to the method of Scatchard. [From Wright et al. (1981).]

respectively, whereas in the absence of molybdate the half-time of receptor inactivation at 4°C was only about 7 h. The enthalpy of receptor inactivation, as measured by Arrhenius analysis, was 21% greater in the presence than in the absence of sodium molybdate (31.9 vs. 26.4 kcal/mol) (Wright et al. 1981). This result is compatible with the postulate that molybdate stabilizes the androgen receptor through direct binding to the protein and thereby protects the receptor against enzymatic cleavage and/or dephosphorylation.

Prevention of receptor degradation is crucial for quantitative receptor assays and, in the case of mouse kidney, this was achieved through the use of 10–20 mM sodium molybdate in the buffers. Equally important, however, is achievement of a complete exchange of endogenously bound steroids with the radioactive ligand used in the measurement. It was, therefore, pertinent to assess the rate of dissociation of ligands from renal androgen receptor. Testosterone was first examined since it is the major physiological androgen in mouse kidney in vivo (Bardin et al. 1973; Bardin and Catterall 1981), and it has been used in most studies in which regulation of renal enzymes was investigated (see below). Androgen receptor–testosterone complex dissociated fairly rapidly at 4°C with a half-time of approximately 6 h (Fig. 2). This

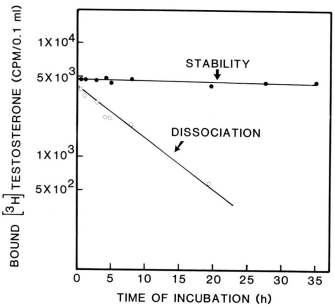

Fig. 2. The rate of dissociation of cytosol androgen receptor–[^{3}H]testosterone complex. Mouse renal cytosol was prepared in TEGD buffer containing 10 mM sodium molybdate and labeled with 15 nM [^{3}H]testosterone for 18 h at 4°C. To start the dissociation rate measurement, a 1000-fold molar excess of nonlabeled testosterone was added to one-half of the sample, whereas the other half was employed for controlling receptor stability. One-tenth milliliter aliquots were removed at timed intervals for measurement of receptor-bound [^{3}H]testosterone by the DE-81 filter technique.

was markedly shorter than the reported dissociation rate of testosterone from rat testicular androgen receptor with a half-time of about 30 h (McLean et al. 1976). 5α-Dihydrotestosterone could not be studied accurately with mouse kidney cytosol as its metabolism under the assay conditions is relatively rapid (Bullock and Bardin 1974; Mowszowicz and Bardin 1974). Another potent androgen, methyltrienolone, dissociated approximately four times more slowly than testosterone, which is in agreement with the four-fold higher affinity of the former steroid to the androgen receptor (Wright et al. 1981). These results were used to develop an exchange assay for measuring total rather than unoccupied androgen receptor concentration in intact or testosterone-treated mice. The routine method requires an 18- to 20-h incubation of the samples at 4°C, a time interval that permitted exchange of at least 90% of the endogenously bound testosterone from the cytosol receptor with [³H]methyltrinenolone (Fig. 2).

III. Validation of Nuclear Androgen Receptor Assays

The measurable androgen receptor concentration in mouse renal nuclei was strikingly influenced by the technique used for nuclear isolation and by the method for extracting the receptors from nuclei.

A. Isolation of Nuclei

In the initial experiments, kidneys were homogenized in the buffer used for cytosol preparation (Fig. 1) and renal nuclei from the 1500-g pellet were then purified by centrifugation through 2 M sucrose. Although appreciable androgen binding was detected in these nuclei, the quantity of receptors recovered accounted for only a minor fraction of that lost from the cytosol after testosterone administration. Since this buffer contained glycerol and sodium molybdate, which are useful in the stabilizing cytosol androgen receptors (Gaubert et al. 1980; Wright et al. 1981), it seemed unlikely that nuclear androgen receptors were degraded in this buffer; rather, loss of receptors from nuclei in later steps of the procedure was considered more plausible. Previous experience in our own (Kokko et al. 1977) and other laboratories (Conn and O'Malley 1975; Hardin et al. 1976; Wray et al. 1977) indicated that hexylene glycol stabilizes nuclear envelope and permits good recovery of many nucleoplasmic proteins that leak out of nuclei prepared in other buffers (Wray et al. 1977). When hexylene glycol-containing buffer was substituted for the cytosol buffer as the initial homogenization medium, a three- to four-fold greater recovery of androgen receptors than those measured without hexylene was obtained (Isomaa et al. 1982). These observations illustrate the importance of the initial homogenization buffer for the subsequent nuclear androgen receptor assays. It remains to be established whether other mechanisms (e.g., inhibition of proteolysis), in addition to stabilization of nuclear envelope, are responsible for the improved receptor recovery with hexylene glycol. Subsequently, this technique has been used successfully in our laboratory to isolate nuclei for other steroid receptor assays from a variety of tissues.

B. Extraction of Nuclear Androgen Receptors

Different concentrations of KCl (Jensen et al. 1968), heparin (Mulder et al. 1980, 1981) and pyridoxal 5'-phosphate (Cidlowski and Thanassi 1978; Dolan et al. 1980; Müller et al. 1980; Isomaa 1981; Mulder et al. 1980, 1981) were studied for their ability to release androgen receptors from renal nuclei. Of these ingredients, pyridoxal 5'-phosphate proved most efficient (Isomaa et al. 1982). Presence of 2.5–5 mM pyridoxal 5'-phosphate in the buffers greatly increased the quantity of androgen receptors recovered in the nuclear extracts (Fig. 3). In subsequent studies, 5 mM pyridoxal 5'-phosphate was routinely utilized.

The efficiency of the pyridoxal 5'-phosphate extraction method for quantitation of renal nuclear androgen receptors was superior to direct exchange assay (Anderson et al. 1972) and KCl extraction (Jensen et al. 1968), two

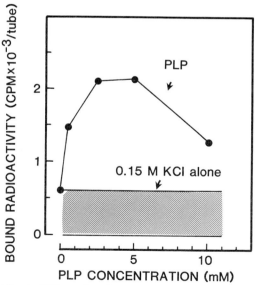

Fig. 3. Effect of pyridoxal 5'-phosphate concentration on the efficiency of extraction of mouse renal androgen receptors from nuclei. Female mice were given 1 mg of testosterone and killed 1 h thereafter. Renal nuclei were isolated with the hexylene glycol technique and extracted using the indicated concentrations of pyridoxal 5'-phosphate in barbital buffer. Extracts were incubated in triplicate with 22.5 nM [³H]methyltrienolone with and without a 1000-fold molar excess of unlabeled testosterone for 18 h at 4°C. Unbound ligand was removed with hydroxylapatite. The cross-hatched area shows nuclear receptors extracted with barbital buffer containing 0.15 M KCl alone. [From Isomaa et al. (1982).]

other techniques widely used in nuclear steroid receptor measurement. The latter two methods yielded nuclear receptor concentrations that were 60–70% of those achieved with the pyridoxal 5'-phosphate extraction technique (Isomaa et al. 1982). These studies were performed with animals treated with nonradioactive testosterone 60 min prior to the nuclear receptor assays and, therefore, were not corrected for unextracted receptors. To determine the efficacy of the pyridoxal 5'-phosphate step, female mice were treated with [³H]methyltrienolone (50 μCi/animal, IP) with and without a 1000-fold excess of the corresponding radioinert steroid. If one assumes that all the radioactive ligand that specifically accumulated in renal nuclei (total radioactivity minus radioactivity not competed for by the radioinert steroid) represented androgen-receptor bound [³H]methyltrienolone, pyridoxal 5'-phosphate had an extraction efficiency of 87% (Isomaa et al. 1982).

C. Characterization of Nuclear Androgen Receptors

Nuclear steroid receptors are usually associated with endogenous ligands that must be exchanged with a [³H]steroid, if total nuclear receptor content is to be measured. It was, therefore, important to assess optimal time and

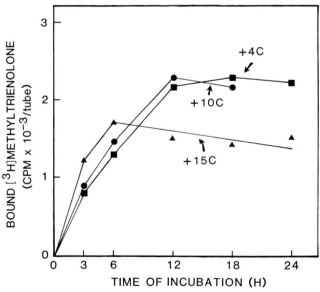

Fig. 4. Effect of incubation time and temperature of [³H]methyltrienolone binding in vitro to renal nuclear androgen receptor. Renal nuclei were isolated from testosterone-treated female mice. Nuclear receptors were extracted with 5 mM pyridoxal 5'-phosphate in barbital buffer. The extract was incubated with 22.5 nM [³H]methyltrienolone in the presence and absence of a 1000-fold molar excess of nonradioactive testosterone at 4, 10, and 15°C. Each point represents specific binding in triplicate incubations. [From Isomaa et al. (1982).]

temperature for [^{3}H]methyltrienolone binding to the renal nuclear androgen receptors. In these studies, the "endogenous" steroid was testosterone, which was administered to animals prior to the isolation of nuclei. The time courses of [^{3}H]methyltrienolone binding at three different temperatures (4°C, 10°C, and 15°C) are illustrated in Fig. 4. The specific binding at 4°C reached a maximum within 12 h and stayed at the same level until 24 h. An almost identical exchange rate was observed at 10°C, whereas at 15°C the rate of [^{3}H]methyltrienolone exchange was initially faster, but the maximal binding was lower by 12 h (Fig. 4). When even higher temperatures (20°C and 25°C) were employed during the exchange reaction, the specifically bound [^{3}H]methyltrienolone concentration was lower than at 15°C, suggesting temperature-dependent receptor degradation. The result that a complete exchange of the endogenously bound testosterone was achieved over an 18-h incubation period at 4°C seems to imply that the binding kinetics of cytosol (Fig. 2) and nuclear (Fig. 4) androgen receptors in mouse kidney are very similar. In addition, the steroid-binding specificities of these two receptor populations are almost indistinguishable (see below).

[^{3}H]Methyltrienolone bound to the pyridoxal 5'-phosphate-extract in a saturable fashion with a K_D of about 5 nM (Fig. 5). The remarkably low nonspecific binding of [^{3}H]methyltrienolone under these assay conditions was of special note; in the highest [^{3}H]methyltrienolone concentration (22.5 nM), which corresponded to about 400,000 cpm/tube, the nonspecific binding accounted for only about 200 cpm/tube. The specific binding of

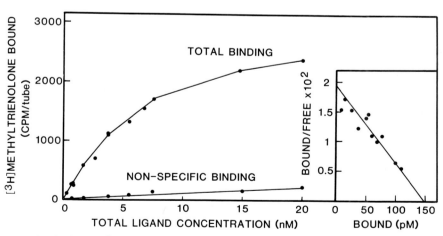

Fig. 5. Binding of [^{3}H]methyltrienolone to pyridoxal 5'-phosphate extracted nuclear androgen receptor. Nuclei were isolated from kidneys of testosterone-treated female mice. The extract was incubated for 18 h at 4°C with indicated concentrations of [^{3}H]methyltrienolone in the presence (nonspecific binding) and absence (total binding) of a 1000-fold molar excess of nonradioactive testosterone. Inset: Specific binding data plotted according to the method of Scatchard. [From Isomaa et al. (1982).]

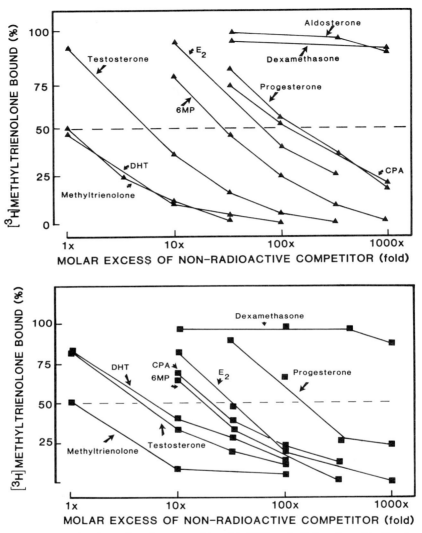

Fig. 6. Competition in vitro of various steroids with [³H]methyltrienolone binding to the nuclear (upper panel) and cytosol (lower panel) androgen receptor from mouse kidney. Nuclear receptors were extracted with 5 mM pyridoxal 5′-phosphate in barbital buffer from kidneys of testosterone-treated female mice. The extracts were incubated with [³H]methyltrienolone (22.5 nM) and increasing concentrations of the indicated nonlabeled steroids at 4°C for 18 h. Bound and free ligands were separated by hydroxylapatite. Renal cytosol was prepared from intact female mice in TEGD buffer (Fig. 1) and incubated with [³H]methyltrienolone and nonradioactive competitors at 4°C for 18 h, after which bound and free radioactive steroids were separated with dextran-coated charcoal. Each point represents the mean of triplicate measurements. The values are normalized to samples incubated with [³H]methyltrienolone alone, which was set at 100%. Abbreviations: DHT, 5α-dihydrotestosterone; 6MP, 6α-methylprogesterone; E$_2$, estradiol-17β, and CPA, cyproterone acetate.

[^{3}H]methyltrienolone to nuclear extracts originating from DNA concentrations of 0–300 μg/sample was linearly related to the nuclear DNA content indicating that the assay was strictly proportional to the amount of receptor in the assay tube. Equally important was the finding that no specific binding of the tracer was detected when buffer alone was substituted for the nuclear extract (Isomaa et al. 1982).

Methyltrienolone is a convenient synthetic steroid for androgen receptor assays, as it binds to the androgen receptor with high affinity, is not readily metabolized under the assay conditions, and does not interact with extracellular androgen-binding proteins (Bonne and Raynaud 1976). This steroid is not, however, androgen receptor specific, as it binds to progesterone receptor with high affinity and to glucocorticoid and mineralocorticoid receptors with moderate affinity (Ojasoo and Raynaud 1978). To ensure that only renal androgen receptor was measured with [^{3}H]methyltrienolone, steroid specificity for the binding was assessed. Of the different classes of steroids tested, methyltrienolone and 5α-dihydrotestosterone were equally potent competitors for [^{3}H]methyltrienolone binding to the nuclear extract, followed by testosterone and 6α-methylprogesterone. Estradiol, cyproterone acetate, and progesterone displayed slight competition, while dexamethasone and aldosterone were inactive (Fig. 6, upper panel). Essentially identical competition curves were generated for different steroids when mouse renal cytosol androgen receptor was utilized instead of the nuclear extract (Fig. 6, lower panel). Poor competition by 5α-dihydrotestosterone in the latter experiments is due to its rapid metabolism in the cytosol (Bullock and Bardin 1974; Mowszowicz and Bardin 1974).

IV. Choosing the Biological Markers of Androgen Action

Of the androgen-responsive proteins in the mouse kidney, β-glucuronidase is most extensively studied due to its relatively large increase following testosterone treatment (Swank et al. 1973, 1978; Bardin et al. 1978; Bardin and Catterall 1981). There is, however, a long lag period from the beginning of treatment until a rise in β-glucuronidase activity (Swank et al. 1978). As a consequence, it was not considered an ideal marker protein for studying acute androgenic responses. In the process of searching for a protein whose activity would increase rapidly with androgen treatment, enzymes involved in polyamine biosynthesis were examined. Of these, ornithine decarboxylase (ODC) was expected to be a possible candidate, since many different anabolic hormones produce an early stimulation of this enzyme in a variety of organs (Jänne et al. 1978). Moreover, chronic androgen administration was known to stimulate ODC activity in mouse kidney (Henningsson et al. 1978). Before adopting this enzyme as the marker for androgen action, we evaluated steroidal specificity of ODC regulation in the mouse kidney and determined the dependency of ODC regulation on functional androgen receptors and pituitary hormones.

V. Hormonal Control of Renal ODC Activity in the Mouse

The basal activity of renal ODC was 12-fold higher in intact male than in female mice (Fig. 7). This enzyme activity decreased rapidly following castration, and reached the female values within two days. On the other hand, castration did not affect the activity of S-adenosylmethionine decarboxylase (AdoMet-DC), another enzyme involved in polyamine biosynthesis, nor was there a sex difference in this enzyme activity. As was expected from the ODC activities, the renal putrescine concentration was significantly higher in male than in female mice. By contrast, other polyamine (spermidine and spermine) concentrations were not affected by the sex or castration of the animals (Pajunen et al. 1982).

To demonstrate which classes of steroids participated in the regulation of mouse renal ODC activity, castrated male mice were treated for five days with 5α-dihydrotestosterone (1 mg/day), estradiol (20 μg/day), progesterone (1 mg/day), cortisol (1 mg/day), and dexamethasone (100 μg/day). In addition, groups of animals were also treated with 5α-dihydrotestosterone combined with either estradiol or progesterone at the preceding doses. Five-day treatment was employed, since a preliminary study indicated that maximal ODC activity was obtained after this period. Treatment with 5α-dihydrotes-

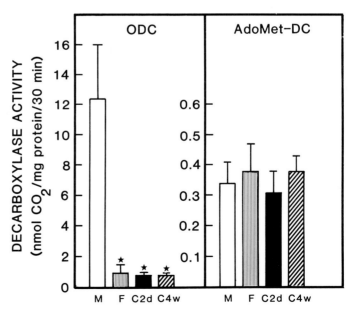

Fig. 7. Activities of renal ODC and AdoMet-DC enzymes in intact and castrated mice. Mice of NCS strain were used in the experiment. Each bar shows the mean ± SD for six animals. Asterisks indicate that values are significantly different from those of intact male mice. Abbreviations: M, intact male; F, intact female; C2d, male mice 2 days after castration; and C4w, male mice 4 weeks after castration. [From Pajunen et al. (1982).]

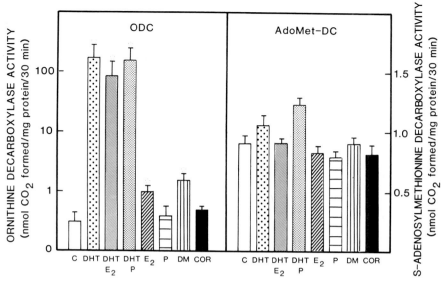

Fig. 8. Effect of different steroids on the activities of ODC and AdoMet-DC in mouse kidney. Castrated male mice of A/J strain were treated for 5 days with daily intraperitoneal injections of 5α-dihydrotestosterone (DHT, 1 mg/day), estradiol (E_2, 20 μg/day), progesterone (P, 1 mg/day), cortisol (COR, 1 mg/day), and dexamethasone (DM, 100 μg/day). The effects of the administration of DHT plus E_2 and DHT plus P were also examined using the same doses. The animals were killed 24 h after the last treatment. Each bar shows the mean $\pm$ SE for 6–10 animals. Asterisks indicate values significantly different from those of vehicle-treated, castrated control mice. The ODC activity from DHT alone was significantly higher than that after DHT plus E_2. [From Pajunen (1982).]

tosterone evoked a 600-fold increase in the renal ODC activity (Fig. 8). A similar increase was achieved with a 5-day testosterone treatment (data not shown). Estradiol administration produced only a two-fold stimulation of enzyme activity. However, when this steroid was given concurrently with 5α-dihydrotestosterone, it had a slight but significant antiandrogenic action. Progesterone did not affect renal ODC activity when given alone or in combination with 5α-dihydrotestosterone (Fig. 8). Dexamethasone increased enzyme activity two-fold; cortisol, which is known to stimulate rat kidney ODC activity (Brandt et al. 1972; Nicholson et al. 1976), had no effect even after 5 days of treatment. None of the preceding steroids elicited a statistically significant change in the renal AdoMet-DC activity.

To determine whether the androgen-induced stimulation of ODC activity was dependent on the presence of functional androgen receptors, 5α-dihydrotestosterone (1 mg/day for 6 days) was administered to Tfm/Y mice, which have a defective androgen receptor (Bullock and Bardin 1974; Attardi and Ohno, 1974). Although Tfm/Y mice had a basal renal ODC activity similar to that in the untreated mice, androgen treatment failed to elicit significant changes in the activity of the enzyme as it did in intact female

Table 1. Induction of Renal ODC Activity by 5α-Dihydrotestosterone Administration: Requiring functional androgen receptors but independent of pituitary hormones[a]

Animal group	Ornithine decarboxylase activity (nmol CO_2 formed/mg protein in 30 min)
Tfm/Y, controls	0.27 ± 0.15 (4)
Tfm/Y, 5α-DHT	0.40 ± 0.21 (4)
Balb/c, hypophysectomized	0.36 ± 0.08 (4)
Balb/c, hypophysectomized + 5α-DHT	47.78 ± 15.04 (4)
Balb/c, intact controls	0.16 ± 0.07 (4)
Balb/c, 5α-DHT	29.90 ± 10.32 (4)

[a] The animals were treated with 5α-dihydrotestosterone (5α-DHT, 1 mg/day for 5 days) and killed 24 h after the last steroid dose. Control mice received the vehicle only. Hypophysectomized mice were used for the experiments 2 weeks after hypophysectomy. (Pajunen AEI, Isomaa VV, Jänne OA, Bardin CW, J Biol Chem 257: 8190–8198, 1982).

Balb/c mice treated in an identical manner (Table 1). These results and those cited above indicate that androgen treatment strikingly stimulates renal ODC activity, and the magnitude of the response is greater than that previously reported for other androgen-regulated renal enzymes (see Bardin and Catterall 1981). Moreover, the stimulation appears to be androgen specific and requires functional androgen receptors.

Hypophysectomy has been shown to dramatically reduce ODC activity in the rat kidney (Nicholson et al. 1976). In the mouse, pituitary hormones were not essential for androgen-stimulated increase in ODC, since this enzyme activity was similar in intact and hypophysectomized animals both before and after treatment with 5α-dihydrotestosterone (Table 1). Thus, mouse kidney ODC is similar to renal alcohol dehydrogenase and D-amino acid oxidase enzymes (Swank et al. 1977), whose activities are controlled by androgens independent of the pituitary. These enzymes are distinctly different from β-glucuronidase, whose androgen responsiveness is partially dependent on pituitary hormones (Swank et al. 1977).

VI. Acute Androgen Action: Receptor Changes and Biological Responses in Mouse Kidney

In these studies, male mice were treated 2 days after castration with single injections of testosterone at four different doses (0.3, 1, 3, and 10 mg). The changes in cytosol and nuclear androgen receptor concentrations and activities of renal ODC and β-glucuronidase enzymes were followed from 0.5 to 216 h after steroid treatment.

A. Receptor Concentrations

Cytosol androgen receptor concentrations decreased immediately following testosterone treatment with a nadir at 30–60 min (Fig. 9). Receptor decline was dose-related in that the concentrations were 28% (0.3 mg), 24% (1 mg),

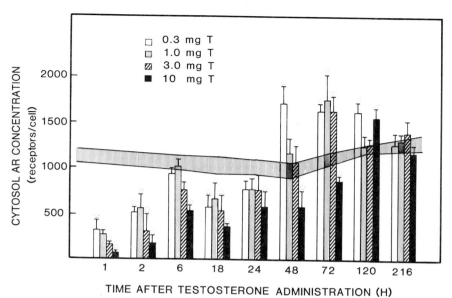

TIME AFTER TESTOSTERONE ADMINISTRATION (H)

Fig. 9. Changes in cytosol androgen receptor concentrations in mouse kidney after administration of a single dose of testosterone. The indicated doses of testosterone were given intraperitoneally to male mice of NCS strain which had been castrated for 2 days. Control animals received vehicle only. Animals were killed at the specified time intervals. The shaded area shows the receptor concentration (mean ± SE range) in vehicle-treated animals, while the bars indicate mean ± SE values in at least 6 animals per experimental group. The values which are significantly different from the control level are marked with asterisks. [From Pajunen et al. (1982).]

14% (3 mg), and 7% (10 mg) of the pretreatment values 60 min after the various testosterone doses. The receptor content returned to control values between 4 to 6 h after 0.3 and 1 mg of testosterone and by 8 h after 3 mg of the steroid. By contrast, the highest dose of testosterone (10 mg) brought about prolonged decline in the cytosol androgen receptor content, which returned to the pretreatment level between 72 and 120 h after steroid injection. The time required for cytosol receptor replenishment was directly dose-dependent whereas the overshoot in the receptor concentrations above those of controls was inversely related to the dose of testosterone (Fig. 8). For example, following the three smallest doses (0.3, 1, and 3 mg), the cytosol androgen receptor content exceeded the control levels by 78%, 60%, and 48%, respectively, between 48 and 72 h, whereas after the largest dose the receptor concentration exceeded the control levels by only 26% at 120 h (Fig. 9).

The concentration of renal nuclear androgen receptors was five times higher than in the vehicle-treated control animals at 30 min after testosterone administration (Fig. 10). Interestingly, at this time nuclear receptor accumulation was not dose-dependent. The androgen receptor concentrations in renal nuclei reached their maxima within one hour at which time there were

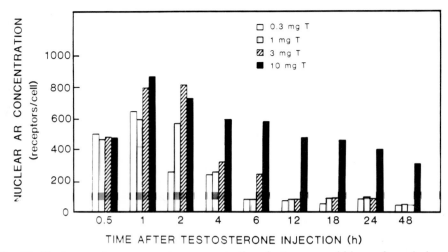

TIME AFTER TESTOSTERONE INJECTION (h)

Fig. 10. Nuclear androgen receptor concentrations in the mouse kidney after administration of testosterone. Two-day castrated male mice of the NCS strain were given single intraperitoneal doses of testosterone (0.3, 1, 3, and 10 mg) and killed at timed intervals. Nuclear androgen receptors were measured from kidneys pooled from four to six animals. The cross-hatched area shows the mean ± SE range for receptor values in vehicle-treated control animals, while each bar represents mean values for two to three separate experiments. [From Pajunen et al. (1982).]

about 600 receptors/cell in animals treated with the smaller doses (0.3 and 1 mg) and about 800–850 receptors/cell with the larger doses (3 and 10 mg). After the first hour, the duration of nuclear androgen receptor residence was related to the dose of testosterone in that the smaller the steroid dose, the faster receptor concentration declined to the level of vehicle-treated animals (Fig. 10). In mice that received 0.3–3 mg testosterone the nuclear androgen receptors returned to control levels within 6–12 h. However, in animals that were treated with 10 mg, receptor concentration was still increased threefold at 48 h posttreatment (Fig. 10). Even at 72 h, it was twice as high as in the controls (250 receptors/cell).

B. Renal ODC and β-Glucuronidase Activities

ODC activity increased rapidly in mouse kidney after a single dose of testosterone (Fig. 11). The stimulation occurred in a dose-dependent manner until 8 h of steroid administration (Fig. 11, inset); thereafter, the largest dose of testosterone produced an increase in ODC activity that was proportionally much greater than those achieved with the other steroid doses. It is of note, however, that the three smallest doses of testosterone increased ODC activity in a dose-dependent manner during all the time intervals. In these groups the peak activities were measured at 12 h after testosterone administration, and were 5–12 times higher than the vehicle-treated animals (Fig. 11, inset). The largest dose of testosterone brought about a striking and long-lasting

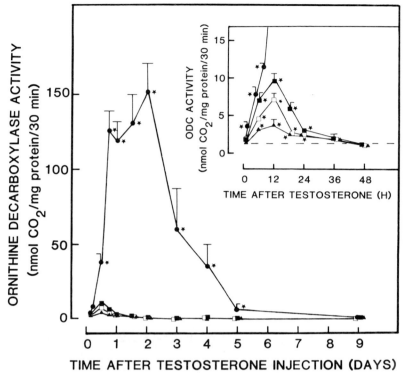

Fig. 11. Mouse renal ODC activity after testosterone administration. Groups of castrated male NCS mice were treated with single doses of testosterone (▲, 0.3 mg; ⊟, 1 mg; ■, 3 mg; ●, 10 mg) and killed at the indicated times. Each symbol with a vertical bar shows the mean ± SE for a given experimental group comprising at least six animals. An absent vertical bar indicates that the SE was less than the width of the symbol. Asterisks indicate significantly different values from the control ODC level, which was 0.83 ± 0.21 nmol CO_2 formed/mg protein in 30 min. Inset: Changes in renal ODC activity during the first 48 h after testoserone administration. Note the five-fold expanded scale in the ordinate. [From Pajunen et al. (1982).]

stimulation of ODC activity, which increased 150-fold at 18 h after testosterone; over the ensuing 30 h, there was a minor additional increase after which ODC activity subsided. Enzyme activity comparable to that in controls was regained between 5 and 9 days after the administration of 10 mg testosterone (Fig. 11). In addition, the lag period required to detect increased ODC activity was shorter (1.5–2 h) after the largest dose than the other doses of testosterone (4–6 h).

The changes in renal β-glucuronidase activity occurred much later than those in the ODC. The three smallest doses of testosterone brought about a slight but dose-dependent stimulation of renal β-glucuronidase activity that was not observed until the fourth posttreatment day (Fig. 12). By contrast, the 10-mg dose of testosterone elicited an earlier and larger increase in β-glucuronidase activity. For example, by the second posttreatment day a

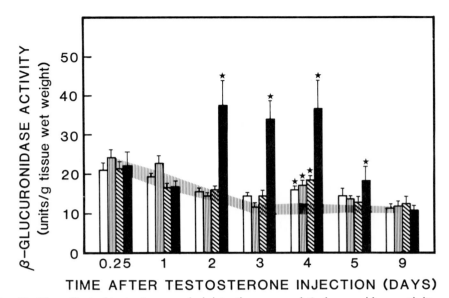

TIME AFTER TESTOSTERONE INJECTION (DAYS)

Fig. 12. The effect of testosterone administration on renal β-glucuronidase activity in castrated male mice. Groups of male mice of NCS strain (castrated 2 days) were administered a single intraperitoneal dose of testosterone ($\square$, 0.3 mg; ▥, 1 mg; ▨, 3 mg; ▪, 10 mg) and killed at the indicated times. Each bar represents the mean ± SE for at least six animals per experimental group. The cross-hatched area shows the mean ± SE range for the vehicle-treated control animals. Asterisks indicate significant differences between the experimental and the corresponding control groups. [From Pajunen et al. (1982).]

three-fold increase occurred that was maintained through day 4 (Fig. 12). These observations suggest that the high and sustained concentration of nuclear androgen receptor produced by the large dose of testosterone was associated with not only a greater stimulation of β-glucuronidase activity but also a shorter lag in the time required for the enzyme activity to increase (Figs. 10 and 12).

It thus appears that the hormonal regulation of ODC and β-glucuronidase activities is markedly dissimilar: 2 to 4 days were required before an increased β-glucuronidase activity was detected, while ODC activity rose only a few hours after testosterone administration. Moreover, the relative stimulation of ODC activity was at least 10 times greater than that of β-glucuronidase. These differences may be inherent in dissimilar turnover rates of the enzymes (Swank et al. 1978; Russell and Snyder 1969; Jänne et al. 1978) or in different accessibility of their genes for androgen receptor interaction. As far as receptor-mediated regulation of β-glucuronidase gene activity is concerned, these data seem to provide the first direct evidence for the hypothesis that both the duration of the lag period in β-glucuronidase induction and the subsequent rate of synthesis of this enzyme are regulated by the number of androgen receptors associated with "β-glucuronidase chromatin" (Watson et al. 1981). Thus, the higher and longer-lasting nuclear receptor accumulation resulted in a shorter lag period and a larger increase in β-glucuronidase activity (Fig. 12). In the case of ODC, a similar situation seems to exist, except for a much shorter lag period in the induction of this latter enzyme (Pajunen et al. 1982).

C. Androgen Receptors as Regulators of Renal ODC and β-Glucuronidase Activities

The results in this report indicate that androgens elicit a rapid and profound stimulation of ODC activity in mouse renal cells despite the relatively low amount of nuclear receptors in these cells. This latter point is illustrated by the fact that rat prostate contains over 10,000 androgen receptors/nucleus after testosterone administration (Van Doorn and Bruchovsky 1978), while less than 1000 receptors/nucleus can be maximally translocated in the renal cells of the mouse (Isomaa et al. 1982). That even a relatively small nuclear androgen receptor content is capable of maintaining defined androgenic functions in the mouse kidney is demonstrated by the physiological nuclear receptor concentrations in the female and male mice: about 300 androgen receptors/cell in the male animals were associated with a 12-fold higher ODC activity than in the female mice, whose renal cells possessed about 100 receptors/nucleus (Isomaa et al. 1982). Although this sex difference might suggest that it is the nuclear receptor content per se that regulates the magnitude of a long-term androgenic response, the results of experiments on acute testosterone responses indicate that there is a more complex relationship between the nuclear receptors and biological response. In the mouse kidney, as in estrogen-responsive tissues (Clark and Peck 1976; Eriksson et al. 1980), it is not only the amount of nuclear receptors initially accumulated that regulates the extent of biological response; the residence time of these receptors in the nuclei also seems to be an important phenomenon. If the nuclear androgen receptors are true regulators of ODC and β-glucuronidase gene activities, we may hypothesize that a certain threshold nuclear receptor concentration must be present for an extended period of time before full stimulation of these genes is acquired. Thus, the extent of stimulation is some function of receptor content and receptor residence time. In addition, it would appear from these preliminary studies that the activated stage of the genome is maintained as long as the threshold receptor level is exceeded.

VII. Summary

To relate androgen receptor activity to biological response, reliable methods for receptor measurements were established. Cytosol receptors from mouse kidney were assayed by taking advantage of the stabilizing effect of sodium molybdate, while pyridoxal 5'-phosphate was used to extract androgen receptors from nuclei isolated with the hexylene glycol technique. This extraction method yielded about two-fold higher receptor concentrations than KCl extraction or direct nuclear exchange reaction; the use of hexylene glycol increased the recovery of nuclear androgen receptors by three-fold as compared with other techniques. The nuclear receptor was androgen specific [5α-dihydrotestosterone = methyltrienolone > testosterone > estradiol > progesterone $\gg$ dexamethasone = aldosterone], had a K_D of 5 nM for methyltrienolone, and had a sedimentation rate of 4 S. The nuclear androgen receptor concentrations in intact male and female mice were 280 ± 100 and 90 ± 20 receptors/cell, while the corresponding figures for the cytosol receptors were 1070 ± 80 and 1470 ± 380 receptors/cell. Renal ODC and β-glucuronidase enzymes were used as marker proteins in studies where receptor changes were related to biological androgen action. Treatment of castrated male mice with different single doses of testosterone (0.3, 1, 3, and 10 mg) produced a comparable initial increase in nuclear androgen receptor content at 30–60 min (600–850 receptors/nucleus), while the time of receptor residence in nuclei was dose-related. The activity of ODC increased in a dose-dependent manner with peak responses 12–18 h after testosterone. The nuclear androgen receptor residence time but not the initial receptor content was an important regulator for ODC stimulation: prolonged residence time was accompanied by a great and sustained increase in ODC and a shortening of the lag period required to detect increased enzyme activity. β-Glucuronidase was induced more slowly than ODC in that its activity did not increase until 2 days after testosterone administration. In the case of β-glucuronidase, a high and prolonged nuclear androgen receptor accumulation was also associated with a greater stimulation and shorter lag period of enzyme induction.

Acknowledgments. This work was supported by grants from NIH (HD-13541 and RR-05860). The skillful technical assistance of Mr. Paul Van Duyne is gratefully acknowledged.

References

Anderson JN, Clark JH, Peck EJ Jr (1972) Biochem J 126: 561–567
Anderson KM, Phelan J, Marogil M, Hendrickson C, Economou S (1980) Steroids 35: 273–280
Attardi B, Ohno S (1974) Cell 2: 205–212
Bardin CW, Bullock LP, Sherins RJ, Mowszowicz I, Blackburn WR (1973) Rec Prog Horm Res 29: 65–109
Bardin CW, Brown TR, Mills NC, Gupta C, Bullock LP (1978) Biol Reprod 18: 74–83
Bardin CW, Catterall JF (1981) Science 211: 1285–1294
Bonne C, Raynaud J-P (1976) Steroids 27: 497–507
Brandt JT, Pierce DA, Fausto N (1972) Biochim Biophys Acta 279: 184–193
Brown TR, Bullock LP, Bardin CW (1979) Endocrinology 105: 1281–1287
Bullock LP, Bardin CW (1974) Endocrinology 94: 746–756
Bullock LP, Bardin CW (1975) Steroids 25: 107–119
Cidlowski JA, Thanassi JW (1978) Biochem Biophys Res Commun 82: 1140–1146
Clark JH, Peck EJ Jr (1976) Nature 260: 635–637
Conn PM, O'Malley BW (1975) Biochem Biophys Res Commun 64: 740–746
Dolan KP, Diaz-Gil JJ, Litwack G (1980) Arch Biochem Biophys 201: 476–485

Eriksson H, Hardin JW, Markavarich B, Upchurch S, Clark JH (1980) J Steroid Biochem 12: 121–130

Gaubert CM, Tremblay RR, Dube JY (1980) J Steroid Biochem 13: 931–937

Hardin JW, Clark JH, Glasser SR, Peck EJ Jr (1976) Biochemistry 15: 1370–1374

Henningsson S, Persson L, Rosengren E (1978) Acta Physiol Scand 105: 385–393

Isomaa V (1981) Biochim Biophys Acta 675: 9–16

Isomaa V, Pajunen AEI, Bardin CW, Jänne OA (1982) Endocrinology 111: 833–843

Jänne J, Pösö H, Raina A (1978) Biochim Biophys Acta 473: 241–293

Jänne O, Bullock LP, Bardin CW, Jacob ST (1976) Biochim Biophys Acta 418: 330–343

Jensen EV, Suzuki T, Kawashima T, Stumpf WE, Jungblut PW, DeSombre ER (1968) Proc Natl Acad Sci USA 59: 632–637

Kokko E, Isomaa V, Jänne O (1977) Biochim Biophys Acta 479: 354–366

McBlain WA, Shyamala G (1980) J Biol Chem 255: 3884–3891

McLean WS, Smith AA, Hansson V, Naess O, Nayfeh SN, French FS (1976) Mol Cell Endocrinol 4: 239–255

Mowszowicz I, Bardin CW (1974) Steroids 23: 793–807

Mulder E, Vrij L, Foekens JA (1980) Steroids 36: 633–645

Mulder E, Vrij L, Foekens JA (1981) Mol Cell Endocrinol 223: 283–296

Müller RE, Traish A, Wotiz HH (1980) J Biol Chem 255: 4062–4067

Nicholson WE, Levine JH, Orth DN (1976) Endocrinology 98: 123–128

Nielsen CJ, Sando JJ, Vogel WM, Pratt WB (1977) J Biol Chem 252: 7568–7578

Noma K, Nakao K, Sato B, Nishizawa Y, Matsumoto K, Yamamura Y (1980) Endocrinology 107: 1205–1211

Ojasoo T, Raynaud J-P (1978) Cancer Res 38: 4186–4198

Pajunen AEI, Isomaa VV, Jänne OA, Bardin CW (1982) J Biol Chem 257: 8190–8198

Russell DH, Snyder SH (1969) Mol Pharmacol 5: 253–262

Swank RT, Davey R, Joyce L, Reid P (1977) Endocrinology 100: 473–480

Swank RT, Paigen K, Davey R, Chapman V, Labarca C, Watson G, Ganshow R, Brandt EJ, Novak E (1978) Rec Prog Horm Res 34: 401–436

Swank RT, Paigen K, Ganshow RE (1973) J Mol Biol 81: 225–243

Toft D, Nishigori H (1979) J Steroid Biochem 11: 413–416

Van Doorn E, Bruchovsky N (1978) Biochem J 174: 9–16

Watson G, Davey RA, Labarca C, Paigen K (1981) J Biol Chem 256: 3005–3011

Wray W, Conn PM, Wray VP (1977) Meth Cell Biol XVI: 69–86

Wright WW, Chan KC, Bardin CW (1981) Endocrinology 108: 2210–2216

Discussion of the Paper Presented by O. Jänne

THOMPSON: Two questions. One, you gave us the number of nuclear androgen receptor sites per cell. I understand that the proximal tubule cells are the sole site of action of androgens in the kidney. Since those numbers you gave us for sites per cell are not corrected for the proportion of cells represented by the proximal tubules, your number of sites per cell is vastly underestimated.

JÄNNE: I would not say that the numbers are vastly underestimated.

THOMPSON: What percentage of the total cells are represented by the proximal tubule cells acted upon by androgens?

JÄNNE: The information available in the literature indicates that about 70% of the renal cells respond to androgens.

THOMPSON: So, 70% of the kidney is proximal tubule. Is that right, Wayne (Bardin)?

BARDIN: Yes, 70%. But I don't think anybody has ever looked to see if androgen receptors are only in that cell. That's the most androgen-responsive part of the kidney.

THOMPSON: Well it's of interest, because there are only a couple of dozen or so genes which are influenced by androgens in the kidney, according to the work from Paigen's group in Buffalo. So if you only found a few hundred receptors in the nucleus, that's pretty efficient compared to the other steroid receptors in their sites of action. That's why it's an important question to answer. Second point; in the studies where you pointed out the dose–response relationship between steroid and ODC induction, did you say that at the lower doses you gave, the response was converse to the amount of nuclear receptors?

JÄNNE: No. The induction was not related to the initial nuclear androgen receptor accumulation after any of the testosterone doses. It was related to the disappearance rate of the nuclear receptors, or the time of residence of the nuclear receptors.

THOMPSON: That was clear for the megadose but is that also true for the lower doses?

JÄNNE: Yes.

MOUDGIL: In one of your slides with molybdate effect on the percentage binding of control. I believe it's for 10 mM molybdate where you get the maximum binding and it sharply decreases afterwards. You have used as much as 100 mM molybdate to show that it inhibits. Could you comment on this?

JÄNNE: It is possible that even 100 mM sodium molybdate does not decrease the stability of the androgen receptor. In this particular experiment, we utilized DEAE filter technique to separate free and bound ligand and therefore, 100 mM molybdate may have resulted in an ionic strength that was high enough to prevent receptor binding to the DEAE filters.

MOUDGIL: My other question is related to the work of Cidlowski and Thanassi published a few years ago. They looked at the pyridoxal phosphate extraction of nuclear glucocorticoid receptor and I believe that they found that about 50 mM would be required to extract something like 40–50% nuclear receptor. In your case you were able to extract the receptor with 5 mM pyridoxal phosphate, which is very nice. Why is there a disparity between that concentration you used. Is it that because you have 0.15 M KCl with pyridoxal phosphate?

JÄNNE: No, we can use 5 mM pyridoxal phosphate alone to extract nuclear androgen receptors. However, we usually include 0.15 M KCl in the buffer to decrease nonspecific binding of the tracer during the hydroxylapatite separation step.

MOUDGIL: How does the dose–response curve look if you use 10 mM pyridoxal phosphate? Would it increase?

JÄNNE: It goes down a little bit. If we add 0.4 M KCl to pyridoxal phosphate it does not improve the extraction efficiency. It thus appears that pyridoxal phosphate and KCl extract the same type of receptors, but pyridoxal phosphate is more efficient.

MOUDGIL: So your receptor system is probably more sensitive than rat liver glucocorticoid receptor system. Because in other systems 5 mM of pyridoxal phosphate won't even inhibit complete binding to the DNA or nuclei and it would be very surprising to have everything extracted by that concentration. It is very clear that your system is more sensitive to it.

JÄNNE: 10 mM pyridoxal phosphate has been used to elute glucocorticoid receptor from a DNA matrix (e.g., *J. Biol. Chem.* 254: 9284–9290, 1979), and we have similar experience with the progesterone receptor. We can also extract progesterone receptor from uterine nuclei with 5–10 mM pyridoxal phosphate without any additional salt. It is possible that the method we use for isolation of nuclei may influence the extractability of the receptors by pyridoxal phosphate.

MOUDGIL: In your androgen administration experiment, where cytosol receptor is moving into nucleus by 18 h, is that an actinomycin D desensitive step?

JÄNNE: We have not performed this experiment.

LITWACK: Just on the point of pyridoxal phosphate, we've been able to get as much as 80% out of nuclei with 5 mM pyridoxal phosphate in agreement with your data and if you go much above that, certainly at 10 mM you begin to dissolve the nuclei.

JÄNNE: It is possibly one of the reasons we saw a decline in the efficiency.

BROWN: I was wondering whether in the bookkeeping experiments that you did with cytosol and nuclei, comparing these values to the total amount, whether in fact the determinations in cytosol and nuclei were made from the same original preparation. In other words, were the nuclei from the same kidneys from which you measured the cytosol.

JÄNNE: They were from the same animals but from different kidneys, because if we initially homogenize the renal tissue in hexylene glycol containing buffer we cannot measure accurately the androgen receptors from the cytosol for two reasons. First, androgen receptor in the cytosol does not tolerate well hexylene glycol. Second, when we homogenize the tissue for isolation of nuclei we use about 20 volumes of hexylene glycol containing buffer per gram of tissue and therefore, the resulting cytosol is very diluted. Thus, we routinely take one kidney for cytosol receptor measurements and the other for nuclear assays, and pool 4–6 kidneys for a given assay.

BROWN: Have you looked at the effects of DTT and/or molybdate in terms of its ability to extract nuclear receptor?

JÄNNE: We routinely use 5 mM dithiothreitol in the extraction buffer. If we add molybdate, we will not improve the recovery of the androgen receptor. We have not studied the stability of the receptor in the presence or absence of molybdate.

BROWN: Just one other thing. Have you looked at the kinetics of the binding in the Tfm mouse kidney in terms of the effect of molybdate on the binding there?

JÄNNE: I have not.

BROWN: Do you have any comments on what Chung and his co-workers reported at the Federation meetings. First, in terms of being able to measure the receptor in the Tfm, something which people have either ignored or negated in the past, secondly, whether this is in your opinion actually the receptor that they were able to measure.

BARDIN: In the cytosol with molybdate one finds 10% of the androgen receptor in the Tfm as in normal. If you use the buffer without molybdate you get about 5%, which is a measurable amount.

BROWN: Have you looked at the kinetics of the binding? Have you looked at the affinity?

BARDIN: No.

PECK: In your bookkeeping experiment were those single-point assays, i.e., using a single concentration of ligand, or were you using saturation analysis for each of those points.

JÄNNE: For cytosol receptor we use Scatchard-type plotting. For nuclear receptors they were single-point assays.

PECK: In your nuclear assays then, did you run your reactions at different volumes to insure that you don't have interference with endogenous ligand, which in fact can create that kind of pattern.

JÄNNE: Yes, we have performed that experiment. The receptor assay is completely linear for the extract volumes ranging from 0.05 μl to 0.5 μl. In addition, the line goes via the origin.

PECK: No. Given a particular volume of extract, run the ligand binding reaction at different total volumes. Do you still get the same B_{max}? During the time period you are discussing, you have endogenous ligand, that is, the ligand you gave the animal, and this will reduce the radiospecific activity of your labeled ligand.

JÄNNE: You mean in the nuclei?

PECK: Yes, nuclei will carry with them quite a bit of endogenous ligand which is not receptor bound and this will dilute the radiospecificity of your ligand and give you an apparent depression of B_{max}.

JÄNNE: This does not seem to be the case in our assays, since the residual testosterone concentration in the renal nuclei at 60 min after a 1-mg dose of testosterone was less than 10% of that of ^{3}H-methyltrienolone used in the receptor assay. The concentration of testosterone in the nuclear extract was measured by radioimmunoassay (Isomaa V, Pajunen AEI, Bardin CW, Jänne OA, Endocrinology, in press).

PECK: I'm not talking about that free or that bound but both, that which is bound as well as free. There is a small fraction of free, but there is a great deal bound either specifically or nonspecifically. When you effect an exchange reaction that unlabeled ligand is released into the tube and dilutes the radiospecific activity of your ligand. This will give you an apparent depression in that which is measured as bound. With a single-point assay you cannot correct for this unless you run the reaction at different total volumes.

JÄNNE: I agree.

BARDIN: In other experiments with radioactive testosterone over 90% of the steroid isolated with renal nuclei is receptor bound and so if you think of that amount that is in the nucleus and compare it the amount that is added, it really shouldn't make a difference because there is so little in the nucleus to begin with.

PECK: In the hypothalamic system with estrogens we do see that artifact and we can correct it by correcting for the endogenous unlabeled ligand.

JÄNNE: But you have much more lipids around.

PECK: Yes, absolutely.

Discussants: E.B. THOMPSON, O. JÄNNE, C.W. BARDIN, V.K. MOUDGIL, G. LITWACK, T.R. BROWN, and E.J. PECK

Chapter 18

Modulation of Androgen Action in Rat Liver by Thyroid and Peptide Hormones

Arun K. Roy, B. Chatterjee, N.M. Motwani, W.F. Demyan, and T.S. Nath

I. Introduction

The mammalian liver is increasingly being recognized as an important target for male and female sex hormones (Roy and Chatterjee 1983). The androgenic hormones cause a significant increase in hepatic protein synthesis in the rat. Much of this increase is due to an immunochemically related group of low-molecular-weight proteins, called α_{2u}-globulin, which are secreted from the liver, rapidly filtered through the kidneys, and appear as the principal urinary protein in the mature male (Roy et al. 1966). Although α_{2u}-globulin constitutes the most abundant androgen-dependent protein in the rat liver, recent studies in our laboratory have identified several other sex-hormone-regulated hepatic proteins. Figure 1 shows the autoradiographic pattern of the proteins synthesized by hepatocytes obtained from male and female rats. The patterns show a considerable degree of sexual dimorphism for several proteins, with α_{2u}-globulin being the most predominant one. Isoelectric focusing and SDS–polyacrylamide gel electrophoresis can separate α_{2u}-globulin into four to five charge-variant and two molecular-weight-variant forms (Chatterjee et al. 1982b; Roy et al 1983). These polymorphic forms of α_{2u}-globulin are coded by a family of approximately 18 genes (Kurtz 1981). Unlike reproductive tissues, there is no hormone-mediated differentiation and cell proliferation in the liver. Accordingly, we have utilized α_{2u}-globulin as a convenient model for studying androgen-dependent regulation of specific gene expression. α_{2u}-Globulin appears in the male rat at the time of puberty ($\sim$40 days), reaches a peak level at about 100 days of age, and disappears at senescence ($>$750 days) (Roy 1973a; Chatterjee et al. 1981). The age-dependent appearance and disappearance of α_{2u}-globulin have been found to be associated with corresponding changes in the hepatic concentration of a cytoplasmic androgen-binding protein (Roy et al. 1974). Although the correlative studies and the steroid specificity of the cytoplasmic androgen binder indicate its role in the regulation of α_{2u}-globulin synthesis, this cytoplasmic "receptor" does not seem to translocate into the nucleus (Roy 1979). Therefore, its postulated role in the regulation of gene expression is probably an indirect one.

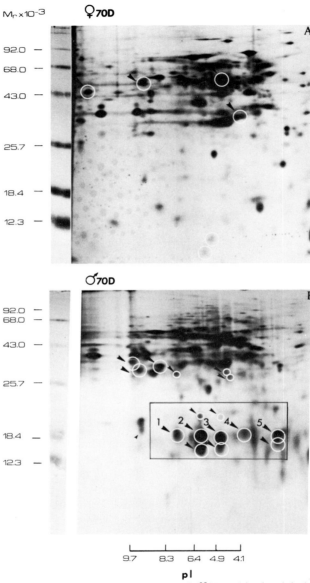

Fig. 1. Autoradiogram of the two-dimensional slab gels of [^{35}S]methionine-labeled proteins synthesized by isolated rat hepatocytes from male (A) and female (B) rats. The sex-specific proteins are circled. The proteins were separated in the first dimension by isoelectric focusing and in the second dimension by SDS–polyacrylamide gel electrophoresis. The positions of the molecular-weight markers are indicated on the left.

The androgen-dependent synthesis of α_{2u}-globulin requires the permissive influence of glucocorticoids, thyroxine, growth hormone, and insulin (Roy 1973b; Roy and Leonard 1973). Multiple hormone interaction in the synthesis of α_{2u}-globulin is best exemplified in the hypophysectomized male rat treated with various hormone combinations. Hypophysectomy of the adult male rat causes cessation of the androgen-dependent synthesis of α_{2u}-globulin. Renewed synthesis of α_{2u}-globulin in the hypophysectomized male rat requires, in addition to androgen, three other hormones, i.e., glucocorticoid, thyroxine, and growth hormone (Roy 1973b). The mechanism of the modulation of androgen action by other hormones, especially the nonsteroid hormones, has been the subject of considerable interest in our laboratory. In this article we review the interactions between growth hormone, thyroxine, and insulin in the androgen-dependent synthesis of α_{2u}-globulin.

II. Growth Hormone

Androgen-dependent synthesis of α_{2u}-globulin requires the influence of growth hormone (Roy 1973b). The effect of growth hormone on the expression of α_{2u}-globulin gene can be conveniently examined in the in vivo "reconstitution" experiments involving hypophysectomized male rats. Loss of α_{2u}-globulin synthesis in the hypophysectomized rats and its reversal after multiple hormone supplementation are associated with corresponding changes in the hepatic concentration of the mRNA for α_{2u}-globulin (Roy and Dowbenko 1977). In these animals α_{2u}-globulin and its mRNA are not induced when growth hormone is omitted from the hormone combination. Essentially the same results are obtained whether the mRNA is examined by translational assay or hybridizational analysis with a cloned cDNA probe (Fig. 2). A short-term culture of rat hepatocytes exposed to growth hormone in vitro but derived from hypophysectomized male rats treated for 8 days with 5α-dihydrotestosterone (DHT), corticosterone, and thyroxine also failed to synthesize α_{2u}-globulin (Fig. 3). These results demonstrate that growth hormone acts to regulate α_{2u}-globulin synthesis at a pretranslational level (Roy et al. 1982). An earlier claim (Kurtz et al. 1978) for translational regulation of α_{2u}-globulin by growth hormone has now been withdrawn (Lynch et al. 1982). Although the pretranslation mode of regulation of α_{2u}-globulin by growth hormone has been clearly established, the precise mechanism of growth hormone action remains to be determined.

One puzzling aspect of growth hormone action on the hepatic synthesis of α_{2u}-globulin is the effect of this hormone on the cytoplasmic androgen-binding protein. Recent experiments in our laboratory performed in collaboration with J.A. Gustafsson and his associates have shown that continuous infusion of a large amount of growth hormone (through osmotic minipumps) causes almost a complete disappearance of the cytoplasmic androgen-binding protein and more than 85% reduction in the hepatic synthesis of α_{2u}-globulin. In the mature male rat, growth hormone is released from the pitui-

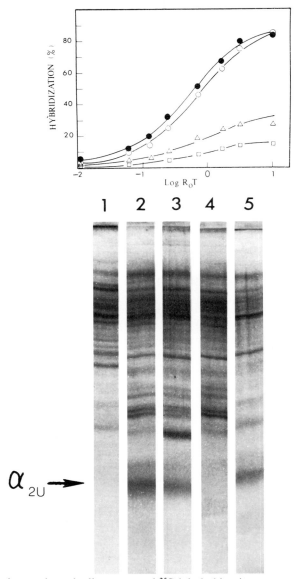

Fig. 2. Autoradiogram of the electrophoretically separated ^{35}S-labeled in vitro translation products of hepatic mRNA from male rats: (1) hypophysectomized without any hormone treatment; (2) normal control; (3) hypophysectomized with dihydrotestosterone, corticosterone, and growth hormone; (4) hypophysectomized with dihydrotestosterone, corticosterone, and thyroxine; (5) hypophysectomized with dihydrotestosterone, corticosterone, thyroxine, and growth hormone.

The inset on the top shows the kinetics of hybridization between a cloned cDNA probe for α_{2u}-globulin and the α_{2u}-mRNA present within total poly(A)-containing hepatic RNA fractions obtained from rats with different endocrine states. □, hypophysectomized without any hormone supplementation; ○, hypophysectomized with growth hormone, corticosterone, and dihydrotestosterone; △, hypophysectomized with thyroxine, corticosterone, and dihydrotestosterone; ●, hypophysectomized with growth hormone, thyroxine, dihydrotestosterone, and corticosterone.

$M_r \times 10^{-3}$

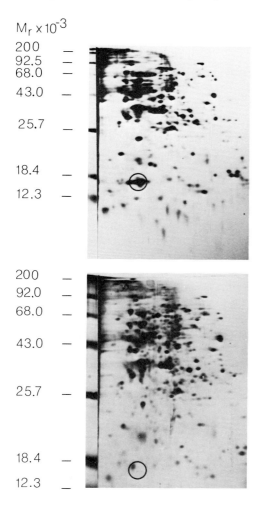

200 —
92.5 —
68.0 —
43.0 —
25.7 —
18.4 —
12.3 —

200 —
92.0 —
68.0 —
43.0 —
25.7 —
18.4 —
12.3 —

Fig. 3. Autoradiogram of the [^{35}S]methionine-labeled proteins synthesized by isolated rat hepatocytes and separated by two-dimensional gel electrophoresis. (A) Proteins synthesized by hepatocytes derived from hypophysectomized rats that received eight daily treatments of DHT, corticosterone, thyroxine, and growth hormone. (B) Proteins synthesized by hepatocytes derived from hypophysectomized rats that received eight daily treatments of DHT, corticosterone, and thyroxine. In both cases the hepatocytes were cultured in vitro in the presence of DHT, corticosterone, thyroxine, and growth hormone. The position of pure α_{2u}-globulin in the two-dimensional gel is circled. Nonequilibrium pH gradient gel electrophoresis (NEPGE) was first performed from left (acidic) to right (basic). SDS–polyacrylamide slab gel electrophoresis (SDS–PAGE) was then run from top to bottom. Positions of the molecular-weight-marker proteins are indicated on the left.

tary gland in a pulsatile fashion. Hourly assays of plasma growth hormone in the male rat show several daily "spikes," each followed by more than 10-fold lower "dips." In the case of the female, growth hormone is released in a more continuous and sustained fashion (Eden 1979). Continuous infusion of growth hormone in the normal male rat is known to cause alteration of hepatic steroid metabolism similar to that found in female animals (Kramer and Colby 1976; Mode et al. 1981). These observations have led to the hypothesis that sex differences in hepatic steroid metabolism may be due to episodic or continuous mode of growth hormone secretion from the pituitary gland (Mode et al. 1981). Since under normal physiological conditions the pituitary gland functions under maximum inhibitory influence of the hypothalamic hormone, supplementation of additional amount of growth hormone to normal rats is not expected to cause feedback readjustment of growth hormone secretion. Therefore, in addition to the dampening of the large fluctuations in its circulating concentrations, the supraphysiological level of growth hormone may indirectly influence hepatic function through other metabolic alterations.

III. Thyroxine

Thyroidectomy results in about a 90% reduction in the hepatic synthesis of α_{2u}-globulin (Roy 1973b). In addition to its effect on α_{2u}-globulin, thyroid hormone is also required for the synthesis and secretion of pituitary growth hormone (Hervas et al. 1975; Martial et al. 1977). In a normal physiological situation, these two hormones, therefore, interact to regulate target gene expression. Initial studies have shown that the reduced synthesis of α_{2u}-globulin in hypothyroid rats and its reversal with thyroid hormone are associated with corresponding changes in the hepatic concentration of the mRNA for this protein (Roy et al. 1976; Kurtz et al. 1976). This observation and the discovery of nuclear thyroid hormone receptors (Oppenheimer et al. 1974) have led to the idea that α_{2u}-globulin can serve as a possible model system for studying thyroxine-mediated regulation of hepatic gene transcription. However, recent results from our laboratory have indicated that the effect of the thyroid hormone on the hepatic concentration of the messenger RNA for α_{2u}-globulin is an indirect one and is mediated through pituitary growth hormone (Roy 1983; Chatterjee et al. 1982a). Figure 4 shows that the decreased hepatic concentration of the mRNA for α_{2u}-globulin in the hypothyroid rat can be brought back to normal not only with thyroxine but also with an optimized dose of growth hormone (0.2 IU/100 g). In addition, treatment of the hypophysectomized male rat with only DHT, corticosterone, and growth hormone (0.2 IU/100 g) can induce α_{2u}-globulin mRNA to the level of normal animals, whereas treatment of these animals with DHT, corticosterone, and thyroxine failed to induce α_{2u}-globulin mRNA. Thus, both in the case of hypophysectomized and thyroidectomized rats, growth hormone but not thyroxine is capable of inducing α_{2u}-globulin mRNA.

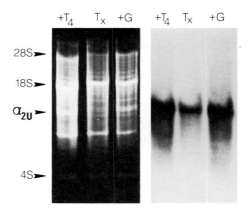

Fig. 4. Effect of growth hormone and thyroxine on the hepatic mRNA for α_{2u}- globulin. The picture on the left shows electrophoretic pattern of the in vitro translation products of male rat liver mRNA obtained from thyroidectomized (T_x), thyroidectomized treated with thyroxine (+T), and thyroidectomized treated with growth hormone (+G) animals.

The right frame shows electrophoretic autoradiogram of the corresponding hepatic mRNA hybridized with ^{32}P-labeled cloned α_{2u}-globulin cDNA probe.

In addition to its indirect effect on the hepatic concentration of α_{2u}-globulin mRNA, thyroxine also influences the hepatocytes directly to promote the synthesis of α_{2u}-globulin. This is indicated by both in vivo in the thyroidectomized rat and in vitro studies with cultured hepatocytes (Roy 1973b; Motwani et al. 1980). Although growth hormone alone can restore the normal hepatic level of the mRNA for α_{2u}-globulin, the synthesis of this protein (as indicated by in vivo pulse labeling and its hepatic concentration) is raised to only 50% of the normal level (Fig. 5). A similar disproportionate relationship between the hepatic concentration of the albumin mRNA and albumin synthesis in thyroidectomized rats has also been reported (Peavy et al. 1981). Both of these observations may be explained on the basis of thyroxine-mediated membrane proliferation in the liver cell. Earlier studies have shown that thyroxine-induced metamorphosis of the bullfrog tadpole is associated with a coordinated and sequential increase in the rate of synthesis of RNA, membrane phospholipid, and serum albumin (Tata 1967). The increased membrane synthesis is reflected by a marked proliferation of the rough endoplasmic reticulum. We have also found a paucity of rough endo-

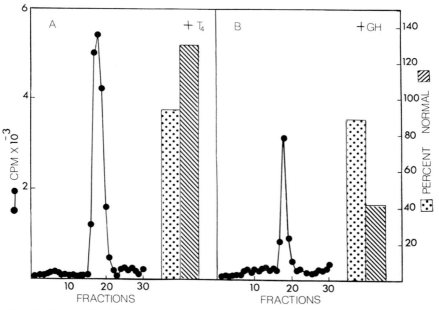

Fig. 5. Correlation between the hepatic content of α_{2u}-globulin and its mRNA and incorporation of [^{35}S]methionine into α_{2u}-globulin after an in vivo pulse. Thyroidectomized rats were treated with either thyroxine (A) or growth hormone (B) as indicated in the respective frames. The histograms on the right of each frame show the hepatic content of α_{2u}-globulin (▨) and α_{2u}-globulin mRNA sequences (▦) as percent of the normal male. The distribution of the in vivo pulse labeled α_{2u}-globulin radioactivity after SDS–polyacrylamide gel electrophoresis is shown on the left (●——●).

plasmic reticulum in the hepatocytes of the thyroidectomized rat as compared to the normal control. Treatment of thyroidectomized animals with thyroxine but not with growth hormone was found to correct this ultrastructural abnormality (Roy 1983). Thus, thyroxine may be involved in modulating the androgen-dependent synthesis of α_{2u}-globulin in at least two different ways, i.e., indirectly influencing the concentration of α_{2u}-globulin mRNA through growth hormone and directly acting on the liver cells to provide the appropriate organizational framework for the synthesis and processing of the secretory proteins. Substantiation of this working hypothesis will require further investigation.

IV. Thyroxine and Insulin

Hypoinsulinimic diabetes is known to cause decreased protein synthesis in various tissues, including the liver. The synthesis of certain proteins is more sensitive to insulin deficiency than are other proteins. A mild form of experimental diabetes (with blood sugar levels between 250 and 350 mg/dl) induced by streptozotocin causes only a 10% shift of the hepatic polysomes to smaller (<five ribosome) aggregates and a slight dispersion in the stacking of the rough endoplasmic reticulum. Despite this small organizational change in the protein-synthesizing apparatus, more than an 80% reduction in the hepatic synthesis of α_{2u}-globulin was noted. The decreased synthesis of α_{2u}-globulin in the diabetic rat can be accounted for by corresponding changes in the hepatic level of the mRNA for this protein (Roy et al. 1980). Interestingly, treatment of diabetic animals with thyroxine resulted in an increased synthesis of α_{2u}-globulin through a corresponding increase in the level of α_{2u}-globulin mRNA (Fig. 6). Studies from several laboratories have shown that diabetes is associated with a 30% decrease in plasma level of protein-bound iodine (PBI) through corresponding decrease in the output of hypothalamic TRH (Kumaresan and Turner 1966; Gonzalez et al. 1980). However, such a small decrease in the level of circulating PBI may not be sufficient to explain an 80% reduction in the hepatic synthesis of α_{2u}-globulin. Streptozotocin treatment and the resulting hyperglycemia increase both the synthesis and secretion of somatostatin (Patel and Weir 1976). The decreased level of thyroxine and the elevated level of somatostatin would cause a reduction in the pituitary output of growth hormone. Another possible contributing factor in this complex multihormonal interaction is the reported observation of decreased thyroid hormone receptor in the liver cells of the diabetic rat (Wiersinga et al. 1982). Thus, the androgenic hormones operate on the liver cell under a subset of control mechanisms involving a number of peptide hormones and thyroxine, and the optimum effect of the androgen at any physiological situation is dependent on the contributions from each of these hormonal influences.

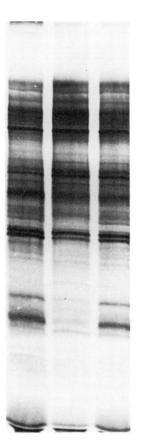

Fig. 6. Autoradiogram of the electrophoretically separated in vitro translation products of hepatic messenger RNA isolated from rats of various endocrine status. Normal male (N); diabetic male without any hormone treatment (D); diabetic male treated for seven days with thyroxine (D + T).

IV. Summary and Conclusion

The liver shows a considerable degree of sexual dimorphism. Sex difference in the hepatic tissue is mediated through altered gene expression by sex steroids. Synthesis of α_{2u}-globulin and its mRNA serves as a model system for the study of androgen-dependent gene expression in rat liver. The ability of the androgen to stimulate α_{2u}-globulin synthesis is restricted by the availability of several other hormones, such as glucocorticoids, thyroxine, growth hormone, and insulin. The influence of thyroxine, growth hormone, and insulin on the androgen-dependent synthesis of α_{2u}-globulin is reviewed in this article. Both growth hormone and insulin influence α_{2u}-globulin synthesis by increasing the hepatic concentration of the mRNA for this protein. Thyroxine promotes α_{2u}-globulin synthesis indirectly through its effect on pituitary growth hormone and by acting directly on the hapatocytes to regulate a yet unidentified step involving translation–processing. In addition to insulin deficiency, complex multihormonal interaction in diabetes may be due to altered levels of circulating thyroid hormones, somatostatin, and hepatic thyroid hormone receptors. Although in this article the permissive effect of the peptide and thyroid hormones on the androgen-dependent synthesis of only α_{2u}-globulin is emphasized, it is believed that such multihormonal modulation of steroid hormone action is a common phenomenon. Under physiological conditions, the multihormonal regulation of steroid hormone is integrated through both systemic and cellular changes. The modifying influence at the cellular level is primarily exerted through changes in hormone receptivity, gene transcription, and protein synthesis.

References

Chatterjee B, Demyan WF, Roy AK (1982a) J Biol Chem (in press)
Chatterjee B, Motwani NM, Roy AK (1982b) Biochim Biophys Acta 698: 22
Chatterjee B, Nath TS, Roy AK (1981) J Biol Chem 256: 5939
Eden S (1979) Endocrinology 105: 555
Gonzalez C, Montoya E, Jolin T (1980) Endocrinology 107: 2099
Hervas F, Escobar GM, Escobar del Rey RF (1975) Endocrinology 97: 91
Kramer RE, Colby HD (1976) Endocrinology 71: 449
Kumaresan P, Turner CW (1966) Endocrinology 79: 828
Kurtz DT (1981) J Mol Appl Genetics 1: 29
Kurtz DT, Chan KM, Feigelson P (1978) Cell 15: 743
Kurtz DT, Sippel AE, Feigelson P (1976) Biochemistry 15: 1031
Lynch KR, Dolan KP, Nakhashi HL, Unterman R, Feigelson P (1982) Cell 28: 185
Martial JA, Baxter JD, Goodman HM, Seeburg PH (1977) Proc Natl Acad Sci USA 74: 1816
Mode A, Norstedt G, Branimir S, Eneroth P, Gustafsson JA (1981) Endocrinology 108: 2103
Motwani NM, Unakar NJ, Roy AK (1980) Endocrinology 107: 1606
Oppenheimer JH, Koerner D, Surks MJ, Schwartz HL (1974) J Clin Invest 53: 768
Patel YC, Weir GC (1976) Clin Endocrinol 5: 19
Peavy DE, Taylor JM, Jefferson LS (1981) Biochem J 198: 289
Roy AK (1973a) Endocrinology 92: 957
Roy AK (1973b) J Endocrinol 56: 295
Roy AK (1979) In: Litwack G (ed) Biochemical Actions of Hormones. Vol. 6, Academic Press, New York, p 481
Roy AK (1983) In: Oppenheimer JH, Samuels HH (eds) Molecular Basis of Thyroid Hormone Action. Academic Press, New York (in press)
Roy AK, Chatterjee B (1983) Ann Rev Physiol (in press)
Roy AK, Dowbenko DJ (1977) Biochemistry 16: 3918
Roy AK, Leonard S (1973) J Endocrinol 57: 327
Roy AK, Milin BS, McMinn DM (1974) Biochim Biophys Acta 354: 213
Roy AK, Neuhaus OW, Harmison CR (1966) Biochim Biophys Acta 127: 72
Roy AK, Schiop MJ, Dowbenko DJ (1976) FEBS Lett 64: 396
Roy AK, Chatterjee B, Prasad MSK, Unakar NJ (1980) J Biol Chem 255: 11614
Roy AK, Chatterjee B, Demyan W, Nath TS, Motwani NM (1982) J Biol Chem 257: 7834
Roy AK, Chatterjee B, Demyan W, Milin B, Motwani N, Nath T, Schiop M (1983) In: Greep RO (ed) Recent Progress in Hormone Research. Vol 39, Academic Press, New York (in press)
Tata JR (1967) Biochem J 104: 1
Wiersinger WM, Frank HJL, Chopra IJ, Solomon DH (1982) Acta Endocrinol 99: 79

Discussion of Paper Presented by A. K. Roy

LIAO: The results and conclusion concerning the mechanism of growth hormone action that you have presented are quite different from another group who work on this system.

ROY: I believe the contradictory results will soon be retracted by the laboratory where it originated.

LIAO: Can you eliminate the question of processing?

ROY: I carefully listened to what John (Baxter) had to say this morning and I guess the results that we have do not contradict his but I could reinterpret John's (Baxter) results on the basis of our EM data. If the mRNA is not bound to the endoplasmic reticulum, they will probably undergo a more rapid degradation and they might lose the poly(A) that he was showing. So I guess none of us has the answer, but these results are not mutually exclusive.

PECK: This may be a very dumb question, but is the effect of thyroid hormone on the organization of the reticulum common to all cell types—do you see it anyplace else or is this the only place it occurs? What's the story there? Have you looked elsewhere?

ROY: The story goes back to about 10 years ago when Tata did a very detailed analysis of the bullfrog tadpole and he found a very similar pattern as I showed. He presented some data on the rat in that paper, which he published in 1969 or 1968 in *Nature*. Those were normal rats that were treated with either thyroxine or growth hormone, and he looked into the synthesis of ER by phospholipid labeling and the results were very similar.

LIAO: Do you have any data on the nuclear androgen receptor complex?

ROY: The binding entity that I described does not seem to translocate into the nucleus. However, since the correlation between the presence of this binder and the androgen-dependent synthesis of α_{2u}-globulin is very strong, I feel that it is somehow involved in androgen action. It might be acting as a sponge to soak up the androgen and then give it to another protein.

LIAO: You don't see any complex inside the nucleus.

ROY: That is right.

KORNEL: Do you have any data to show whether the action of growth hormone is mediated by somatomedin?

ROY: It's possible, but I don't know. I don't have any data on that.

KORNEL: Do you think there is a possibility that because thyroxine and growth hormone are acting in concert, prolactin would also have something to do with modulating the synthesis of α_{2u}-globulin.

ROY: When we examined the effect of multiple hormone in the hypophysectomized rat, we used all different kinds of hormone combinations, TSH, ACTH, prolactin, etc., we did not find any effect of prolactin on this particular gene expression. Dr. Gustafsson may want to say something about somatomedin.

GUSTAFSSON: I don't really have so much to say. We have been studying a similar system, i.e., the sex differentiation of rat liver over the years, and we have been able to show that this sex differentiation is mediated by growth hormone solely, so that if you hypophysectomize, adrenelectomize, thyroidectomize, and castrate the rat and then you obtain essentially a masculine type of liver metabolism and then if you infuse growth hormone (human growth hormone or rat growth hormone) into such a rat, you can obtain a complete feminization, and with this I mean, for instance, induction of prolactin receptors from practically zero to the normal female-level induction of the cytochrome-P450-dependent 15β-hydroxylase to at least 3000 times and so on. In other words, we can turn on specific proteins and we can also turn off specific proteins by giving growth hormone. So this is in a way a more simple system than the one you are describing, because we don't need various other hormones.

There is no multihormone control here. But again I'm not sure whether somatomedin is involved or not. I don't see any reason why somatomedin has to be involved. It could well be a direct effect of growth hormone in the system and we are in the process of isolating some of these proteins and raising antibodies against them, and we hope that it may provide a system to study the mechanism action of a growth hormone.

Discussants: A.K. ROY, S. LIAO, E.J. PECK, L. KORNEL, and J.A. GUSTAFSSON.

Chapter 19

A New Class of 5α-Reductase Inhibitors

G.H. RASMUSSON, T. LIANG, AND J.R. BROOKS

I. Development of Steroidal Inhibitors of Androgen Action

Over the past 20 years several approaches to the control of androgen action have been investigated by research teams at Merck & Co., Inc. Basically, these approaches can be described as those that would cause chemical interference at certain critical stages where hormone development or action takes place (Fig. 1). These studies were initiated largely through the efforts of the late Dr. Glen E. Arth, the Steroid Chemistry group leader in the Department of Synthetic Chemical Research (1957–1975).

The first approach investigated was that of direct interference with the conversion of pregnanes into androstanes. The enzyme involved, 17,20-desmolase or -lyase, irreversibly converts 17α-hydroxyprogesterone or 17α-hydroxypregnenolone into the corresponding 17-oxoandrostane. A subsequent step, a steroid 17-oxido-reductase, converts these products into testosterone, the major circulating androgenic hormone.

To achieve specificity a variety of steroidal compounds was prepared and tested for the ability to inhibit the rat testicular 17,20-lyase (Arth et al. 1971). A series of 17β-acylaminoandrostanes was found to be very selective in their action on this enzyme in vitro and certain compounds (e.g., Fig. 2) were active when given acutely in vivo to the rat. Chronic in vivo studies, however, showed a rebound of testicular testosterone levels, presumably because of increased LH levels resulting from an initial absence of the normal negative feedback effect of testosterone (Patanelli et al.). Although not investigated, interference with the 17,20-lyase might also have an effect on estrogen biosynthesis as estrogens must be formed in a subsequent step.

A second approach was stimulated by the discovery of cyproterone acetate and its ability to directly antagonize androgen action in hormonally sensitive tissues (Neumann et al. 1969). Such antagonism (boxed-in area, Fig. 1) is the result of interference by a compound with the binding of testosterone (T) or 5α-dihydrotestosterone (DHT) to the intracellular receptor, an action that is necessary for a hormonal effect to occur in these tissues.

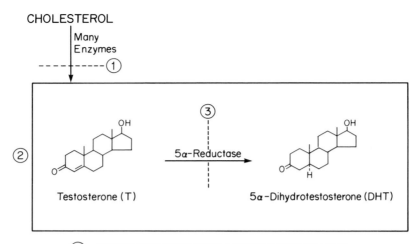

① TESTOSTERONE SYNTHESIS INHIBITORS
② ANDROGEN ANTAGONISTS (Ā)
   Peripheral antagonists all androgens
③ 5α-REDUCTASE INHIBITORS

Fig. 1. Androgen-depriving mechanisms.

The aldosterone antagonist, spironolactone, was also found to be a good androgen antagonist (Steelman et al. 1969). As an offshoot of a study of similar aldosterone antagonists, a series of $2',3'-\alpha$-tetrahydrofuran-2'-spiro-17-(1,2α-methylene-4-androsten-3-ones) was prepared and tested for the ability to reverse the effects of administered testosterone enanthate on castrate rats (Rasmusson et al. 1972). One of these compounds (Fig. 3) was found to be orally effective in reducing the size of the dog prostate (Brooks

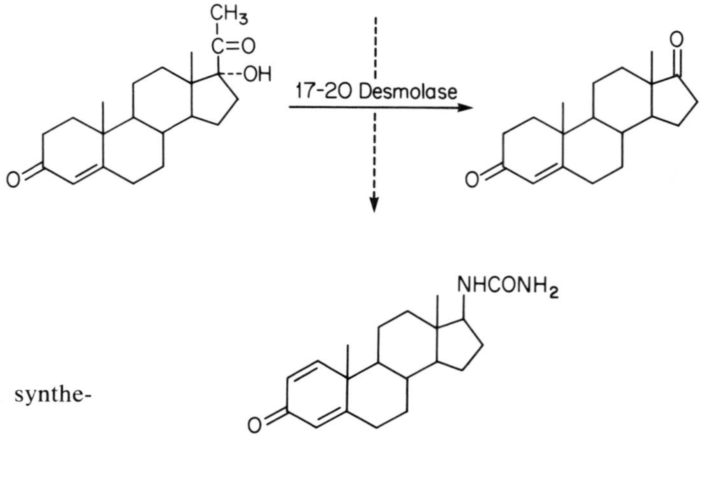

Fig. 2. Androgen synthesis inhibition.

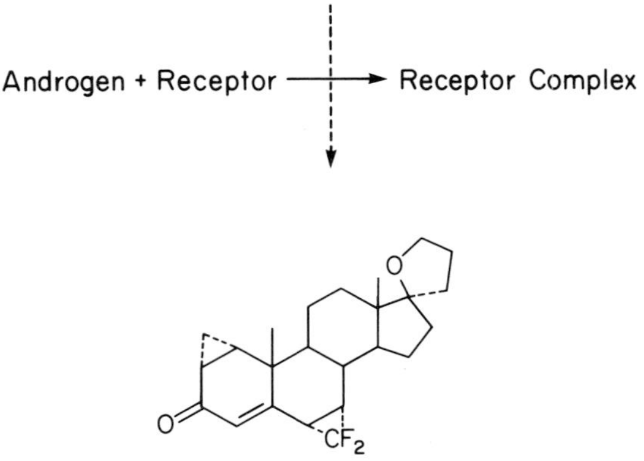

Androgen + Receptor ⟶ Receptor Complex

Fig. 3. Androgen receptor antagonist.

et al. 1973). Although relatively free of other hormonal effects, this compound was somewhat less potent than cyproterone acetate or the nonsteroid antagonist, flutamide, as an antiandrogen.

The third approach, and the one to which the remainder of the chapter will be addressed, is the inhibition of steroid 5α-reductase. In 1968 two laboratories reported that DHT is the major androgen present in the nucleus of rat prostate cells (Bruchovsky and Wilson 1968b; Anderson and Liao 1968). Since the nucleus is the locus of action of steroids, it became clear that DHT was the true trophic agent in the prostate and probably other androgen-responsive tissues. Thus, it was of interest to investigate the effects of compounds that would interfere with the conversion of T to DHT. Medical application of such agents might be envisioned for the treatment of benign prostatic hypertrophy and certain skin–hair conditions, such as acne, female hirsutism, and male pattern baldness. Each of these conditions has with its occurrence a demonstrated local excess of DHT over that found in normal tissue (Siiteri and Wilson 1970; Price 1975).

II. Bioassay of Steroids as 5α-Reductase Inhibitors

Initial in vitro studies in our laboratories utilized a crude rat prostatic 5α-reductase enzyme preparation (Ham et al.). Some 120 samples from our steroid collection were assayed as inhibitors of this enzyme. It was found that the best inhibition of the enzymic conversion of T to DHT was obtained with 3-keto-Δ^4 steroids. A variety of substituents at C-17 was tolerated by the enzyme; the most effective inhibitors were not androgen derivatives but compounds such as progesterone and 20-dihydroprogesterone. Hydroxyl substitution at C-6β, C-11α, and C-11β reduced the enzyme inhibitory potency of parent compounds, as did C-6α and C-7α methyl substitution. 19-Nor compounds showed an inhibitory potency about equivalent to the parent compounds. The spiroether, L-586,659, was chosen for further study because of its potency and lack of other endocrine effects. When applied to chick combs, it was more effective in preventing T-stimulated growth than would have been predicted from its modest androgen antagonist activity in the rat.

L-586,659

In 1973 Voigt and Hsia described the treatment of the androgen-stimulated hamster flank organ with the 5α-reductase inhibitors, 4-androsten-3-one-17β-carboxylic acid (17βC) and its corresponding methyl ester (17βME). The fact that these nonhormonal steroids could influence the action of T but not DHT on this sebaceous gland led investigators at Merck to prepare new steroids for tests using this animal model as a readout (Johnston et al.). About 140 new steroidal analogs were prepared and first tested as 5α-reductase inhibitors in the rat prostatic enzyme assay. The more active compounds in this assay were tested topically on the testosterone-stimulated hamster flank organ. Although structural variations were performed on all the steroid rings and side chain, the most effective agent prepared was the diethyl amide, L-631,810, of 17βC. This material was consistently more active than either 17βC or 17βME in the hamster gland assay, and it showed selective activity only on the gland to which it was administered, indicating a lack of a systemic effect. Such a systemic effect is found on topical treatment of the gland with the androgen antagonist, flutamide.

CO_2R

$CON(C_2H_5)_2$

R = H (17βC)
   = CH_3 (17βME)

R^1= H (L-631,810)
    = CN (L-633,649)

The best 5α-reductase inhibitor prepared was the 4-cyano derivative, L-633,649. However, it was less effective topically on the hamster flank organ than its parent compound (L-631,810), indicating that features in addition to the enzyme inhibitory activity are necessary for optimal activity.

III. Development of In Vivo Active 5α-Reductase Inhibitors

In the course of this work a paper by Imperato-McGinley et al. (1974) appeared describing the characteristics of people with a genetic deficiency of steroid 5α-reductase. Although the absence of DHT dramatically reduces the development of the male sex accessory glands and external genitalia in utero, the postpubertal action of a DHT-deficiency would appear to have no serious medical implications. Adult men with this disorder achieve normal male build and psychosexual orientation but have only scanty beard growth and no acne; the prostate is so small that it cannot be palpated.

Our interest then turned to developing an agent that would be systemically effective in preventing DHT formation. The 5α-reductase inhibitors observed up to this point were uniformly inactive when tested for reducing DHT levels in vivo. It was felt that all these agents were substrates for reductase enzymes and were quickly reduced to inactive 4,5-dihydro forms before adequate concentrations could accumulate to have an effect in the prostate. Thus, our objective was to find an agent that could block the 5α-reductase without being deactivated peripherally by it or related enzymes.

The proposed mechanism of reduction of 3-keto-Δ^4-steroids by steroid 5α-reductase is depicted in Fig. 4. The enzyme must bind the steroid and the nicotinamide portion of NADPH in close proximity such that pro-S hydrogen transfer (as hydride) from the cofactor to the steroid can take place (Suzuki and Tamaoki 1974). This results in a transition state (still enzyme-bound through electrophile E) in which the steroid is held in its enolic form. Ketonization by water protonation at the C-4 position results in a conformational flip of the steroid A-ring and debinding from the enzyme. Energy for debinding from the enzyme could be derived from the ketonization process as the A–B *trans* ring juncture of steroids experiences increased strain by unsaturation between C-3 and C-4. Relief of this strain should provide energy adequate to change the conformation from that required for binding to the enzyme.

This rationale for the mechanism is in line with the high in vitro activity of the cyano compound, L-633,649. If this material is reduced by the enzyme to the enolic intermediate, it is stabilized by conjugative overlap of the resulting

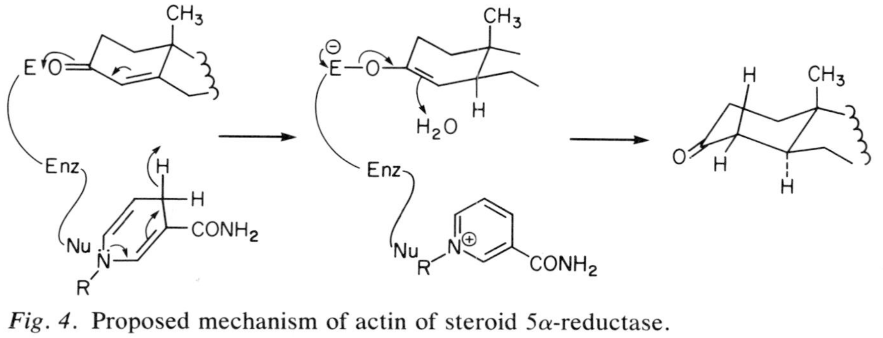

Fig. 4. Proposed mechanism of actin of steroid 5α-reductase.

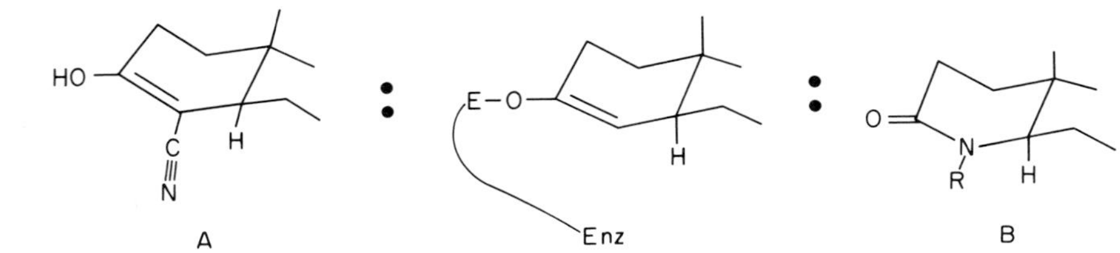

Fig. 5. Enzymic intermediate and isosteres.

C-4,5-double bond with the C-4 nitrile group. Such stabilization would retard the ketonization step and thus attenuate the turnover rate of the steroid on the enzyme.

Our studies thus led us to a new class of inhibitors in which a stable array of ring atoms at the 2-, 3-, 4-, and 5-positions of a 5α-reduced steroid A-ring would lie in a plane similar to that found for the enolic form of reduced L-633,649 (Fig. 5A). A highly active series of compounds was found that met these criteria, namely, steroids containing the 4-aza modification shown in Fig. 5B. The conformation of this lactam ring system should very closely approximate that of the proposed 3-enol form of the steroid on the enzyme. The irreversible inhibitor (Fig. 6B) recently described by Metcalf et al. (1980) should also adopt this conformation.

As in the case of the carbocyclic 3-keto-Δ^4-steroids, various semipolar side chains at C-17 of these 4-azasteroids permitted high 5α-reductase inhibition (Fig. 6A). Interestingly, the 4-methyl derivatives were somewhat better inhibitors than the corresponding unsubstituted derivatives.

Examples of highly active inhibitors are shown in Fig. 7. The IC_{50} values are for inhibition of rat prostatic 5α-reductase in a cell-free system. The variation of structure of the side chain from the lipophilic character of L-642,318 to that of the ionic L-642,022 is indeed striking. The compound L-636,028 (4-MA) has been chosen as a candidate for further study because of its high in vivo activity, its activity versus the human prostatic enzyme, and its lack of other endocrine effects. A detailed account of studies with this material follows.

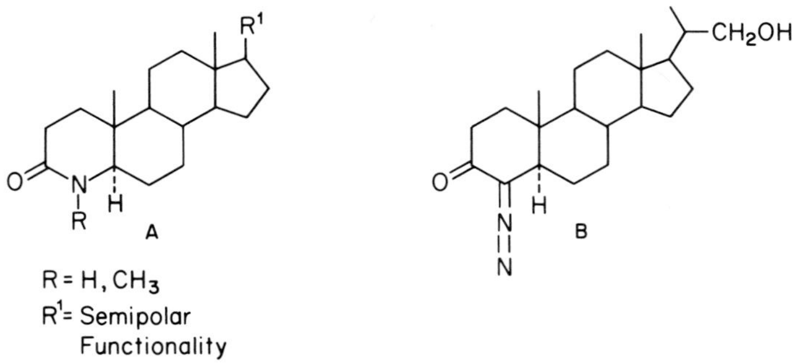

R = H, CH_3
R^1 = Semipolar Functionality

Fig. 6. A–B trans 5α-reductase inhibitors.

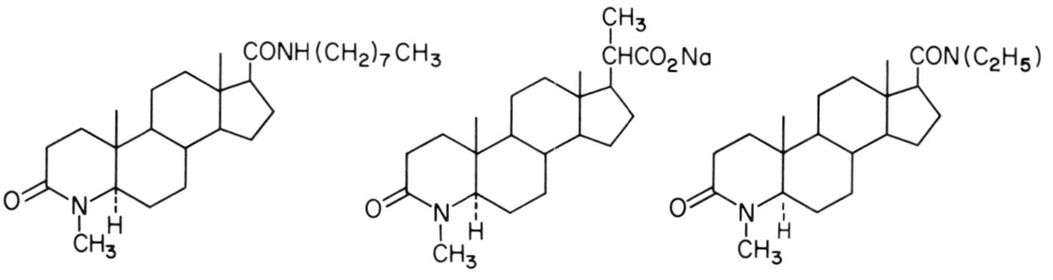

$$L\text{-}642,318$$
$$IC_{50} = 2 \times 10^{-10}$$

$$L\text{-}642,022$$
$$IC_{50} = 2 \times 10^{-9}$$

$$L\text{-}636,028\,(4\text{-}MA)$$
$$IC_{50} = 1 \times 10^{-8}$$

Fig. 7. Potent 5α-reductase inhibitors.

IV. In Vitro Evaluation of 4-MA

The current view of androgen action in many tissues, such as the prostate and the seminal vesicles, involves (1) the conversion of testosterone, the major circulating androgen in adult males, to 5α-dihydrotestosterone by the enzyme 5α-reductase (in the extranuclear compartment); (2) the binding of 5α-dihydrotestosterone to the cytosol receptor protein, followed by a temperature-dependent transformation of the 5α-dihydrotestosterone-receptor complex; (3) the translocation of the transformed 5α-dihydrotestosterone-receptor complex into the cell nucleus (Liao 1977); and (4) the stimulation of nuclear RNA synthesis through mechanisms yet to be elucidated. We investigated the effect of 4-MA on the first three of these steps of androgen action in the rat prostate.

V. Inhibition of 5α-Reductase by 4-MA in Rat Prostate Cell-Free Systems

In rat ventral prostate homogenate, 5α-reductase activity, measured by the conversion of radioactive T to DHT, is distributed in the nuclear (63%), mitochondrial (22%), and microsomal (15%) fractions. Cytosol contains virtually no 5α-reductase activity. The subcellular distribution of 5α-reductase that we observed is consistent with that reported by others (Frederiksen and Wilson 1971; Nozu and Tamaoki 1974). Since the enzyme activities in the nuclear fraction and in the cytoplasmic membrane fractions have been shown to exhibit the same catalytic properties, such as pH profile and K_m for T (Frederiksen and Wilson 1971), we combined the nuclear, mitochondrial, and microsomal fractions as the enzyme source in our studies. Although the enzyme preparations are rather crude, they exhibit an absolute requirement for exogenous NADPH for activity. In addition, after incubation with radioactive T, DHT accounts for more than 99% of the radioactive products.

In the presence of T (1 μM), NADPH (50 μM), and the enzyme preparation (1 mg protein), DHT formation at 37°C is linear for at least 30 min. The

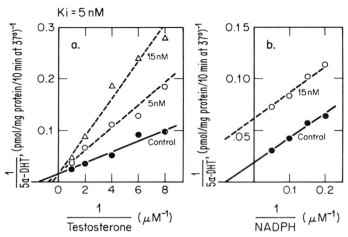

Fig. 8. Lineweaver–Burk plots of the inhibition by 4-MA of rat prostate 5α-reductase catalyzed conversion of testosterone (T) to 5α-dihydrotestosterone (DHT): (a) Inhibition by 4-MA at varying concentrations of T. The reactions were carried out in the absence (●, control) and in the presence of 5 nM (○) and 15 nM (△) 4-MA. A saturating level of NADPH (50 μM) was used in the reaction. (b) Inhibition by 4-MA at varying concentrations of NADPH. The reactions were carried out with 10^{-6} M T in the absence (●, control) and in the presence of 15 nM 4-MA (○). [From Liang and Heiss (1981); 4-MA was previously abbreviated as DMAA).]

reaction is inhibited by 4-MA in a dose-dependent manner and the inhibition reaches 96% at 1 μM of 4-MA, indicating that 4-MA is a very potent 5α-reductase inhibitor.

Figure 8 shows Lineweaver–Burk plots of 4-MA inhibition of 5α-reductase at varying concentrations of T and NADPH. As expected, 4-MA inhibition of 5α-reductase is competitive with testosterone and is uncompetitive with NADPH. The apparent K_i for 4-MA is $5.3 \pm 0.6 \times 10^{-9}$ M (mean ±SD, 4 experiments). The apparent K_m is $7.9 \pm 2.8 \times 10^{-7}$ M for T and is $2.1 \pm 0.3 \times 10^{-5}$ M for NADPH. Thus, 5α-reductase has a much higher affinity for 4-MA than for T.

VI. Inhibition of the Androgen Receptor Binding of 5α-Dihydrotestosterone by 4-MA

Since the action of DHT requires that it first bind to its receptor protein (Liao 1977), 4-MA has been evaluated for its affinity for the androgen receptor of rat prostate cytosol. To demonstrate [³H]DHT binding to the androgen receptor, prostate cytosol from rats castrated overnight was incubated with [³H]DHT (1 nM) at 0°C for 90 min. The resulting [³H]DHT receptor complex was precipitated with ammonium sulfate (35% saturation). After redissolving the pellet, the solution was either analyzed by sucrose gradient centrifugation or treated with dextran-coated charcoal. The amount of the [³H]DHT

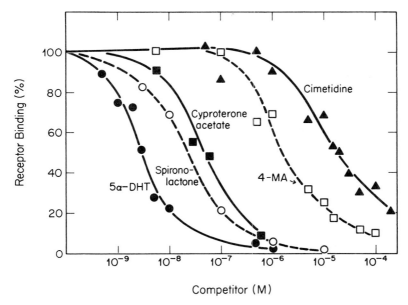

Competitor (M)

Fig. 9. Inhibition of [³H]DHT binding to the rat prostate cytosol androgen receptor by selected compounds. [³H]DHT (1 nM) was incubated with rat prostate cytosol (30 mg protein) in the presence and absence of a competing steroid at 0°C for 90 min. Ammonium sulfate was then added to 35% saturation. After a centrifugation, the resulting pellet was redissolved and the solution was treated with dextran-coated charcoal. The radioactivity in the supernatant represents receptor binding. The receptor binding in the absence of a competitor is taken at 100%. [From Liang and Heiss (1981).]

receptor complex present was determined from the characteristic radioactive peak in the sucrose gradient or from the supernatant after charcoal treatment.

The [³H]DHT receptor complex sediments as a 3–4 S peak in the sucrose gradient containing 0.4 M KCl as reported by Fang and Liao (1971). Inclusion of a 10-fold (10 nM) excess of nonradioactive DHT in the binding incubation inhibited the receptor binding of [³H]DHT by 80%, and the same concentration of 5β-dihydrotestosterone, cortisol, 17β-estradiol or progesterone gave little effect. Thus, the receptor binding is steroid specific.

Figure 9 shows inhibition of [³H]DHT binding to the androgen receptor by 4-MA, DHT, and some antiandrogens. Comparison of the concentration of these steroids required for 50% inhibition indicates that the receptor affinity for 4-MA is 1000-fold, 120-fold, and 40-fold lower than that of DHT, spironolactone, and cyproterone acetate, respectively, and is 7-fold higher than that of cimetidine, a histamine antagonist with weak antiandrogenic activity (Funder and Mercer 1979).

VII. Inhibition of Nuclear Uptake of Androgens by 4-MA in Minced Prostate Incubations

We also evaluated the effect of 4-MA on the nuclear uptake of androgens. Prostates from rats castrated for 20–40 h were minced and then incubated with [^{3}H]DHT or [^{3}H]T in the presence or absence of 4-MA. The nuclei were purified from the tissue homogenate by centrifugation through a 2.2 M sucrose/MgCl$_2$ solution followed by washing of the resulting nuclear pellet with a Triton X-100 solution to remove residual cytoplasmic contaminants. Examination of a nuclear preparation by electron microscopy indicated that the nuclei were mostly morphologically intact with little contamination by cytoplasmic membranes.

[^{3}H]Steroids in the nuclei were extracted with ethyl acetate and the extracts were analyzed with high-pressure liquid chromatography (HPLC). Table 1 summarizes results of one of these studies. When [^{3}H]DHT was used in the tissue incubation, all radioactivity extracted from the nuclei co-chromatographed with DHT regardless of whether or not 4-MA was present in the tissue incubation. However, 4-MA decreased the nuclear uptake of DHT. This effect of 4-MA may be due to its inhibition of the androgen receptor binding of [^{3}H]DHT as described earlier. When [^{3}H]T was used in the tissue incubation, 96% of the radioactivity extracted from the nuclei co-chromatographed with DHT and only 4% as unmetabolized [^{3}H]T. Inclusion of 4-MA in the tissue incubation decreased the nuclear concentration of DHT and increased that of T. The total radioactivity in the nuclei was

Table 1. The Effect of 4-MA on the Nuclear Uptake of [^{3}H]DHT and [^{3}H]T in Minced Prostate Incubations[a]

| | Nuclear extract | | |
| | | % Radioactivity | |
Tissue incubation	cpm/100 μg Protein	DHT	T
[^{3}H]DHT (10^{-8} M)	10552 $\pm$ 121[b]	100	0
[^{3}H]DHT + 4-MA 5 $\times$ 10^{-8} M	9488	100	0
[^{3}H]DHT + 4-MA 1 $\times$ 10^{-7} M	8886	100	0
[^{3}H]DHT + 4-MA 3 $\times$ 10^{-7} M	5548	100	0
[^{3}H]DHT + 4-MA 1 $\times$ 10^{-6} M	4870	100	0
[^{3}H]T (1.1 $\times$ 10^{-8} M)	11307 $\pm$ 3051[b]	96	4
[^{3}H]T + 4-MA 5 $\times$ 10^{-8} M	8241	18	82
[^{3}H]T + 4-MA 1 $\times$ 10^{-7} M	7701	11	89
[^{3}H]T + 4-MA 3 $\times$ 10^{-7} M	5981	5	95
[^{3}H]T + 4-MA 1 $\times$ 10^{-6} M	4927	2	98

[a] Liang and Heiss 1981.
[b] Duplicate incubation. Ventral prostates from rats castrated for 20 h were minced and then incubated with 10 μCi of [^{3}H]DHT (200 Ci/mmol) or [^{3}H]T (180 Ci/mmol) in the presence or absence of 4-MA in cell culture medium 199 at 37°C for 40 min. Nuclei were purified from these tissues and then extracted with a solution containing 0.6 M KCl, 1 mM EDTA and 20 mM Tris · HCl, pH 7.5. Aliquots of the KCl extracts were taken for determining the radioactivity and protein concentrations and for HPLC analysis of [^{3}H] steroids.

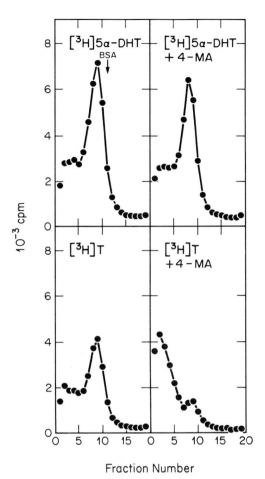

Fraction Number

Fig. 10. Sucrose gradient centrifugation of KCl nuclear extracts of prostates incubated with either [³H]DHT or [³H]T in the presence or absence of 4-MA. Ventral prostates from rats castrated 20 h previously were incubated with 8 μCi of [³H]DHT (200 Ci/mmol, 8×10^{-9} M) or [³H]T (152 Ci/mmol, 1×10^{-8} M) in the presence or absence of 4-MA (10^{-7} M). Nuclei were purified from these tissues and the [³H]steroid–receptor complexes in the nuclei were extracted with a solution containing 0.6 M KCl, 1 mM EDTA, and 20 mM EDTA, and 20 mM Tris · HCl, pH 7.5. The nuclear extract (1.1 mg protein in 0.1 ml) was layered on a 5–18% sucrose gradient containing 0.4 M KCl, 1 mM EDTA, and 20 mM Tris · HCl, pH 7.5. The centrifugation was at 300,000 g for 18 h at 0–2°. The sedimentation was from left to right. Bovine serum albumin (BSA) was layered on a gradient containing no nuclear extract and was centrifuged in the same rotor. The peak position of BSA is indicated in the figure. [From Liang and Heiss (1981).]

decreased by 4-MA. However, this latter effect of 4-MA was not as drastic as its effect on the ratio of DHT/T. These results indicate that 5α-reductase is more sensitive to 4-MA than the receptor binding of androgens.

Figure 10 shows the sucrose gradient centrifugation of 0.6 M KCl extracts of the nuclei. When [³H]DHT was used in the tissue incubation, most of the radioactivity sedimented as a 3.3 S peak, whether 4-MA was present or not. When [³H]T was incubated in the absence of 4-MA, a similar 3.3 S peak was also demonstrated. However, when 4-MA was present, this peak was drastically decreased. In a number of experiments, we observed that the decrease in this peak was associated with a decrease in the proportion of DHT in the nuclei. The disappearance of the peak may be due to dissociation of [³H]T from the receptor during the sucrose gradient centrifugation. To demonstrate this we prebound [³H]DHT and [³H]T to the prostate cytosol receptor, partially purified the [³H] steroid receptor complexes by ammonium sulfate precipitation, and analyzed the complexes by sucrose gradient centrifugation. We found that [^{3}H]T gradually dissociated from the receptor during the centrifugation, whereas [^{3}H]DHT remained associated with the receptor (Liang and Heiss 1981). The higher rate of dissociation of the prostate cytoplasmic T receptor complex than that of the DHT receptor complex has been previously reported (Liao et al. 1973; Wilson and French 1976).

VIII. Absence of 4-MA Binding to Human and Rat Serum Proteins

In vivo interaction of 4-MA with serum proteins would affect the availability of the steroid to tissues. We therefore investigated whether 4-MA binds to serum proteins. Incubation of [1,2-^{3}H]4-MA (16 Ci/mmol) with serum (160× and 800× dilutions) from a mature male rat and a mature male human followed by treatment with charcoal showed no binding of [^{3}H]4-MA to serum proteins of either species. Under the same assay conditions, we observed good binding of [^{3}H]progesterone, [^{3}H]T, and [^{3}H]DHT to human serum proteins and little binding of these steroids to rat serum proteins. This is consistent with previous reports that human serum contains a corticosteroid-binding globulin, which also binds progesterone (Milgrom 1978), and a sex steroid-binding globulin that binds both T and DHT (Rosner and Smith 1975).

IX. In Vivo Evaluation of 4-MA: Effect on Concentration of Prostatic T and DHT in Intact Male Rats

DHT is the androgen present in highest amount in the nuclei of prostate cells (Bruchovsky and Wilson 1968a; Anderson and Liao 1968). Since its formation from T in the cytoplasm is catalyzed by 5α-reductase (Tomkins 1957), it was reasoned that the administration of a 5α-reductase inhibitor should lead to a perturbation in the prostatic concentrations of T and DHT. That this does, in fact, occur was shown by Brooks et al. (1981). In acute experiments, intact 200 g male rats were treated with one of a number of compounds and sacrificed 4 h later. The ventral prostate was removed, homogenized, and extracted with ethyl acetate. The extract was subjected to column chromatography to separate T and DHT and the concentrations of the two androgens determined by radioimmunoassay (RIA). The compound, 17β-N,N-diethylcarbamoyl-4-methyl-4-aza-5α-androstan-3-one (4-MA), which had proved to be a highly active inhibitor of 5α-reductase in vitro, was used as a reference standard at a subcutaneous level of 1 mg/rat. Treatment with 4-MA caused a consistent increase in the concentration of prostatic T and a decrease in that of DHT (Table 2). The total of T plus DHT was similar to that noted in control prostates. Although the potent antiandrogens cyproterone acetate and flutamide lowered tissue levels of DHT, they also reduced the combined total of T plus DHT. The latter result is considered

Table 2. Effect of Subcutaneous Administration of Various Compounds on the Concentration of T and DHT in the Ventral Prostate of Intact 200 g Male Rats[a]

Group	Dose	T ng/100 mg tissue ± SEM	DHT ng/100 mg tissue ± SEM
Control	1 ml oil	0.12 ± 0.02	1.68 ± 0.07
4-MA	1 mg	0.58 ± 0.06[d]	1.11 ± 0.04[d]
Progesterone	10 mg	0.25 ± 0.03	1.95 ± 0.16
Cyproterone acetate	10 mg	0.10 ± 0.01	1.01 ± 0.06[e]
17βC[b]	10 mg	0.15 ± 0.01	1.84 ± 0.11
17βME[c]	10 mg	0.18 ± 0.06	1.74 ± 0.14
Megestrol acetate	10 mg	0.14 ± 0.01	0.82 ± 0.14[d]
Medrogestone	10 mg	0.09 ± 0.02	1.61 ± 0.09
Flutamide	10 mg	0.13 ± 0.02	1.17 ± 0.13[f]

[a] Brooks et al. 1981.
[b] Androst-4-en-3-one-17β-carboxylic acid.
[c] Androst-4-en-3-one-17β-carboxylic acid methyl ester.
[d] $P < 0.001$.
[e] $P < 0.01$.
[f] $P < 0.05$.

indicative of interference with the binding of androgen to receptor protein. It was of interest that several compounds that had been reported to inhibit 5α-reductase in vivo did not reduce DHT concentration in the assay system described earlier. These included progesterone (Mauvais-Jarvis et al. 1974), androst-4-en-3-one-17β-carboxylic acid (17βC) and its methyl ester (17βME) (Voigt and Hsia 1973), megestrol acetate (Geller et al. 1976), and medrogestone (Tan et al. 1974). Since these are all Δ^4-steroids, it can be speculated that they had been reduced and rendered inactive as 5α-reductase inhibitors.

X. 5α-Reductase Inhibition in Castrated Male Rats

Young adult (200 g) male rats were castrated 24 h prior to treatment. They were given a subcutaneous (s.c.) injection of 1 or 10 mg of 4-MA and 2 h later an injection of either 200 μg testosterone propionate (TP) or 400 μg dihydrotestosterone propionate (DHTP). Two hours after receiving androgen, the rats were sacrificed and the ventral prostate was removed for determination of the concentrations of T and DHT.

The results of these experiments (Table 3) showed that pretreatment with 1 mg of 4-MA caused a significant reduction in the amount of DHT subsequently accumulated by the prostate of rats given TP. A concomitant increase in prostatic T level was also noted. In rats treated with DHTP, 1 mg of 4-MA did not affect the amount of DHT found in the prostate, nor did it reduce the total amount of androgen in the gland. Taken together, these data are consistent with the hypothesis that 4-MA acts by inhibiting 5α-reductase and thus the concentration of prostatic DHT.

Table 3. Influence of 4-MA on the Accumulation of T and DHT in Ventral Prostate after the Administration of TP or DHTP to Castrated Rats[a]

Treatment	T (ng/100 mg tissue ± SE)	DHT (ng/100 mg tissue ± SE)
Exp. 1		
Oil	0.01 ± 0.00	0.02 ± 0.02
TP	0.04 ± 0.01	1.39 ± 0.02
TP + 10 mg 4-MA	0.27 ± 0.05^{b}	0.01 ± 0.01^{c}
10 mg 4-MA	0	0
Exp. 2		
Oil	0.01 ± 0.01	0.01 ± 0.01
DHTP	0.03 ± 0.02	0.94 ± 0.20
DHTP + 10 mg 4-MA	0.02 ± 0.01	0.18 ± 0.04^{b}
10 mg 4-MA	0.01 ± 0.01	0.02 ± 0.02
Exp. 3		
Oil	0	0.05 ± 0.05
TP	0 ± 0	1.16 ± 0.21
TP + 1 mg 4-MA	0.87 ± 0.23^{b}	0.42 ± 0.11
DHTP	0.06 ± 0.04	1.16 ± 0.30
DHTP + 1 mg 4-MA	0.07 ± 0.07	1.22 ± 0.30
1 mg 4-MA	0.08 ± 0.02	0.13 ± 0.07

[a] Brooks et al. 1981.
[b] $P < 0.01$.
[c] $P < 0.001$.

In contrast to the results just outlined, the administration of 10 mg of 4-MA lowered not only the concentration of DHT but also the total androgen content of the prostate in both TP- and DHTP-treated rats. This latter result suggests that a high dosage level of 4-MA probably interferes not only with the conversion of T to DHT but also with the binding of androgen to receptor.

XI. Effect of 4-MA in Antiandrogenic Assays

The ability of 4-MA to counter the growth-promoting action of T or TP and DHT or DHTP on the ventral prostate and seminal vesicles of castrate immature male rats was assessed using standard methods. The animals were treated s.c. for 7 days. On the eighth day, they were sacrificed and sex accessory glands were weighed. The most striking finding of these tests was that 4-MA antagonized the action of T or TP more effectively than it did that of DHT or DHTP (Table 4 and Fig. 11). This is a result that might be anticipated if it were assumed that the primary action of 4-MA was to inhibit 5α-reductase rather than to block androgen binding to receptor. In any event, the selective antiandrogenicity exhibited by 4-MA is very unusual and, in fact, may not have been observed previously with any other compound.

Table 4. Effect of Subcutaneous 4-MA and Flutamide on TP or DHTP-Induced Growth of the Ventral Prostate (VP) and Seminal Vesicles (SV) of Immature Castrate Male Rats

| Group | Daily dose | N | Avg. organ wgts. (mg) ± SD | |
			VP	SV
Control	0.20 ml sesame oil	6	12 ± 3	8 ± 1
TP	20 μg	6	50 ± 8	26 ± 6
TP + flutamide	20 μg + 33 μg	6	36 ± 7^b	18 ± 2^a
TP + flutamide	20 μg + 300 μg	6	17 ± 2^c	8 ± 0.4^c
TP + 4-MA	20 μg + 0.5 mg	6	34 ± 11^a	17 ± 2^a
TP + 4-MA	20 μg + 1.5 mg	6	32 ± 9^b	12 ± 4^b
TP + 4-MA	20 μg + 4.5 mg	6	17 ± 5^c	9 ± 2^c
DHTP	40 μg	6	54 ± 9	20 ± 4
DHTP + flutamide	40 μg + 33 μg	6	61 ± 6	20 ± 4
DHTP + flutamide	40 μg + 300 μg	6	31 ± 6^b	10 ± 1^b
DHTP + 4-MA	40 μg + 0.5 mg	6	60 ± 7	22 ± 2
DHTP + 4-MA	40 μg + 1.5 mg	6	57 ± 5	20 ± 3
DHTP + 4-MA	40 μg + 4.5 mg	6	47 ± 6	19 ± 6

[a] $P < 0.05$.
[b] $P < 0.01$.
[c] $P < 0.001$.

XII. Effect of 4-MA on Organ Weights and Fertility in Male Rats

Mature male rats of proven fertility were given either 10 or 50 mg/kg per day of 4-MA by s.c. injection over a 67-day period. Twice a week these males were caged overnight with a proestrous female. Vaginal smears were taken to determine whether mating had occurred and mated females were autopsied 17–20 days later. The males were sacrificed the day after their final injection, day 68 of the test, and the weight of various organs was recorded.

Results obtained in this experiment are shown in Table 5. Highly significant reductions in weight of the ventral prostate and seminal vesicles were found in both treated groups. The significant increase in epididymal weight noted in the group given 10 mg/kg per day is not considered to be of physiological importance. Body weight and weights of the kidneys, adrenals, thymus, thyroid and pituitary were not significantly altered by treatment with 4-MA.

It was of considerable interest that males that received 4-MA continued to mate and remained fertile throughout the course of the experiment. Both treatment levels of 4-MA were above the level required, as shown in the earlier tests, to inhibit DHT formation in the prostate. This would appear to suggest that in the rat DHT is either not required or is needed in minor amounts for maturation of spermatozoa. The period of treatment was longer than the time needed to complete an entire spermatogenic cycle in the rat.

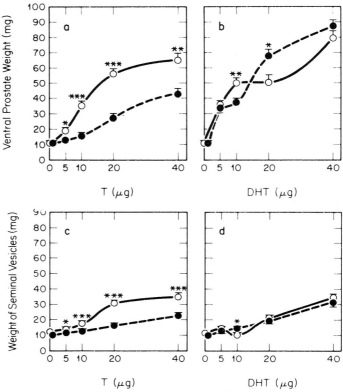

Fig. 11. Effect of subcutaneous 4-MA on growth of the ventral prostate and seminal vesicles in rats treated with increasing s.c. levels of testosterone (T) or dihydrotestosterone (DHT). Open circles (O——O) indicate no 4-MA, closed circles (●--●) indicate rats received 1 mg 4-MA per day. Values are means ± SE. (a) Ventral prostate weights in rats treated with T in the presence or absence of 4-MA. (b) Ventral prostate weights in rats treated with DHT in the presence or absence of 4-MA. (c) Weight of seminal vesicles in rats that received T in the presence or absence of 4-MA. (d) Weight of seminal vesicles in rats that received DHT in the presence or absence of 4-MA. *$P < 0.05$; **$P < 0.01$; ***$P < 0.001$: level of statistical significance from group receiving a corresponding level of T or DHT. [From Brooks (1981).]

XIII. General Effect of High Levels of 4-MA in Young Adult Male Rats

An experiment was conducted to determine the effect of high treatment levels of 4-MA on organ weights and viability of male rats. Dosages of 50, 150, and 450 mg/kg per day were administered by s.c. injection for 14 days. These levels were up to about 100 times higher than that which causes a lowering of prostatic DHT concentration. The rats were sacrificed on the fifteenth day and various weights were taken (Table 6).

Table 5. Response of Mature Male Rats to 67 Days Treatment with Subcutaneously Administered 4-MA[a]

Group	Dose kg/day	BW Changes g ± SEM	Avg. organ wts. (mg) ± SD					
			Testes	Prostate	Sem. ves.	Epididymis	Adrenal	Thymus
Control	1 ml Sesame oil (10% ETOH)	139 ± 12	3272 ± 302	617 ± 302	798 ± 108	1070 ± 40	47 ± 3	209 ± 32
4-MA	10 mg	114 ± 12	3488 ± 281	396 ± 89[b]	530 ± 81[b]	1155 ± 57[c]	48 ± 4	221 ± 29
4-MA	50 mg	111 ± 8	3529 ± 202	271 ± 34[b]	270 ± 39[b]	1055 ± 77	50 ± 4	256 ± 58

[a] Brooks et al. 1982b. Six rats were in each group and all remained fertile throughout the experiment.
[b] $P < 0.001$.
[c] $P < 0.05$.

Table 6. Effect of Daily Subcutaneous Administration of 4-MA for 14 Days on Body and Organ Weights of 200–300 Gram Male Rats[a]

Group	Dose mg/kg/day	N	BW change g ± SEM	Organ wts. (mg) ± SEM					
				Testes	Prostate	Sem. ves.	Epididymis	Thymus	Adrenals
Control	4 ml Sesame oil (10% ETOH)	7	90 ± 5	3026 ± 353	262 ± 20	224 ± 16	785 ± 63	385 ± 28	49 ± 3
4-MA	50	8	84 ± 3	3233 ± 66	173 ± 10[c]	122 ± 9[c]	764 ± 33	450 ± 35	52 ± 3
4-MA	150	8	60 ± 7[b]	3259 ± 53	92 ± 8[c]	82 ± 7[c]	638 ± 22[b]	373 ± 26	52 ± 2
4-MA	450	3	-3 ± 3	2573 ± 298	52 ± 11	54 ± 6	397 ± 98	133 ± 81	67 ± 10

[a] Brooks 1982b.
[b] $P < 0.01$.
[c] $P < 0.001$.

Ventral prostate and seminal vesicle weights were significantly reduced in rats given 50 or 150 mg/kg per day, but testis, thymus, and adrenal weights were not different from control values. In the group of rats treated with 450 mg/kg per day, toxicity was apparent. Five of the eight animals died before the experiment was over and the survivors exhibited both an increase in adrenal weight and a decrease in thymus weight. These latter changes are indicative of stress.

XIV. Action of 4-MA on the Dog Prostate

Intact mature male beagles of at least 3 years of age were used in a test designed to determine the effect of chronic administration of 4-MA on prostate size and prostatic and testicular histology (Brooks et al. 1982a). A pretreatment laparotomy was done, and prostate length, width, and depth were determined. The dogs were allowed about 10 days for recovery. They were then treated orally twice daily for 4 weeks with 1 or 3 mg of 4-MA per kg per day.

When the dogs were sacrificed, prostate dimensions were recorded and sections of the prostate and testis removed for histological purposes. Both treatment levels of 4-MA caused a significant reduction in prostate size (Table 7). This was coupled with a dose-related atrophy of the prostatic epithelium. Testicular changes were limited to two dogs in the 3 mg/kg per day group. In one animal there was a very slight degeneration of the spermatogenic epithelium and the second showed slight Leydig cell hyperplasia. These results indicate that the minimal oral dosage level of 4-MA needed to cause a decrease in prostate size is about 1 mg/kg per day. This level for dogs compares very well with the effective treatment levels of the highly potent antiandrogens flutamide (Neri and Monahan 1972) and cyproterone acetate (Neri et al. 1968).

XV. General Bioassays with 4-MA

A number of standard tests were conducted to more fully establish the endocrinological spectrum of 4-MA. These included the uterotropic and vaginal cornification assays of estrogenicity, progestational assay, gonadotropin inhibition assay utilizing the parabiotic rat, and a test to determine whether the compound induced fetal feminization of the male rat. A detailed description of these assays and the results obtained has been reported (Brooks et al. 1982b). Briefly, 4-MA exhibited no estrogenic, progestational or gonadotropin-inhibiting activity. It did, however, feminize male fetuses carried by females treated s.c. over days 13–21 of pregnancy with 3 or 9 mg/ rat per day.

Table 7. Effect of 42 Days Treatment with Orally Administered 4-MA on Prostate Volume and Histology of the Prostate and Testis of the Dog[a]

Treatment[b] and dog no.	Prostate Volume (cm³)		Histologic findings	
	Pretreatment	Posttreatment	Prostate	Testis
Control				
1	7.7	8.8	Normal	Normal
2	11.6	11.5	Normal	Normal
3	10.8	14.7	Normal	Normal
4	9.4	11.9	Normal	Normal
Avg. ± SEM	9.9 ± 0.9	11.7 ± 1.2		
4-MA (1.0 mg/kg/day)				
1	18.2	8.6	Slight atrophy	Normal
2	17.4	9.9	Normal	Normal
3	14.3	10.8	Normal	Normal
4	14.0	4.5	Moderate atrophy	Normal
Avg. ± SEM	16.0 ± 1.1	8.5 ± 1.4[c]		
4-MA (3.0 mg/kg/day)				
1	17.9	4.5	Moderate atrophy	Normal
2	11.9	3.1	Marked atrophy	Normal
3	16.7	5.2	Moderate atrophy	Slight Leydig cell hyperplasia
4	25.4	5.9	Marked atrophy	Very slight atrophy
Avg. ± SEM	18.0 ± 2.8	4.7 ± 0.6[c]		

[a] Brooks et al. 1982a
[b] The dogs were randomly assigned to groups and treated twice daily.
[c] Significant within-group difference between pretreatment and posttreatment prostate volume ($P < 0.01$).

XVI. Summary

4-MA $(17\beta\text{-}N,N\text{-diethylcarbamoyl-4-methyl-4-aza-5}\alpha\text{-androstan-3-one})$ is an example of a new class of highly active reversible inhibitors of 5α-reductase. These steroids attenuate the conversion of T to DHT both in vitro and in vivo. The in vitro inhibition by 4-MA is competitive with T (apparent K_i of $5.3 \pm 0.6 \times 10^{-9}\ M$) but uncompetitive with NADPH. 4-MA has moderate affinity for rat prostate cytosol androgen receptor. This affinity is 1000-fold, 120-fold, and 40-fold lower than that of DHT, spironolactone, and cyproterone acetate, respectively, but 7-fold higher than that of cimetidine. In tissue incubates of minced rat prostate, 4-MA decreases the nuclear concentration of $[^3H]DHT$ but increases that of $[^3H]T$. Total radioactivity is reduced. However, this latter reduction is slight in comparison to the effect of 4-MA on the ratio of nuclear $[^3H]DHT$ to $[^3H]T$. These latter results indicate that 5α-reductase is affected more readily by 4-MA than is the receptor binding of androgens. 4-MA failed to bind to serum proteins of either the male rat or human, whereas under the same conditions progesterone, T, and DHT all bound well to human serum proteins but not to those of the rat.

In intact 200-g male rats, treatment with 4-MA caused an increase in the concentration of T in the ventral prostate and a concomitant decrease in the concentration of DHT. These effects indicate that 4-MA inhibits the 5α-reductase catalyzed conversion of endogenous T to DHT.

Although active in in vitro tests, progesterone, androst-4-en-3-one-17β-carboxylic acid and its methyl ester did not inhibit 5α-reductase in vivo. The potent antiandrogens, flutamide and cyproterone acetate, reduced prostatic concentrations of both T and DHT, a result that probably reflects their ability to interfere with the binding of androgen to androgen receptor protein.

In the 24-h castrate male rat, pretreatment with 1 mg of 4-MA resulted in an attenuation of the conversion of injected TP to DHT as judged by the concentrations of T and DHT recovered in the ventral prostate. Under similar conditions, 4-MA did not affect the prostatic concentrations of DHT in rats injected with DHTP. It was noted that 10 mg of 4-MA reduced the total concentration of prostatic androgens. That suggests that high treatment levels probably influence androgen–receptor binding.

In standard assays for antiandrogenicity, 4-MA exhibited more activity against T or TP than DHT or DHTP. This differential effect has not been noted with classical antiandrogens and is thought to be due to the primary mode of action being that of 5α-reductase inhibition.

At oral treatment levels of 1 or 3 mg/kg per day, 4-MA caused marked regression in size of the dog prostate. In a 67-day test, 10 or 50 mg/kg per day of 4-MA sharply reduced sex accessory gland weights in mature male rats. These chronically treated rats remained fertile throughout the treatment period. That suggests that DHT may not play a major role in sperm maturation in the rat.

4-MA showed no activity in assays for estrogenic, progestational, or gonadotropin-inhibiting activity. It did cause feminization of male fetuses carried by treated female rats, a result consistent with the hypothesis that DHT is required for normal development of the external genitalia in males.

References

Anderson KM, Liao S (1968) Nature 219: 277

Arth GE, Patchett AA, Jefropoulus T, Bugianesi TL, Peterson LH, Ham EA, Kuehl FA Jr, Brink NG (1971) J Med Chem 14: 675

Brooks JR, Baptista EM, Berman C, Ham EA, Hichens M, Johnston DBR, Primka RL, Rasmusson GH, Reynolds GF, Schmitt SM, Arth GE (1981) Endocrinology 109: 830

Brooks JR, Berman C, Glitzer MS, Gordon LR, Primka RL, Reynolds GF, Rasmusson GH (1982a) The Prostate 3: 35

Brooks JR, Berman C, Hichens M, Primka RL, Reynolds GF, Rasmusson GH (1982b) Proc Soc Exp Biol Med 169: 67

Brooks JR, Busch RD, Patanelli DJ, Steelman SL (1973) Proc Soc Exp Biol Med 143: 647

Bruchovsky N, Wilson JD (1968a) J Biol Chem 243: 2012

Bruchovsky N, Wilson JD (1968b) J Biol Chem 243: 5953

Fang S, Liao S (1971) J Biol Chem 246: 16

Frederiksen DW, Wilson JD (1971) J Biol Chem 246: 2584

Funder JW, Mercer JE (1979) J Clin Endocrinol Metab 48: 189

Geller J, Albert J, Geller S, Lopez D, Cantor T, Yen S (1976) J Clin Endocrinol Metab 43: 1000

Ham EA, Kuehl FA, Arth GE, unpublished results

Imperato-McGinley J, Guerrero L, Guatier T, Peterson RE (1974) Science 186: 1213

Johnston DBR, Schmitt SM, Arth GE, Brooks JR, Berman C, Primka RL, unpublished results

Liang T, Heiss CE (1981) J Biol Chem 256: 7998

Liao S (1977) In: Litwick G (ed) Biochemical Actions of Hormones. Academic Press, New York, pp 351–406

Liao S, Liang T, Fang S, Castaneda E, Shao T-C (1973) J Biol Chem 248: 6156

Mauvais-Jarvis P, Kuttenn F, Baudot N (1974) J Clin Endocrinol Metab 38: 142

Metcalf BW, Jund K, Burkhart JP (1980) Tetrahedron Lett 21: 15

Milgrom E (1978) In: O'Malley WO, Birnbaumer L (eds) Receptors and Hormone Action. Vol II, Academic Press, New York, pp 473–490

Neri RO, Casmer C, Zeman WV, Fielder F, Tabachnik IIA (1968) Endocrinology 82: 311

Neri RO, Monahan M (1972) Invest Urol 10: 123

Neumann F, von Berswordt-Wallrabe R, Elger W, Steinbeck H, Hahn JD, Kramer M (1969) Recent Prog Horm Res 26: 337

Nozu K, Tamaoki B-I (1974) Biochim Biophy Acta 348: 321

Patanelli DJ, Berman C, Primka R, Steelman SL, unpublished results

Price VH (1975) Arch Dermatol 111: 1496

Rasmusson GH, Chen A, Reynolds GF, Patanelli DJ, Patchett AA, Arth GE (1972) J Med Chem 15:1165

Rosner W, Smith RN (1975) Biochemistry 14: 4813

Steelman SL, Brooks JR, Morgan ER, Patanelli DJ (1969) Steroids 14: 449

Siiteri PK, Wilson JD (1970) J Clin Invest 49: 1737

Suzuki K, Tamaoki B-I (1974) J Steroid Biochem 5: 249

Tan SY, Antonipillai I, Murphy BEP (1974) J Clin Endocrinol Metab 39: 936

Tomkins GM (1957) J Biol Chem 225: 13

Voigt W, Hsia SL (1973) Endocrinology 92: 1216

Wilson EM, French FS (1976) J Biol Chem 251: 5620

Discussion of Paper Presented by G.H. Rasmusson, T. Liang, and J.R. Brooks

MEYER: When you gave very massive doses of the compound what did the pathology of those animals look like?

BROOKS: You're talking about the animals that were given the compound over a 14-day period. The veterinarian did look at those rats and said it looked like generalized stress, whatever that means.

LIAO: Does the inhibitor work for 5β-reductase?

BROOKS: Yes, it also inhibits 5β-reductase. We've seen that in the liver.

LIAO: So it may alter the metabolic pattern of the androgen.

BROOKS: I think it could probably alter the metabolic pattern. Those 5β compounds are mainly excretory products and I don't think in the economy of the animal they are going to be important.

BARDIN: Have you ever done an experiment where you have held the ratio of testosterone to inhibitor as a constant and then done the same experiment where you held the ratio of dihydrotestosterone to inhibitor constant. In other words, can testosterone alone stimulate prostate without being converted to dihydrotestosterone?

BROOKS: We've done an antiandrogen assay where the animals were given a constant dose of inhibitor with increasing levels of either T or DHT. The compound inhibited the effect of T on prostate weight in these castrated animals, whereas it did not affect the ability of DHT to induce an increased prostate weight.

BARDIN: Let me draw a theoretical experiment. If this is a dose response to testosterone and this is on the prostate, theoretically a large amount of this is due to conversion of DHT and both testosterone and DHT both bind to the androgen receptor. But in the prostate there is something different about the androgen receptor and on sucrose gradients, testosterone dissociates. So the question is does testosterone per se have a biological activity. So the question is what does this shift the response of testosterone to the right, will testosterone by itself when this conversion is blocked will have it's effect.

BROOKS: With an antiandrogen such as flutamide or cyproterone acetate, you can get an inhibition of the testosterone or testosterone propionate induced change in ventral prostate weight that you cannot see when you give 028. You can't reduce ventral prostate weight to as low a point with L-636, 028 as you can with those potent antiandrogens and on that basis I believe that testosterone itself may have some ability to cause an increase in weight of the ventral prostate.

KORNEL: I was intrigued by one of your slides where you have shown the testicular weights and you pointed out that there was no significant increase in the weight, but still the control weights were in the order of 3200 of whatever units you had, and they went up to 3600 with your compound, and what comes to mind is whether this compound may indirectly influence spermatogenesis. My question to any one of the experts here pertains to possible differences between DHT and testosterone: Which one has a higher affinity to the hypothalamic receptors for the feedback inhibition of LH?

BARDIN: As far as I know dihydrotestosterone has a higher affinity for almost all androgen receptors regardless of where you are measuring. The way androgen receptors differ in different parts of the body is in their dissociation rate for testosterone. Dr. Liao showed many years ago that he ran a sucrose gradient on the androgen receptor of prostate that testosterone was dissociated. You just confirmed that study and that's been shown by many other people. Yet for the pituitary, the hypothalamus, the kidney, the salivary glands, when you use the androgen receptor in these organs, you get beautiful gradients with testosterone and by using charcoal assays you can show that the rate of dissociation is different. So there is a difference in affinity but in the relative competition study you always can show a higher affinity for

DHT than testosterone in almost every assay. So the receptors are different in some unusual way.

KORNEL: Now if this be so that the DHT has a higher affinity for the hypothalamic receptor, it may well be that the administration of your compound results in only partial inhibition of LH secretion and as a result of that the secretion rate of testosterone will be higher, and what is important the intratesticular levels of testosterone will be much higher and then there will be stimulation of spermatogenesis. If this is really so, then, as an offshoot of the whole thing, you may consider using this compound to improve spermatogenesis whenever this is desired.

BROOKS: You're saying that you think possibly there is an increase in LH with this compound?

KORNEL: Yes. You haven't measured the levels directly.

BROOKS: No, we haven't measured them directly. The point you're basing your comments on is the increase in testicular weights. Now that was not a significant increase. We haven't really seen a trend toward increased testicular weight with this compound. We've seen that it is not a gonadotropin inhibitor. If LH is increased, there should be a short-term increase in testosterone and also in estrogen if aromatase is unaffected. Increased estrogen and testosterone would exert a negative feedback effect and you should end up right back where you started really, I would think. With no real change in gonadotropin levels.

LIAO: I have a question, probably the most important one. Any plan on clinical tests?

BROOKS: If all goes well, the compound should be in the clinic in a few months.

MEYER: Do you have any data as far as the compound's effect on the red blood system is concerned?

BROOKS: I think I understand why you're asking that question, but we've done relatively little on this. In the 67-day-treated rats, we did do red blood cell counts and they were normal.

MEYER: Do you think that indicates that the androgen that is important for the red blood cells system is testosterone rather than DHT?

BROOKS: I can't really give you an answer to that because I'm not that familiar with the mechanism of hematopoiesis, so I'd rather not speculate on it.

KORNEL: With regard to your prospective intended clinical trials, I would like to mention here some very poorly known facts or findings in some female patients with hirsutism or mild virilization due to (presumed) excessive adrenal androgen production. We have investigated very carefully a large number of such patients (the lab tests included free testosterone, bound testosterone, etc., and all those indices) and we have seen three patients in whom the only significantly increased level of androgen was that of DHT.

BROOKS: Well, as you know, there is a lot of 5α-reductase in the skin (Kuttenn et al. (1977) J Endocr 75, 83–91), and I think a lot of the testosterone is going to be converted to DHT there. If the target organ sensitivity in an individual is high, then I would think the DHT present might well bring about hirsutism.

KORNEL: I don't think that in those patients I'm referring to the high levels of DHT came from the compound escaping from the skin. I think, in those patients (by the way, all these patients were only mildly hirsute, were young, had menstrual irregularities, couldn't get pregnant, etc.) the activity of hepatic 5α-reductase was significantly increased, and that is something you may really want to look into.

BROOKS: Dr. Liang could answer your question on hepatic 5α-reductase as related to androgen levels.

LIANG: Dr. Kornel, your observation on the increase of hepatic 5α-reductase activity and the elevation of the DHT levels in the blood of some hirsute women is interesting. If they are correlated, a 5α-reductase inhibitor may be useful for treating these patients. With regard to the relationship between hepatic 5α-reductase and androgen levels, it was first reported in 1958 (Yates, FE, Herbst AL, and Urquhart J (1958) Endocrinology 63 887–902) that in mature rats, the hepatic 5α-reductase activity is 3–10 times higher in females than in males, and the hepatic level of 5α-reductase of male rats increased after castration. Hepatic activity of 5α-reductase is also increased by thyroxin treatment (McGuire JS, Tomkins GM (1959) J Biol Chem 234 791–794). It is not known whether the increase of the hepatic level of 5α-reductase by thyroxin treatment causes an increase of the DHT level in the blood.

Discussants: W.J. MEYER, J.R. BROOKS, S. LIAO, C.W. BARDIN, T. LIANG AND L. KORNEL

Chapter 20

Comments on Steroid Hormone Action

MODERATOR: A.K. ROY

ROY: For many years we have been hearing about the acceptor sites and acceptor proteins that may have some role in steroid hormone action. However, Dr. Gustafsson presented his results showing specific interaction between the glucocorticoid receptor and the naked DNA. Maybe somebody could comment on the implication of these results for the general area of steroid hormone action.

CLARK: I don't necessarily wish to discuss the implications but I would like to make a few comments. The question of specific nuclear acceptors has been a difficult one. One reason for this is the existence of many nonspecific sites to which RE complexes will bind. Jack Gorski and I first demonstrated this in 1969 by showing that RE complexes will bind any charged surface, including glass. Therefore, in order to demonstrate specific sites in the midst of large numbers of nonspecific sites one must do competition experiments with unlabeled RE complexes. Since this is rarely done, most work on nuclear acceptor sites does not hold up to the criterion of specificity and therefore is not valid. Dr. Gustafsson's work is an exception to this and was done in an entirely different manner.

With respect to the location of nuclear acceptor sites, as all of you know every conceivable nuclear component has been proposed as a nuclear acceptor site for receptors. Perhaps there are many sites involved in such complex regulation. One can imagine that acceptor sites on DNA and nuclear matrix are involved and may even interact in the steps that lead to DNA and RNA synthesis.

ROY: Still, this is not very clear to me. I still don't understand what is the role of acceptor proteins or "sites," whatever you want to call them, when free DNA can bind to the purified receptor in a sequence-specific manner.

CLARK: The point is the DNA-binding experiments have not been done properly because they have not been shown to be specific.

PECK: I am taking the microphone away from him. Seldom has an experiment been done with an activated receptor that has appropriately used a competitor to prove specificity of that interaction.

CLARK: That's what I just said.

PECK: But you didn't say it right. In fact, activated estrogen receptor, the one that I speak for best, will stick to anything and you cannot simply study binding of receptor to things and say that it is specific.

ROY: Let us hear from Dr. Gustafsson.

GUSTAFSSON: With regard to the specificity, I think that it was quite clearly shown that it was binding specifically to just certain pieces of DNA. With regard to competition experiments, these have been performed, and if you add five times excess of the MMTV genome (of unlabeled genome), you can compete away the P^{32}-labeled DNA from the filters; however, if you add other types of DNA to which the glucocorticoid receptor does not bind, then you get no competition. I indicated there seem to be three different places along the PMMTV genome that specifically bind the glucocorticoid receptors, whereas perhaps only one of them is really functional, and maybe you need proteins to shield the other parts so that they don't interact with the glucocorticoid receptors. I don't know, but there might well be a place for protein to direct the glucocorticoid receptor to the functional place.

CLARK: Jan-Åke, yours is the one experiment out of the 100 that's different, because you have a different assessment of specificity. Your experiments are very clear and the specificity lies in the gene fragments to recognize the receptor hormone complex.

ANDERSON: Dr. Gustafsson, in your recent paper didn't you people use a 10,000-fold excess of calf thymus DNA, showing it had no competitive effect. It was stated in there; I don't think the data was shown.

GUSTAFSSON: The competition experiments were carried out with the vector genome. They were added in five times excess and that didn't have any effect. In the paper you referred to calf thymus DNA was not used. The plasmid vector DNA was used in this study and that did not displace.

ANDERSON: Wouldn't you expect that if indeed there is specific binding the calf thymus DNA *would* give some competitive effect because the DNA would be expected to have some glucocorticoid binding sites.

GUSTAFSSON: These results will be described in detail in the forthcoming PNAS paper.

ANDERSON: Wasn't it said in the paper that 10,000-fold excess of calf thymus DNA did not exhibit competition.

GUSTAFSSON: No, but if you wait another month you will see the PNAS paper and you can check that.

SHAPIRO: I'd like to make one comment about this question also. I think that the idea that a receptor may bind specifically to a piece of naked nonchromatin DNA may be more reasonable for a gene such as the MMTV gene, which has two properties that make it a better candidate than some others. One is that the gene is usually on at a very low level in the latent cells, so there is already an equilibrium between the receptor and whatever site it binds to, and the other is that the response is extraordinarily rapid in cells, raising the possibility that the way you can achieve such a rapid response is to have a system where you really don't have to displace various chromatin proteins from sites on the genome. These proteins could be modulating in some fairly direct and rapid way accessibility to promoter sites or something like that. The second thing I'd like to point out is that when you have a gene that responds very rapidly to a hormone–receptor complex, then you have to ask how the hormone receptor finds the gene this quickly, and the possibility is that in addition to the actual, perhaps tight binding sites at which hormone–receptor complex may interact with DNA that there is a significantly larger region of DNA that has some affinity for receptor. Perhaps the receptor can rapidly slide up and down this DNA until it reaches its ultimate binding site. As time goes on and this site is perhaps localized in a smaller fragment of DNA, it may become more clear whether you are looking at a process in which the receptor is actually in a region of DNA for which it has some affinity and it is sliding as it searches for a correct binding site, or whether the receptor is exclusively interacting with one specific class of DNA sequence.

MOUDGIL: Drs. Munck and Litwack's labs have clearly shown that there is some physiological involvement of activation process. Second, I want to make a comment about the activated receptor sticking to everything. If we take, for example, activated hormone–receptor complex and look at its DNA-cellulose binding and receptor binding to just Cellex 410, it doesn't bind.

KORNEL: I would like to move to the other topic discussed during this conference, namely, the heterogeneity of the cytoplasmic receptors, particularly that discussed by Dr. Peck and by Dr. Litwack. We have studied the arterial mineralocorticoid and glucocorticoid receptors and have found that in the cytosolic fraction of arterial blood vessels there are at least three different binders of glucocorticoids, one with charcteristics of a classical dexamethasone glucocorticoid receptor that has high affinity for dexamethasone and is translocating to nucleus and binding to chromatin; the other is quite specific for cortisol, has rather low affinity for dexamethasone, and possesses all those properties of receptors, including the translocation to the nucleus and binding to chromatin; and then a third one that is present in rather high concentrations is a very specific cortisol binder with transcortinlike properties, but we know this is not transcortin. So I would like to ask Dr. Peck or Dr. Litwack, or both of them, if they would give us a brief prophetic look into the future of what you think eventually will determine the possible importance of those multiple receptors to one class of steroids.

LITWACK: It's gratifying that this audience has such simple questions: I think we have some data to suggest that there may or may not be two receptors. The resolution of that problem, I think, depends on having highly specific probes like monoclonal antibodies, both to the traditional receptor that Dr. Gustafsson has described and to what we call IB. If it should turn out that IB is, in fact, a piece of the traditional receptor, then we are hard put to explain some of our data that I mentioned in my talk. If, on the other hand, it is a separate gene product, which has not been determined as yet, then it seems possible to me that IB could be similar to the receptor in glucocorticoid-resistant P1798 lymphosarcoma and also the resistant S49 cell that John Baxter mentioned briefly, having molecular weights and other properties in the case of John Stevens' studies rather similar to IB. Since the resistant cell lives in the presence of the hormone and the sensitive cell dies, it would be tantalizing to consider that IB receptor would be the *anabolic* receptor and the binder II would be the *catabolic* receptor. We know from analysis of hepatocytes and liver-type cells using two-dimensional gels (Phelps and Litwack) that not only do you see proteins that are induced, you also see simultaneous repressions in response to glucocorticoids. I think this is probably more of a commonplace than not, depending on what tissues are looked at. These are some of the ideas that we're batting around but we have to wait for the direct evidence on that.

ROY: Dr. Litwack, are you still leaving the possibility open that the IB could be a proteolytic fragment of the receptor?

LITWACK: If it is, it will not square with the antibody data we presented. So my view at the moment is that it is not. I have to entertain the view that it's a separate gene product. But at the same time I think we have to have better reagents than we have, to clarify this point.

THOMPSON: Well here's a prophetic model for you: I suggest that receptors are going to be a multigene family in which the steroid-binding site (region) of the receptor is relatively constant. Consequently, we can't tell them apart on that basis, with the possible exception of a few cases like yours or like the ones in the kidney. On the other hand, the DNA-binding regions (or if you like, the acceptor-binding regions) are represented by a widely variable set of sequences in the genomic DNA. The two are put together to provide appropriate specificity for appropriate organs. This is a real problem, particularly with the case of glucocorticoids, where one sees all kinds of different responses in different organs. So if there is a single monolithic glucocorticoid receptor protein, specificity must occur at some other level. That could occur by blocking off genes so that the receptor complex can't get to them. Indeed, that is suggested by our data from the somatic cell hybrids. But when one selects for variant cells that show noncoordinate control of a set of responses, as we and a couple of other labs have done, then you can't use that explanation anymore. Specificity must be provided at other levels, one of which could be the receptor itself (E.B. Thompson et al. (1979) Cold Spring Harbor Conf Cell Biol 6:339). So I think it would be really interesting to see if this particular prophecy holds true in the next few years.

GUSTAFSSON: If we just speak about what we really know, we know that there is one receptor, whether IB represents a protolytic fragment or not remains to be shown. Then there is something called transcortin that has often been mistaken as an intracellular receptor for the simple fact that if you don't perfuse your liver well enough, you will get the impression that it occurs on the cytosol. However, if you perfuse it well, there is no transcortin present. I should also point out that when you use dexamethasone as the ligand, it will bind to both the glucocorticoid receptor and the mineralocorticoid receptor which makes up quite a substantial part of the specific dexamethasone binding in various tissues. You may know that Raynaud has published some studies where he has used an unlabeled specific glucocorticoid named RU6988. With that he can block specifically the glucocorticoid receptor-binding sites so that what is left, so to say, is the mineralocorticoid-binding sites. If you run your experiments in the presence of RU60988, then you can see that what you label with dexamethasone is something else than the glucocorticoid receptor. It has a different isoelectric point, for instance, as shown by Dr. Wrangle in our laboratory. That may explain the heterogeneity as seen by various workers when using dexamethasone to detect glucocorticoid receptor in rat kidney. I would like to ask Dr. Roy about his androgen receptor in the liver. Does this unconventional sort of androgen receptor that you have described occur also in female rats?

ROY: Normally the female rats contain very small amounts of this binding protein; however, it is inducible. When the spayed female rats are treated with androgen one finds a rise of this binding protein as well as an increase in the intensity of the androgen response.

GUSTAFSSON: Do you have any real direct proof that this unconventional androgen receptor is really involved in androgen action. In other words, have you seen translocation of this protein to the cell nucleus.

ROY: No, this hepatic androgen binder does not translocate into the nucleus. At least, we were not able to show that. The reason, we believe, that it may be involved in androgen action is that there is a strong correlation between the presence of this binder and the ability of the liver to synthesize α_{2u}-globulin under androgenic stimulation. Since this binder does not translocate to the nucleus, its role in androgen action may be an indirect one. I may comment that only a few labs (which include yours) have been able to provide results indicating a direct influence of steroid receptors in gene activation. I have a sneaky suspicion that there may still be some postreceptor steps that have not been discovered yet. The concept that the receptor has to bind to the DNA and open it up may be too simplistic a view. I think there are a lot of steps missing, and in years to come the picture may look a bit more complex.

LIAO: I remember, actually I was discussing this with Dr. Shapiro, that antiandrogens like cyproterone acetate do not bind to this hepatic receptor. Is that right?

ROY: Yes, that's another important point that suggests that this binder has something to do with the physiological function of the androgen. Cyproterone acetate does not compete for this binding protein and cyproterone acetate also fails to block α_{2u}-globulin synthesis. Flutamide is also totally ineffective both in inhibiting α_{2u}-globulin and in displacing labeled DHT from the cytoplasmic binding protein. However, as I have said, we do not know how this cytoplasmic androgen binder may influence androgen action. In light of recent findings on type II estrogen receptor and what Dave Shapiro has described as mid-affinity receptor in the liver, Dr. Liao may want to comment on the α-protein of the ventral prostate.

LIAO: As I commented after Dr. Clark's presentation, α-protein binding of steroids in the rat ventral prostate may be very similar to the type II estrogen-binding activity in the rat uterus. I do not feel that we have enough evidence to say what kind of roles α-protein of type II binding protein may have in the target cells. α-Protein, however, has many interesting properties. For example, one of the polypeptide components that does not bind steroid can prevent, in the cell-free systems, the interaction of androgen-receptor complex with DNA or chromatin. The receptor complex bound to nuclei (chromatin or DNA) can also be released by the addition of this polypeptide chain. This model system may indicate that in the target cell the receptor activity in the nuclei can be modulated by certain polypeptides (see Chapter 16).

Another interesting property of α-protein is that this protein, as isolated, has about 1 mol of cholesterol per mole of protein. The protein also binds phospholipid and sediments as a high-density lipoprotein (HDL). I wonder whether this protein can function like a plasma HDL.

ROY: Let's conclude with something from Dr. Don Coffey concerning nuclear matrix.

COFFEY: I want to say one thing about an experiment that Jim Clark just mentioned a while ago. He is able to bring about a very specific estrogenic effect in the rat vagina i.e. cornification by stimulating the surface with a glass rod. And that is important for the following reason. There is a great belief and growing feeling that the structure of the cell can control cell function and that you can induce structural changes in those cells and have a dramatic effect on cell functions; for example many of you are familiar with Folkman's experiment where he showed that changing the shape of the cell by either chemical or mechanical means can alter nuclear function like DNA synthesis. One of the things we've been working on as a basis and background of our hypothesis is that the structural elements are being changed by hormones. A lot of people ask me "Well, if all the hormone is doing is coming in and changing these structural elements that bring about these effects, somebody ought to be able, without a hormone, change the structure of the cell and bring about the hormone's effect." Now there are several examples that I won't get into, but Clark's experiment is one of them where no steroid is involved and no receptor is going into the nucleus and yet you see the whole hormone effect. I think Dr. Liao remembers the experiments of Dr. Huggins, where fibroblasts come up and touch collagen and totally change its differentiation state to that of an osteoclast to form bone. I'll leave you with this thought, because I've used it once or twice in some enzyme kinetics lectures that I give. What is the rate-limiting thing in a Volkswagen that is driving down the street. Now you can go in and measure many events in the Volkswagen and try to say what is the rate-limiting thing that governs this Volkswagen, and you can go to the crank shaft to the spark plugs to the distributor and finally you trace it back and you try to get the earlier and earlier events and you finally end up with the legs of the lady who is driving the car and the ATP level in the muscle and everything else, and then it finally ends up that the rate-limiting thing is the light coming through the windshield. If it's dark or black, it stops, or if it is red it stops; at 50 miles per hour it slows down. In the case of steroid action all we're really seeing is the steroid molecule coming through that windshield like the light and all of these things begin to happen. I think we try to put too much on just any one of these events. I believe it brings about a whole correlated series of events that are orchestrated and so I'm not going to be surprised as Arun Roy was saying that there is a whole battery of those things that might be taking place and as Jim Clark says that a steroid can bind to the DNA or the receptor can, into the nuclear proteins, etc., but whatever it does it brings about dramatic effects on the structure of that cell. Last thing, Gospodarowicz showed that he can mechanically or physically change the shape of the cell and totally change the response of that cell to fibroblast growth factor or epidermal growth factor, based on the shape the cell was in when the hormone comes up and touches it. So I'd just go off the air by saying that I think we have to give a lot of thought to structure in the mechanism of hormone action.

ROY: Well, it looks like we have made good progress and a lot of rewarding surprises may be lying ahead.

Discussants: A.K. ROY, J.H. CLARK, E.J. PECK, J.A. GUSTAFSSON, J.N. ANDERSON, D.J. SHAPIRO, V.K. MOUDGIL, L. KORNEL, G. LITWACK, E.B. THOMPSON, S. LIAO and D.S. COFFEY

Index